PROVING GROUND

PROVING GROUND

A Memoir

W. DAVID TARVER

ebuktu
media llc

Proving Ground

A Memoir

Copyright © 2012 by W. David Tarver

All rights reserved. No portion of this book may be reproduced in whole or in part, by any means whatever, except for passages excerpted for purposes of review, without the prior written permission of the publisher. For information, or to order additional copies, please contact:

Ebuktu Media LLC, Birmingham, Michigan.

www.ebuktumedia.com

Book design by Jacinta Calcut.

Permissions acknowledgments for previously published material can be found beginning on page 473.

Publisher's Cataloging-in-Publication Data

Tarver, W. David (William David)
   Proving ground : a memoir / W. David Tarver.
   p. cm.
   Includes index.
   ISBN 978-0-9847962-6-7 (alk. paper)
   1. Tarver, W. David (William David) 2. Entrepreneurship. 3. Businessmen—United States—Biography
4. African American businesspeople—Biography   I. Title.
   HC102.5.T27A3 2012
   658.4'21'092   2011944372
   QBI11-600230

Printed in the United States of America

10 9 8 7 6 5 4 3 2 1

First Edition

For Pops,

*and for all those other fathers who,
despite achieving little wealth, and no fame,
relentlessly kept the flame alive.*

Scan for more information.
www.provinggroundbook.com

TABLE OF CONTENTS

PROLOGUE..1

PART 1: FLINT
    1. FLINT RIVER..6
    2. SCIENCE FAIR LOSER..................................17
    3. 1963..23
    4. RADIO...32
    5. KID BUSINESS..38
    6. 1968..45
    7. ONE OF A KIND.......................................53
    8. SCIENCE FAIR WINNER.................................63
    9. MOST LIKELY TO......................................71
    10. "GMI IS FREE, BOY!"...............................80
    11. AC AND ME...86
    12. A FORK IN THE ROAD................................97
    13. MOOG MAGIC.......................................104
    14. GOODBYE, POPS....................................107

PART 2: ANN ARBOR
    15. MICHIGAN MAN.....................................118
    16. CHEATED..131
    17. NO-PLEA MCAFEE...................................137
    18. GOTCHA!..142
    19. ALMOST OBSOLETE..................................149
    20. NOT DR. TARVER...................................155

PART 3: NEW JERSEY
    21. MECCA..162
    22. GARDEN STATE.....................................177
    23. A NIGGER IN THE WOODPILE.........................184
    24. BREAKTHROUGH.....................................193
    25. A DREAM DIES.....................................196

26. An Idea is Born..................203
27. Love and Technology............209
28. Movin' on Up....................219
29. Nerd World War..................226
30. Last Stand......................237
31. Basement Brigade................243
32. Leaving Bell Labs...............252

Part 4: Venture
33. Money...........................256
34. The Gauntlet....................262
35. 1984............................268
36. World's Oldest Profession......279
37. The Missing One.................285
38. Southern Surprise...............292
39. Double-Cross....................306
40. Nuts, Misfits, and Rejects.....315
41. California Showdown.............324
42. International...................331
43. World Class.....................341
44. Mandela is Free, Bill...........363
45. Hitting the Wall................371
46. Hard Drive......................377
47. In the Valley...................386
48. Turnaround.....................394
49. Exit Strategy..................400
50. Break Some Eggs................411
51. The Dubious Dance..............416
52. End Game.......................422
53. Free at Last...................427
54. The Long Goodbye...............432

Epilogue............................443
Acknowledgments.....................455
Author's Note Regarding Pseudonyms..457
Index...............................459
Permissions Acknowledgments.........473

Prologue

TWO DAYS BEFORE THANKSGIVING 1995 I emerged from a law office in Red Bank, New Jersey, as a newly minted African American millionaire. Not bad, especially given that my new financial status had nothing to do with professional sports or entertainment. It resulted from the sale, for $30 million, of Telecom Analysis Systems, a technology company I started in the basement of my home.

The achievement was especially gratifying because I had predicted it years earlier, not only to myself but also to the two African American cofounders I'd recruited. The dollars were frankly more than we had expected, but probably much less than we could have gotten had we waited a few years. We weren't concerned about the "what-ifs," though. On that November afternoon, after twelve years of exceedingly hard work, patience, ingenuity, and loyalty, we gladly reaped our reward.

Twenty-five years before that red-letter day, I sat in a high school classroom sketching my imaginary company headquarters, a building in the shape of an interlocking "T" and "E," for a company to be named Tarver Electronics. It was, of course, audacious to peer so precisely into such an uncertain future. The civil rights movement of the 1950s and '60s had opened many doors to African Americans, but the career path I planned to take was largely untested. Starting a black-owned electronics manufacturing company in one's basement was, to say the least, not a typical endeavor.

A semiconductor electronics revolution was beginning to take hold, but twelve years before the advent of the IBM PC, it was impossible to know if my chosen vocation would explode or fizzle. Yet I was determined even then, as a teenager in Flint, Michigan, to start a company making "electronics stuff." A few years later, as a young engineering student, I vowed that I would sell my as-yet-unborn company by age forty and begin a new life as a financially independent person—a "free man." So I felt relieved, fulfilled, and proud as I closed that sale in Red Bank—at the ripe old age of forty-two.

Years before signing those sale papers in the Giordano, Halleran & Ciesla boardroom, I described to my friends and family what I planned to do in my post-business existence in the same words I use today: "study anthropology." That phrase was whimsical shorthand for observing the human condition, getting involved in my community, and seeking interesting and fulfilling experiences. In the years following the sale, I largely fulfilled that mission, but I also felt it was important to document my journey from aspiring dreamer to successful entrepreneur. My generation of African Americans was the first who could plausibly hope to realize such bold dreams. Consequently, the business venture my cofounders and I undertook amounted to a social experiment. The results revealed much about us, but also revealed a lot about the world around us. Ironically, our business venture itself yielded many important "anthropology" lessons.

No lucrative government contract, or angel investor, or venture capital firm, or minority business development program fueled what my cofounders and I achieved. We were black entrepreneurs out on our own, knocking on doors in a mostly white world, with nothing between us and failure except our ability to invent, manufacture, and deliver superior products. The journey was by no means easy. Along the way, we had to develop effective business strategies, keep abreast of rapid technological change, and deftly handle race and cultural relations. The money that came with the brass ring in 1995 was great, and my tuition-paid anthropological studies have been fulfilling, but from the beginning, our struggle has been about much more than making money and buying time. It has been about three young black men proving to themselves and the world that we were prepared enough, and that the world had changed enough, for us to compete and win on a global scale.

I worked incredibly hard and took many career risks to fulfill my childhood dreams. I probably fit the description "self-made man" as well as anyone, but in truth, that term is a misnomer. We all, if wise, absorb every ounce of skill and insight we can from the best and brightest in the places where we work, but I was fortunate to have access to some of the *best* places. I benefited greatly from family, friends, and mentors who helped clear my path to success, but many people are not blessed with such supporters. I was able to launch my dreams on a powerful wave of social and technological progress, but millions of African Americans with similar dreams who preceded me, including my father, were not able to ride that wave.

Looking back, I am amazed at the role luck, or divine intervention, played in my success. If my mother, for example, had not had the audacity to use a WHITES ONLY restroom in the Georgia hospital where she worked, and been severely reprimanded for it, my life would have been completely different. I would not have been raised in the North, and would not have been exposed to amazing opportunities in a city that was a leading engineering and manufacturing center. If my father, a World War II-era U.S. Army Signal Corps veteran, had been able to find a day job in his beloved electronics field, he might not have created a basement workshop with vacuum tubes and transistors and electronics parts of every kind that would lure his youngest son toward creating a company that made "electronics stuff."

My mother *did* use that restroom, and my father *did* fill our basement with electronics stuff, and the world *had* changed enough that when I sketched my company headquarters while sitting in a dull high school class, my own dream did not have to be deferred.

I have been blessed with a unique and rewarding journey that began in Flint and took me to the University of Michigan, Bell Laboratories, and then around the world establishing my own company as a global player. I hope my story will inspire entrepreneurs of all races to endure and to understand that their journey is about much more than making money. I hope it will help people negotiate the complex web of business strategy, technology, and cultural relations that defines modern global business. Above all, I hope this story will help each reader chart his or her own path to success.

Part 1: Flint

CELEBRATING THE FIFTY-MILLIONTH GENERAL MOTORS VEHICLE, DOWNTOWN FLINT, 1954.

courtesy Sloan Museum

1. Flint River

MY MOTHER, MY GRANDMOTHER, AND I STOOD beside the tracks outside the tiny Flint train depot. It was after midnight, and a soft breeze cooled the sultry summer air. We heard the train's faint horn, and seconds later I saw its headlight in the distance. The horn sounded again, much louder, as the enormous engine crawled past the Michigan Lumber Company yard and rumbled to a stop right in front of us. Up close, the train's blinding light and thunderous horn terrified me. I wished I were home in bed.

I was just five years old, due to start kindergarten when the summer ended. I had never traveled beyond the Flint city limits, so the idea of climbing aboard a huge, noisy machine in the middle of the night was frightening. My grandmother was taking me on a trip into the unknown, to visit "our family in Georgia." I had no idea what or where Georgia was, and I didn't know—until my mother and grandmother told me—that I had additional family members in some faraway place.

The conductor helped me climb onto the train car's high first step, and then he helped my grandmother. We found an empty row of seats near the middle of the car, and I scooted to the window and searched for my mother. She was still standing beside the track, and when she saw me, she smiled and waved. I fought tears and waved back.

The only thing that made the trip bearable was my grandmother's presence. I called my grandmother Mimama. She had lived with our family since before I was born, and because my parents both worked full-time, I spent more time with her than with anyone else. I would go anywhere as long as Mimama was with me. The train rumbled, and I saw lights moving outside the window—we were on our way. After a few minutes, I saw fewer and fewer lights, and then there were none. I put my head against Mimama's shoulder and fell asleep.

Fast-forward thirty-eight years. I was sitting on the beach in Runaway Bay, Jamaica, and my grandmother and that childhood trip to Georgia were very much on my mind. Mimama had died less than a year earlier, just before her 101st birthday, and I sorely missed her. When she died I was in the final stages of selling my company, Telecom Analysis Systems Inc., to a British firm called Bowthorpe plc. Several months later, with that exciting but all-consuming stage of my life completed, I had the time and money to organize a family trip in honor of my mother and her three siblings: Aunt Mary, Uncle Bill, and Uncle Emerald. Each had survived difficult childhood circumstances and had lived an interesting, productive, rewarding life. I wanted to honor them while they were still alive, and to honor, by extension, my grandmother. Sixteen family members made the trip, and we had a great time eating, drinking, relaxing, and enjoying each other's company. I could imagine Mimama smiling down on us from heaven, because she had always loved family gatherings.

Lingering at the table after long breakfasts, caressed by warm Caribbean breezes, I listened to my mother and her siblings tell stories of our family's past. Those leisurely history lessons were one of the trip's most gratifying aspects. I had heard bits and pieces over the years, but during that trip, I listened more intently and tried to catalog the events in my mind. I didn't know how many more such trips we would take; I didn't know how much time I would have to get the story straight. With my greatest career goal achieved at last, and my desire to pursue a new life "studying anthropology," the history lessons mattered—a lot. They comprised chapter 1 of my own story.

The furthest back anyone could trace our family was Cassandra "Cassie" Alexander. She was Mimama's great-grandmother, my great-great-great-grandmother. Cassie was born somewhere in Virginia on March 1, 1824. She was the property of a slave owner whose first name has not survived the years, a man known to our family only by his last name—Alexander. According to family legend, Cassie was a resourceful, intelligent woman who could read, write, and manage a household. In plantation parlance, she was a "house Negro," not a "field Negro."

On November 25, 1842, Cassie bore her first child, a boy named Henry. The father was Cassie's master, Alexander. I don't know whether Henry's birth precipitated the events that followed, but soon after his arrival, Cassie and Henry were sold to a new master, one James Newton Askew of Fort Gaines, Georgia. Cassie arrived at the Askew estate with a trunk that contained all her worldly possessions. There must have been special circumstances surrounding Cassie's sale—perhaps Master Alexander felt guilty about her treatment—because it was unusual for a slave to be sold along with offspring or possessions.

CASSANDRA (CASSIE) ALEXANDER, 1824-1900.

Askew's southwest Georgia plantation sat on the east bank of the Chattahoochee River in what is now Clay County. Cassie became the Askew family cook and lived in a cabin behind the big house. James Askew had a wife and children, but he fathered several children by Cassie. Askew's wife and white children knew of the arrangement but could do nothing about it. Askew saw to it that his "colored" children, whom he called his "chaps," learned to read and write, and insisted that they learn a trade. They were allowed to eat at the table in the big house, too. After the white family finished a meal, Askew would stand up and announce to the household help, "Bring the chappies in for dinner!" Askew's white sons resented the treatment accorded the "colored" children, and often remarked that they would "teach them how to be slaves" once the old man was dead.

Cassie bore six children by Askew. The oldest, a girl named Ammie, was born January 1, 1846. She was followed by Walter, Mose, Fannie, Carrie, and Isaac. Between Fannie and Carrie, Cassie bore a daughter fathered by one of Askew's other slaves. For that transgression, Askew beat Cassie so severely that her clothing was seared into the wounds on her back, and her skin became infected. That merciless beating severely strained the relationship between Askew and Cassie, but she later produced two more children with Askew.

When the slaves were freed following the Civil War, Cassie and her children left the Askew plantation. He reportedly gave the family a mule, a wagon, provisions, and some cash—and would have given them more had they chosen to take the name "Askew," but they refused. Cassie's children deeply resented the brutal beating Askew had given their mother. Free to take any name they chose, all elected to take Cassie's last name, "Alexander." To this day, however, some family members refer to Henry, the child sired by Cassie's first master, as the only "true" Alexander, to distinguish the child fathered by slave master Alexander from the subsequent children.

Over time, the Alexanders migrated to Mitchell County, also in southwest Georgia, on the east bank of the Flint River. Cassie's first daughter, Ammie Alexander, worked as a midwife, and it is said that she delivered and took care of most babies, white and black, born in Mitchell County around that time. Ammie also was the cook for a man named Francis Emory Catchings. They eventually married, and in 1870 produced a son, Timothy Titus Catchings. As a young man, Timothy worked for a prominent Mitchell County planter known as Captain Heath. The white plantation owner Heath was so impressed with Timothy's skill and industriousness that he gave young Timothy first option to buy his land. Eventually Timothy Titus Catchings, the enterprising grandson of slave Cassie Alexander, became the prosperous owner of a 170-acre spread one mile east of Camilla.

Timothy married a woman named Alice Brown, and together they produced the remarkable total of sixteen children. Their second child, the eldest girl, was born in 1894. More than a half century later she would come into my life bearing her married name, Elizabeth Bernice Hayden—though she was always known among family and friends as Bernice. To me she was simply "Mimama," and though by age five I was willing to board that enormous, frightening train with her, I could not comprehend at the time how much history I would be visiting when we made that first trip to Georgia.

When Bernice was a young girl, another child challenged her to pick up some smoldering fireplace ashes. Bernice picked up the embers and placed them in her lap. Her clothes caught fire, and she was severely burned over much of her body—so severely that Timothy Catchings did not expect his eldest daughter to reach adulthood. Bernice survived, though, and gradually recovered from her wounds. Over time she assumed responsibility for the household and played a major role in caring for her fourteen younger siblings. Her father, no longer doubting she would survive, doubted any man would ever wish to marry her.

In those days, public schools in the South educated Negro children only through ninth grade, so Bernice's father sent her to boarding school in Atlanta so she could continue her studies. A classmate whose handwriting and grammar were not equal to Bernice's sought to impress a boyfriend in Alabama by enlisting Bernice to write letters for her. The boyfriend showed the letters to one of his friends, William Wise Hayden, who was immediately taken with the quality and style of Bernice's writing. This long-distance, accidental matchmaking bore fruit when young Mr. Hayden showed up to meet Bernice. A courtship ensued, and they eventually married. Hayden served in the army, in France, during World War I. After the war, William and Bernice Hayden migrated to Flint, Michigan.

William Hayden considered himself a master of the dry-cleaning trade, and he reckoned that much money could be made cleaning the clothes of Flint's many automobile factory workers. Unfortunately, William was not only a master dry cleaner but also an alcoholic, a womanizer, and a gambler. All three flaws severely eroded his plans for success in the North. When sober, William was articulate and gregarious. When drunk, he was fighting mad, and often verbally and physically abused Bernice. Despite tough times both at home and at work, the Haydens produced four children: William Wise Hayden Jr., Mary Elizabeth, Claudia Louise, and Emerald Earl. Bernice endured the abuse as long as she could, but in 1934 Cassie's great-granddaughter decided to gather the kids and return to Georgia. Her father drove up from Camilla in his Ford and returned with Bernice and three of the Hayden children. William Jr. elected to stay in Flint with his father.

Back on the family farm in Camilla, life remained difficult for Bernice and the kids. They were regarded as the unsuccessful migrants, the "poor relations," and they felt looked down upon by the family in Camilla. Bernice did domestic work for families in town. Mary, Louise, and Emerald often got jobs picking cotton or pulling peanuts in the hot Georgia sun.

Louise stayed in Camilla long enough to complete her ninth-grade studies at Rockdale High, then left to attend Georgia State. She completed only one semester, though, because she lacked money to continue. Her older sister, Mary, had married and settled in Albany. She and her husband were boarding at the home of an older black couple, the Lees. Mary convinced the Lee family to take Louise in, so Louise moved to Albany and got a job as a maid at Rosenberg's department store.

By 1940, just six years after Bernice fled to Georgia, none of the Hayden children remained in Camilla. Mary and Louise were trying to make a life for themselves in Albany. Emerald, after finishing school at Rockdale, had returned to Flint to see if he could do better there. Meanwhile, William Hayden Jr., who never left Flint, had been convicted of armed robbery and was serving a prison sentence.

William Hayden Sr. wrote to Bernice and implored her to return, and his letter found her feeling lonely and vulnerable. Bernice had grown weary of life in Camilla. Her children were gone; she felt like a failure because she was dependent on her relatives; she disliked the low-paying jobs she performed, providing home care for sick people. Her life in Flint had not been good, but at least it had been *her* life. In 1940, Bernice Hayden packed up her few possessions and returned to Flint. For the first time, Louise and Mary didn't migrate along with their mother—they remained in Albany. That fact led to the meeting that led to my existence.

Mary's husband, Odest Watson, had a job selling insurance in Albany, and he had a friend by the name of Fred Douglas Tarver. One day, Odest brought Fred home and introduced him to Louise. Fred and Louise began a courtship, and on Christmas Eve in 1941, they walked across town to their pastor's house and got married. It was a very small ceremony—the pastor administered the vows, and his wife served as witness.

Fred and Louise did not have much time to enjoy their newlywed status. Just two months later, in February 1942, Louise began nurse training at the Grady Colored School of Nursing in Atlanta, and Fred was inducted into the U.S. Army. Later, he shipped out to the Aleutian Islands to serve as a Signal Corps radio operator. Louise, meanwhile, was attending Grady because it offered a deal that could not be beat: all school expenses paid, plus free housing. Married students were not allowed in the Grady nursing program, so Louise had to keep her matrimonial status secret.

Fred returned to Georgia on furlough in 1943, and his visit resulted in a complication: Louise became pregnant. She left Grady early in 1944, returned to Albany, and resided once again at the Lee home. After settling there, Louise wrote to Grady to explain the circumstances of her departure. In July 1944, Louise bore her first child, Elizabeth Bernice Tarver. The birth took place at home, as did almost all Negro births in the South. As had been the case with the child's maternal grandmother, the family referred to the baby by her middle name—Bernice.

Louise's nurse training at Grady, though truncated, qualified her for a job as a practical nurse at Phoebe-Putney Hospital in Albany. In 1945, with the war over, Fred received an honorary discharge from the army and rejoined Louise in Albany. He began working as a salesman for Pilgrim Life Insurance. In January 1947, the insurance salesman and the practical nurse had a second child, Fred Douglas Tarver Jr. Even though Fred Jr. was a "breech presentation" baby—a potentially life-threatening condition for mother and child—no doctor intervened. He was a Negro baby, so he was born at home. Both mother and son were lucky to survive.

It had been three years since Louise withdrew from Grady, but her strong desire to become a nurse endured. Not long after Fred Jr.'s birth, one of Louise's classmates—Darthulia Caldwell—wrote Louise to inform her that Grady was accepting students who had left without completing the program. Later that year, 1947, Louise Tarver returned to the Grady Colored School of Nursing in Atlanta. She made arrangements for Bernice and Fred Jr. to stay in Atlanta with Fred's mother, Ida Tarver, and his half brother, William David Bell. Given a second chance, Louise diligently pursued her studies, and in 1948 received her nursing degree.

Louise and the children then rejoined Fred in Albany, and Louise returned to Phoebe-Putney Hospital, this time as a registered nurse. She remained there until 1951, when a late-night incident at the hospital changed the course of her life. Like most Southern institutions at the time, Phoebe-Putney maintained separate facilities for whites and blacks. The hospital complex consisted of a new brick and mortar building for white patients and an old wooden structure for "coloreds." A passageway connected the white and black wings of the hospital to facilitate movement of personnel and supplies. The passageway also contained restrooms, but they were marked FOR WHITES ONLY. One night, Louise was on duty in the black patients' wing, near the passageway. A sudden need to use the toilet led her to quickly and discreetly use the "white" restroom in the passageway rather than go all the way back into the bowels of the black wing. A white employee spotted her slipping out of the restroom and reported her to supervision. Louise was severely reprimanded. She had suffered indignities at the hands of whites all her life, but this was the final straw. She and Fred decided to leave Albany and move north—to Flint. They had additional reasons for making the move. Louise's father, William Hayden Sr., had died in 1950, leaving Bernice—my grandmother—to fend for herself in Flint. Fred wasn't exactly prospering in the Albany insurance business, and good-paying jobs seemed plentiful in Flint. Louise had landed a job opportunity at Flint's Hurley Hospital that offered $200 per month, twice her Phoebe salary.

In 1951, Fred and Louise Tarver, together with their two children, Elizabeth Bernice and Fred Douglas Jr., boarded the train at Albany for the long trip north. For Louise, it was a return to the city of her birth and childhood. For Fred, it was a journey into the unknown, and he was apprehensive. The family joined Bernice in the apartment above William Hayden's old dry cleaning shop. Bernice was renting the apartment and was having trouble making ends meet. Sharing the space with Fred and Louise and the children put the family on a better financial footing.

Fortunately, the Tarvers arrived in Flint amid a booming economy. General Motors employed one in four Flint residents and was constantly hiring. GM had originated in Flint, and was home to Buick, AC Spark Plug, and numerous other large GM operations.

Louise began her nursing job at Hurley Hospital. Fred started selling life insurance again. The family had no car, so Louise took the trolley to work and Fred walked or caught a ride.

The Tarver/Hayden apartment was at the corner of Dartmouth Street and Industrial Avenue, directly across from the Buick auto factory entrance. At night the assembly line noises and oily smells drifted out of the open factory windows and into the little apartment.

MY OLDER SIBLINGS IN 1954: FRED (7) AND BERNICE (10).

It wasn't the best life, but it was better than Albany, Georgia. It was progress. After a few months managing the Pilgrim Life office in Flint, Fred landed a job at Buick, but he hated factory work, so he quit after only a few days and got a job as a letter carrier at the Flint branch of the U.S. Post Office.

In 1952, the family, including my grandmother, moved across town to a house on Jasmine Street. This too was a rental, but it was much bigger than the apartment, and it was in a residential neighborhood rather than the heart of Flint's industrial zone. Jasmine Street was a better spot for a family, a better place for kids to grow up. A few months after the move, the landlord decided she wanted to rent the place to someone else, so the Tarvers moved across Jasmine Street to yet another rented home.

Later that year, Louise became pregnant again. On May 22, 1953, at Hurley Hospital, the Tarvers' third and last child was born. Fred and Louise named their youngest after Fred's half-brother in Atlanta. I, William David Tarver, am that child. My folks call me David.

My mother and father were not content to live in rented homes, so they started saving to buy a house. My mother took on a second job

ME, 1954.

as a private-duty nurse to add to the family's savings. My parents looked around town for a suitable home, and in 1955 they found it: a white two-story house in a middle-class neighborhood a few blocks southeast of downtown. The house had a spacious front yard, with a large oak tree on one side and a maple on the other. Our new home was at 813 East Sixth Street, and my parents purchased it for the princely sum of $10,000. Because we moved there when I was only two years old, it is the first home I remember.

My friends and I affectionately refer to our old neighborhood as The Block. It actually comprised several city blocks, bordered by Lapeer Road on the west and Avon Street on the east. The north end of The Block started with Fifth Street and proceeded south to Sixth, Seventh, Eighth, Ninth, Belmont, Wellington, and Kennelworth. The Block was already in the midst of a rapid demographic transition when we arrived. We were one of only two black families on Sixth Street in 1955, but when I started kindergarten three years later only three white households remained, and no white kids. The white families had departed for the suburbs or for the still-white neighborhoods of the city.

The Flint of my youth was one of the leading communities in the entire country. General Motors was the engine that drove Flint's economy, and it was firing on all cylinders. Flint enjoyed essentially full employment, great public schools, and outstanding civic institutions. The city was blessed to have Charles Stewart Mott, an early GM director and preeminent philanthropist, as a leading citizen. The Mott Foundation he established provided Flint with world-class institutions in the arts, education, and recreation. The combination of a great community, expanding civil rights, and a technology explosion produced a fertile learning environment and seemingly boundless opportunity. Through hard work, perseverance, wisdom, skill, luck, and divine intervention, my folks had placed me on the best possible launching pad.

Sitting on that beach in Jamaica I had an extended opportunity to contemplate our family history and its effect on my own development. I was able to ponder intangibles of aspiration and determination and perseverance—qualities that had somehow filtered down to me through a lineage that began with an enslaved woman who managed a plantation household, and whose children, at freedom's first blush, refused to take the name "Askew." I was proud to have established an excellent company in a complex and competitive high-tech arena, and to host the entire extended family at a sun-drenched Jamaican resort. But just how far removed was my path from the one my great-grandfather Timothy Catchings trod, as he ran an exemplary farm in the Deep South not many years after the Civil War? How far was my path from the one on which my mother refused to endure any longer a place that would countenance a WHITES ONLY toilet? How far did my path stray from my father's, on which he, in his quiet way, pursued an electronics passion that led him to an Alaskan island during World War II—a passion that no doubt had immeasurable impact on my success? In Jamaica, I came to understand better (and continue to understand better every day) that all of us, from the "Alexanders" to the Tarvers, were treading different sections of the same path, and engaging in similar struggles along the way.

My success as an entrepreneur was special, no doubt. But gaining skills? Learning to interact effectively with other people? Learning, beyond that, the special skills required to recognize and deal with the nuances of race and culture? African Americans have been doing these things since they first arrived, beaten down and shackled, on American shores. What has changed is the society, the law, and now—beyond anything my ancestors could have imagined—the technology. My success as an inventor and manufacturer and entrepreneur wasn't new. The scale and the technology were, but learning a marketable skill and negotiating one's way through a white world are integral, and essential, parts of the African American experience.

It is purely coincidental, but a body of water called the Flint River greeted my Georgia ancestors when they first tasted freedom. That same Flint River flowed through Albany the night my mother threw off her figurative shackles and decided to leave the South. When she and my father arrived in Flint, Michigan, and took up residence in the little apartment on Industrial Avenue, another Flint River flowed nearby. The name "Flint River" courses through our family history, but it isn't the name alone that resonates. It is the strength, the progress, the surging inevitability of the stream.

2. Science Fair Loser

THE SCIENCE FAIR...MORE PRECISELY, *The Seventh Annual Flint Area Science Fair.* The mere mention of it was enough to raise goose bumps. A few years earlier, the Russians had launched something called Sputnik into space, and suddenly the whole country was crazy about science. I didn't know much about the Russians, except that they were the bad guys, and that when you didn't like someone in our neighborhood, that's what you were supposed to call them, as in "You jive Russian!" The adjective "jive" was totally redundant—no kid ever called someone a Russian without calling him jive. I think every kid on Sixth Street was called a "jive Russian" at some time or other.

Oh, yeah—the science fair. In Flint, in 1963, the science fair was a really big deal. Every student was strongly encouraged to participate. Even fourth-graders like me. Even the kids in Mr. Mullally's class. Mr. Mullally taught special education, but most kids simply called it "special class." Special class included learning-disabled kids and kids who exhibited behavioral problems. Black kids seemed to comprise a disproportionate share of special-class students, including some neighborhood kids who seemed pretty normal to me.

The science fair was supposed to teach kids something called the scientific method. Basically, that meant you had to (a) pose a question, (b) form a hypothesis, (c) design an experiment to test the hypothesis, (d) perform the experiment to prove or disprove the hypothesis, and (e) draw conclusions. For example, your question might be "Do plants grow better in blue light or yellow light?" You might hypothesize that plants grow better in yellow light, and you would design an experiment to test your hypothesis. The plant/light experiment was probably the most popular (and most basic) science fair project ever, the kind of thing you would find in chapter 1 of a science book.

The project I remember best from the 1963 Flint Area Science Fair, other than my own, was Frank Stank's. Frank was in the special class, and he was called Frank Stank by the neighborhood kids because, well, let's just say he had a hygiene problem. Aside from that issue, the dysfunction that landed Frank in special class seemed to be his propensity to precede every statement with the speech fragment "Ossa Jack man." For example, if Frank invited someone to the store, he would say, "*Ossa Jack man* let's go to the store." When Frank addressed me, he would say, "*Ossa Jack man* David." No one in the neighborhood knew where the phrase "Ossa Jack man" came from or why Frank used it. Over time we just accepted it. Sometimes we jokingly used it ourselves.

Frank's science fair question was "Does Coca-Cola rot your teeth?" His hypothesis was "Coca Cola rots your teeth." For his experiment, Frank got some previously owned teeth from a local dentist. He drilled a hole through each tooth and ran a wire through several of them to form something that looked like a cannibal's necklace. He made two of these contraptions. He put one tooth necklace in a cup of water, and the other in a cup of Coca-Cola. Occasionally, he would shake each cup to make the liquid slosh over all the teeth.

Well, I wasn't going to be outdone by anyone, least of all someone in the special class. I thought all this scientific method stuff was pretty obvious and dull, intended for kids who couldn't come up with their own project ideas. I had my own thoughts about what would make a great science fair project.

I had been experimenting with a new gadget called a photocell. My big brother, Fred, introduced me to photocells only a few weeks before science fair time rolled around, and I thought they were really cool. A photocell's intriguing property was that it generated an electric current when exposed to light. The current could be used to control a relay or a motor or some other electrical device.

My idea for the perfect science fair project was to build a light-activated robot. I was sure that would dazzle everyone at the science fair and prove I was the undisputed King of All Things Scientific. I shopped at Radio Tube Merchandising Company to purchase a photocell and the other electronic bits—transistors, resistors, and such—needed to control the robot's motor. Then I visited a local supplier and bought sheets of aluminum to construct the body. For the next few weeks, I spent every spare moment building my robot and designing and testing its control circuitry. All this activity kept me up well into the night. Fred gave me electronic circuit design tips and helped me put everything together.

The night before the science fair I stayed up until 2:00 a.m. finishing my robot. It was a gleaming silver thing that resembled a stack of bean cans with the labels removed. An aluminum cylinder served as the torso. Two smaller cylinders were the legs. Atop it all, an *actual* pork 'n' beans can with the label removed was my robot's head. The photocell was located in the center of the robot's back. It connected to circuitry and a battery inside the torso, and wires ran down one of the legs to an electric motor at the base. The legs stood on the rectangular base, which had a wheel at each corner. The electric motor drove the rear wheels.

I was bleary-eyed and beyond tired as I applied the last pieces of duct tape to seal everything together. Fred stayed up to help (he was a real night person anyway) and we finally got the thing done. He got me a flashlight and I pointed it at the photocell. The robot lurched forward a few inches, got caught on a rug, and fell over. I was ecstatic, certain that science fair victory would be mine!

Kids from all over Flint and many surrounding towns participated in the science fair, which was staged at the huge Industrial Mutual Association (IMA) Auditorium. Each entrant had a "booth"—actually a table space about four feet wide and three feet deep—for his or her project. Contestants were supposed to assemble a display that explained their project, plus any apparatus (in my case, the robot), and a written report. I had spent so much time on the robot itself that my display was really skimpy—basically a folded piece of cardboard with papers attached. My report was skimpy too, but basically told it like it was: "Here is the world's first light-controlled robot!" What more did

I need to say? I finished setting up my booth in no time and was ready to return home when some kids from my class stopped by to look at my robot. They were impressed. So were the teachers who stopped by. No one had ever heard of a fourth-grader building a real working robot. I was certain that I was a shoo-in for first prize.

Flint Area Science Fair winners were accorded celebrity status, especially in the junior high school and high school divisions. Finalists were notified by phone, and were driven to the IMA Auditorium in the pride of Flint's Buick Motor Division—a shiny new Electra 225. A placard on each side of the car proclaimed "Science Fair Winner," and kids in the winner's neighborhood would point and gawk as the car went by. The top winners' photos even appeared on the front page of the *Flint Journal*.

Winners in the elementary school division were not celebrated quite as much. In our age category, winners were broken down into first, second, third, and fourth divisions. Fifty winners filled each of the first three divisions. Fourth division included everyone who didn't make the first three. Basically, fourth division meant, "Thanks for coming." All winners received a ribbon—blue for First Division, red for Second Division, and so on—and blue ribbon winners had their names printed in the *Journal*.

I attended the awards program with my mother and my brother Fred. Blue ribbon winners were announced from the stage, and I listened intently as the names were called in alphabetical order. When the T's passed without my name being mentioned, my heart sank. It had to be a mistake. Emilie won a blue ribbon with her plant/light experiment. *Surely that couldn't beat the world's first light-controlled robot,* I thought. Adding insult to injury, Frank Stank's tooth-decay project won a blue ribbon. I thought I would die.

After the program, I rushed over to my booth to see what had happened. Maybe someone had stolen my robot. Maybe there had been a misunderstanding, or maybe the judges had been confused about what age category my project was supposed to be in. Maybe the announcer had overlooked my name. When I got to my booth, I found a white ribbon taped to my glorious robot. My project was in third division.

I was inconsolable. No one had a project as complex as mine! I didn't even *see* another project that involved electronics in the elementary school category. Surely something was wrong. Fred theorized that the judges didn't believe a fourth-grader could build a working robot. That made me feel even worse—I was being penalized for being *too good!*

One by one, classmates came by my booth to see how my robot had done. Most brought their awards with them—a first-place ribbon here, a second-place ribbon there. Some teased me about my third-division finish. That was payback, because in the days leading up to the science fair I had acted as if my project was superior to all the others. None of my friends could say with a straight face that their project was better than mine. Some, like Mike (Dookey-Hookey) Thompson, echoed the opinion that I couldn't have built the robot by myself, and therefore deserved third division. That was bogus, because I had received less help on my project than most other kids had. Even kids who did the "What type of light makes plants grow best?" experiment received a lot of help from teachers and family members. The whole *idea* for Frank Stank's project probably came from Mr. Mullaly. No, there had to be some other reason for my disappointing finish.

The judges were circulating through the exhibit hall, and one of them stopped in front of my booth to talk with me. He said, "That's a nice robot, son."

I was *really* confused.

"Yes, I spent a lot of time building it," I said, hopefully. "It uses a photocell to switch itself on and off."

The judge looked closely at the back of the robot. *Now maybe we'll get somewhere*, I thought. *Maybe the judges missed the whole point of my exhibit. Maybe they thought I just produced a stack of cans and called it a robot. Maybe they didn't know about the photocell and the circuitry and the motors and all the things that made my robot really special.*

"Yes, I know son. You did a really good job putting this together. I haven't seen anything like it before."

I was biting my lip, but I just had to ask the question. I ended up blurting it out: *"Then why did I only get third division?"*

My words came out a lot more forcefully than I intended, and the judge seemed taken aback. He collected himself and said, "You know, what you have done here is to build a snazzy robot, but you didn't use the scientific method. You didn't pose a hypothesis and then set out to prove or disprove it. You just made a robot. It's a very nice robot, and not many people could build something like this, but that's not the point. Next time, you should be sure to use the scientific method in your project. If you do, I'm sure you'll do much better."

I felt pretty bad, and I still didn't believe the judge. Back at home, Fred and I stayed up late into the night talking about it.

"They just didn't believe I built the robot," I muttered.

"Yeah, that's right!" Fred said.

We watched TV for a while and went to bed.

3. 1963

I started school in kindergarten, but my *education* really began in fourth grade. That's when I started to understand that a world existed beyond my Sixth Street cocoon, a world where things didn't always go as I expected. My bitter science fair loss was just one example of the awakening I experienced during that school year.

My fourth-grade teacher, Miss Hamilton, was known as the toughest teacher at Walker Elementary School. She was an Air Force Reserve officer and she looked every bit the part. Her deep chestnut complexion, her taut, erect frame, and her crisp, no-nonsense voice created an imposing classroom presence. Some days she wore her air force uniform to class, and on those occasions she was even more intimidating.

LIEUTENANT COLONEL THERESA HAMILTON.

Miss Hamilton didn't take any "stuff" from any student, and she didn't hesitate to use her thick ruler or her big wooden paddle to quell any insubordination. She brandished the ruler for minor offenses like talking during class or using a slang term such as "ain't" (for "am not") or "norf" (for "north"). Miss Hamilton would call the offending student to the front of the class and whack the ruler against the kid's open palm. The big paddle was for more serious offenses like talking back or sticking chewing gum on a desk. Some teachers sent students to the principal's office to be disciplined. Not Miss Hamilton. She handled such matters on the spot.

Even before I entered fourth grade, the older neighborhood kids had passed The Legend of Miss Hamilton down to my friends and me. Their Miss Hamilton stories were laced with the same fear and reverence accorded the hermit at Mott Camp. Both Miss Hamilton and the mysterious hermit, the older kids said, were to be avoided at all costs. My friends and I were terrified at the prospect of being assigned to her class. Unfortunately, that is exactly where I found myself in the autumn of 1962.

Miss Hamilton's classroom had quite a mix of students. There were black kids from my neighborhood and white kids from adjoining ones. That was one aspect of the mix. The other was that our class consisted of fourth-, fifth-, and sixth-graders, all in one large room with retractable dividers. On the first day of class, Miss Hamilton said we were part of an experiment, and that being selected for such a special class was an honor. It might have seemed like an honor to us fourth-graders, but I wondered how the sixth-graders felt about being lumped in with younger students. In any case, the "honor" of being in an experimental classroom didn't relieve the terror of being subjected to Miss Hamilton.

By the first day of school, I had formulated a strategy for surviving Miss Hamilton's class, and I executed that strategy to a fault. I did so out of fear as much as a desire to excel. I was attentive, always did my homework, made sure to use "proper English," and participated enthusiastically in class discussions. That wasn't much of a stretch, because my parents and my grandmother had always expected that behavior. When Miss Hamilton asked our class a question, my attention-craving hand was usually one of the first to shoot into the air. I was an eager beaver, wanting to show Miss Hamilton how smart I was, wanting to impress her most. As a result, I was among the elite group who rarely experienced Miss Hamilton's ruler and never felt her paddle.

Before fourth grade, I hadn't paid much attention to the white kids at our school. They had always been there, but none of them lived on The Block, so we had rarely interacted. In fourth grade, though, extracurricular activities began. One such activity was a weekly after-school roller skating party in the gym, open to kids in grades four through six. We skated freestyle to the latest pop music, clowning and showing off our moves, except during "moonlight drive" songs. During those songs the DJ dimmed the regular gym lights, turned on a few colored spotlights, and encouraged boys and girls to hold hands and skate together. Each party featured several moonlight drive songs, and I took advantage of each one to get close to the girls in my class. I skated with the black girls from my neighborhood, like Shirley Hampton and Rhonda Wilson and Cathy Peaks, but I also skated with the white girls, like Emilie Hauser and Becky Simon and Martha Cross. I especially liked skating with Martha, because I had a crush on her. Most boys from my neighborhood didn't skate with the white girls. Some simply sat out the moonlight drive songs.

Instrumental music class began in fourth grade, providing yet more opportunities for getting to know kids from other neighborhoods. The music teacher was Mr. Swindell, a thirtyish man with dark skin and wavy black hair. Mr. Swindell was always well dressed—he wore neatly pressed dress slacks, a crisp white shirt, and a nice tie every day. He wasn't as strict or as imposing as Miss Hamilton, but he tolerated no fooling around. Mr. Swindell started me on trumpet, but the small mouthpiece didn't agree with my big lips. He switched me to trombone, and I took to it immediately.

Many students didn't take music class because of the practice time required, or because their parents had to pay to rent an instrument, or because they simply weren't interested. To reach a wider group, Mr. Swindell started the Bell Ringers Club. Bell Ringers required no practice outside club meetings, no fees, and no talent other than to be able to ring a hand bell at the appropriate time. I enjoyed Bell Ringers almost as much as instrumental music class because the group was larger and more varied.

The science fair wound up being my favorite and most time-consuming fourth-grade activity. I thoroughly enjoyed designing and building my project. Even though my robot didn't do nearly as well as I had hoped, I got to know my classmates on yet another level and to participate for the first time in a city-wide event.

During fourth grade, I made friends with some of the white kids in my class. By the time the science fair ended, I had a variety of friends at school—black and white, boys and girls. One of my best friends outside the neighborhood was Jimmy Bolinsky. Jimmy had lots of good toys, and we enjoyed playing with them together. I often visited his house after school, and often after we played for a while, his mom served us spaghetti and meatballs.

A few weeks after the science fair ended, we had "autograph day" in Miss Hamilton's classroom. Each student brought an autograph book, and all the other kids were supposed to write a short message and sign it. Some of my classmates' autograph books were plain dime-store pads, but mine was different. It had white vinyl covers, and the word "Autographs" was printed in gold on the front. The pages alternated in color—green, blue, yellow, pink, and so on. When Miss Hamilton gave the signal, we passed our autograph books around the room. At the end of the exercise, each student's book was full of short messages and signatures from classmates. As soon as my book came back to me, I carefully scanned each page. I couldn't help smiling as I read the first few messages:

"To David: A very smart boy. Karen"
"To David: Thanks for being so nice. Emilie"
"To David: You're kind of cute. Stay nice. Shirley"
"To David: I like to skate with you! Martha"

I kept flipping through the book and smiling to myself until I reached a page that contained an unexpected message. The page was pink, and the message on it was written in large, uneven letters. It looked more like a ransom note than an autograph. The message said:

"To David: A boy me and Reggie don't like. Mike"

I snatched my eyes from the offensive message and looked across the classroom at Mike. He was snickering and pointing in my direction. Then I looked over at Reggie. He seemed oblivious to Mike's shenanigans.

I had never cared much for Mike. He didn't live in our immediate neighborhood, and he wasn't popular at school. Still, it bothered me that he would record his dislike for me in my autograph book. It was unnecessarily cruel. It bothered me even more that he would say Reggie didn't like me, because I *did* care how Reggie felt. Reggie lived in my neighborhood—his house was on Seventh Street, and mine was on Sixth. Reggie was athletic and popular, so adding his name gave more weight to Mike's insult. It seemed as if Mike was saying, "A lot of us don't like you!"

The more I thought about Mike's message, the more it bothered me. *Why doesn't Mike like me?* That question pounded inside my head for the rest of the day.

After a few days, I stopped worrying about Mike and his cruel message. I convinced myself that he spoke only for himself, and I didn't care if he liked me or not. Then, a few days later, I was unexpectedly confronted by another kid in our class—a kid named Clyde.

I had hardly ever spoken to Clyde, and didn't know much about him. I knew he lived around the block, on Fifth Street. Clyde hardly ever spoke in class, and was one of the few kids who hadn't signed my autograph book. We were walking to gym class when Clyde bumped my shoulder and said, "Hey, punk." I turned to look at him. Clyde was a few inches shorter than me, but that didn't make him any less menacing. His skin was black as night, and an ugly scowl was etched on his face. Clyde said, "You think you somethin', don't you? Don't let me catch you walkin' home by yourself, punk. I'm gonna beat the mess outta you!"

I was terrified. I had never been in a fight. I had played with the kids on Sixth Street all my life and had never even *seen* a fight. Now Clyde was threatening to beat me up, and I didn't know what to do. I couldn't tell Miss Hamilton, because then I really *would* be a punk. I couldn't tell my folks, either. I certainly couldn't expect them to visit Clyde's parents and tell them I was afraid of their son. My folks didn't even *know* Clyde's parents, and anyway, parents didn't get involved in that kind of stuff.

As I tried to make sense of what was happening, I realized that Mike and Clyde didn't see me as one of their group, one of the "gang." Sure, we were all black, and we shared some neighborhood friends, but I acted different. To them, I was a teacher-pleasing, trombone-playing, bell-ringing, white-girl-liking, science-loving traitor. I didn't know what to do. I couldn't change who I was, so I just did my best to avoid Mike and especially Clyde.

A few weeks later, the school year was drawing to a close and I was walking home along Avon Street with my friend Wally Ross. Like me, Wally was a good student. He was tall and handsome, with light brown skin and wavy black hair. His father owned a construction business, his mother was a librarian, and they lived in a nice house on Fifth Street. Although both Wally and Clyde were black, and both lived just one street over, they seemed about as different as two kids could be.

I was enjoying the walk home, and my worries about Clyde's threats had subsided. I had convinced myself that Clyde had forgotten his promise to beat me up, or that he hadn't been serious, or that perhaps he had just been having a bad day. Wally and I were laughing and talking as we crossed Court Street, only a block or so from home. Just then, Clyde appeared out of nowhere and blocked the sidewalk in front of us.

Clyde sneered, "Come on, punk. I told you I was gonna beat you up, didn't I? I'm gonna beat your punk friend up, too!"

Before I could react, Wally caught Clyde with a punch in the gut. When Clyde doubled over, Wally hit him in the head, and then proceeded to punch him all over—chest, back, shoulder, jaw—wherever he could hit him. While Wally was punching, he was taunting Clyde: "You little black punk! You think you can beat me up? Take that home to your mama!"

I had never seen that side of Wally. He had always seemed like a mild-mannered guy, but in that moment he seemed possessed. And Clyde! He just stood there and took his beating. When it was over, he ran away and slithered around the corner at Fifth Street.

Wally asked, "Why did you let that little punk talk to you like that?" I didn't have an answer.

I said goodbye to Wally at the corner of Fifth and Avon. I was ashamed that I had let him fight my battle for me, but I was relieved, too. Clyde didn't bother me again, and before I knew it, the school year was over.

A few weeks later, during summer vacation, I was up the street from my house playing marbles with my friends. We were playing on the packed dirt in the front yard of the apartment house where a kid named Charles lived. I got into an argument with Charles during the game, and he said the dreaded words:

"I'm gonna beat you up."

I said, "I don't want to fight you, Charles," and ran home. My neighbors, the Thompson brothers, followed me. They couldn't believe what they had heard.

Mike Thompson, AKA Dookey-Hookey, was the first to speak up: "You gonna let him talk to you like that, Turtle? You better git over there and kick his butt! I'm gonna tell Freddy what you did—he'll make you go and fight!" (The kids on Sixth Street called me Turtle because I was the slowest runner. They called Mike Thompson Dookey-Hookey for a reason I never understood.)

Fred heard the commotion and came out onto the front stoop where Dookey-Hookey gave him the news: "Your brother is a punk, Freddy! Charles made him run home and hide! Tell him to go back up there and fight!"

Fred looked at me, and right away he could see I was scared. He didn't hesitate to respond.

"Go on up there and fight. You can beat that skinny Charles. I'll go with you to make sure nobody else jumps in."

At that point, Dookey-Hookey said, "Yeah, we'll go too, Turtle. Just go up there and kick his butt."

I had no choice. I had to go back up the street and fight Charles. When we reached his apartment house, Charles was sitting on the front porch, his long legs propped up on the banister. He saw me coming and ran down into the yard.

"So, you came back!" Charles sneered. I got somethin' for you."

Charles pulled a small fingernail file from his pocket and brandished it like a knife. I hadn't expected him to have a weapon, but I couldn't back down. My brother said, "David, here!" and threw me a broom handle he had brought with him. I caught it and swung, catching Charles on the wrist. He dropped the nail file onto the hard dirt, and I dropped the broom handle and lit into him with my fists. In just a few moments the fight was over. Charles had a bloody lip, and he turned and stormed back to his porch. I took my time sauntering out of the yard, and then strolled triumphantly back to my house, Fred and the Thompson boys trailing me down the narrow sidewalk.

"That was good, Turtle," Dookey-Hookey said. "Don't you never run from no punk like that again or else I'll beat you up myself!"

Even though I had won the fight, I was nonetheless troubled by it. Did Charles resent me too, like Mike and Clyde? Had our fight really been about a marbles game, or was it about me being a "traitor," or a good student, or living in a nicer house?

That summer, I noticed that people on TV were constantly talking about "Negroes" and "race" and "civil rights." President Kennedy gave a big speech, and in it he said every child should have an equal chance at success. Then, in August, something called the March on Washington took place. A lot of people from Flint rode a bus more than 650 miles to participate. No one from my family went, but Marsha Taylor, my sister Bernice's friend, did. I was particularly struck when, watching the march on TV, I heard Dr. Martin Luther King say, "little black boys and black girls will be able to join hands with little white boys and white girls as sisters and brothers." That made me think about the after-school roller-skating parties at Walker School. In my mind's eye, I saw Martha Cross and me holding hands and skating around the Walker School gym. Then I thought about the times Mimama had taken me down South to see our relatives in Georgia, where there were signs that said colored and white on the drinking fountains and bathrooms we saw around town. The overtly segregated life in Georgia and the events I was witnessing on TV didn't seem to be a part of my world in Flint, yet for the first time I started to sense a kind of barrier between black and white, even in my town, even at my school.

School started again in September. My fifth-grade teacher was Mrs. Saltman, a graying white lady who seemed much older than Miss Hamilton. There was nothing intimidating about Mrs. Saltman, but she seemed less engaged and less accessible than Miss Hamilton had been.

A few days after the beginning of the school year, a church in Birmingham, Alabama, was bombed, and four little black girls were killed. That news hit close to home, because those girls were close to my age. That tragic bombing, and the images of the dead and injured children being pulled from the church rubble, really brought home the fact that there were some white people, perhaps a lot of white people, who didn't want little black boys holding hands with little white girls.

On November 22, I was sitting in class when someone came to the door and summoned Mrs. Saltman. She came back a few minutes later looking wounded. She said, "I've just been told that President Kennedy has been shot. I know we're not supposed to do this, but I think we need to pray." Mrs. Saltman seemed to be fighting back tears as she led us in a short prayer for President Kennedy and for the country. Then she said class was dismissed early and that we should all go home.

I was devastated. I really liked and admired President John F. Kennedy, because he had spoken out forcefully against segregation, and because he had seemed to support Dr. Martin Luther King. President Kennedy had also solved the Cuban missile crisis and thereby saved the world. I remembered how, just a year earlier, my friend Johnny Reaves's father had said, "If we go to war with Russia, we won't actually *go* anywhere. Each side will fire nuclear missiles at the other, and that will be the end of the world." The president had delivered us from that terror, and because of that, he was a hero to me and to many of my friends. Though he had loomed larger than life, he also had a tangible connection to Flint. When he was campaigning for state candidates that October, he gave a speech at the city hall rotunda, right at the end of Sixth Street.

The events of 1963 made me aware for the first time that the differences between black people and white people might go beyond skin color. I became aware that a bitter struggle was taking place between people who supported full rights for black people and those who opposed those rights. I realized that in the course of that struggle, some people were ready to kill, and others were ready to be killed. I didn't recognize evidence of the struggle in my neighborhood or in my school, but I knew it existed just the same, and I wondered when and how it would touch my life. I thought that maybe there really were two different worlds, and that I might need to be careful about straying too far from my own.

4. Radio

PORTABLE TRANSISTOR RADIOS WERE THE RAGE in the early 1960s, and they were everywhere—perched on desktops, built into car dashboards, carried in shirt pockets. The compact, battery-powered "solid-state" gadgets replaced vacuum-tube radios, which could be bigger than a breadbox and needed to be plugged into a wall socket. Fred kept up with and sometimes led the latest technological fads in our neighborhood, so it was no surprise when I saw him and his friend Leroy Nesbit spend most of a Saturday afternoon building their own solid-state crystal radio. Hobbyists had been building "crystal sets" practically since the late-1800s invention of radio, but the "crystal" in Fred's and Leroy's radio—a tiny germanium semiconductor diode—was a brand-new product of the transistor age.

Fred and Leroy had obtained their components—the diode, a coil, a variable capacitor, and an earphone—from the parts trove my father kept in our basement. They wired the components together and then connected the "ground wire" to a cold-water pipe in my sister's bedroom. They took turns listening to the earphone and playing with the capacitor to see if they could pull in WAMM, Flint's Top Forty AM station. I watched them grow more and more excited as they heard static coming from their little radio. When Fred popped his fingers and started singing, I knew they had succeeded. Fred and Leroy had made their little radio with their bare hands, and it actually worked. They were just sixteen years old, and I was ten. I had no idea where they got the knowledge to make their own radio, but at that moment they were my heroes.

Fred's interest in electronics was inspired by my dad—not directly, but by osmosis. My father had returned from his U.S. Army Signal Corps service in the Aleutian Islands passionate about radio and TV and electronics in general. He never found a job in the electronics field—he spent most of his working life as a maintenance technician at the post office—but he created his own electronics workshop in our basement. He had a side business repairing radios and TVs, and he had all manner of gadgets to help in his work. He collected stacks of electronics books and subscribed to all the popular electronics magazines. Then there were the parts. My father had electronics parts in drawers, on his worktable, and stuffed into stacks of El Producto and Dutch Masters and Banker's Choice cigar boxes. My father's workshop was his haven, his personal electronics laboratory.

A couple of years after Fred and Leroy's radio breakthrough, I noticed my father fiddling with a device I had never seen. It looked like a reel-to-reel tape recorder inside an army-green metal box with reinforced corners. Instead of magnetic recording tape, this contraption's reels contained what looked like white paper tape, which was imprinted with a series of black marks. When my father turned the machine on and the tape started to feed from reel to reel, a series of long and short beeps began to play. My father would listen intently to the machine, and as he listened he would write down a sequence of letters. CHDIL BCRTD HJPLM ASERV NBRDY. The machine made no sense to me, and the letters even less. I asked my father what he was doing, and he said he was practicing his Morse code. He explained that Morse code was how many people communicated with each other by radio, and that he had used "the code" when he was in the army.

I was fascinated. My father went on to explain that people could actually send and receive messages over the radio. He said radio wasn't just something you listened to; radio could be used for two-way communication. What's more, he explained, I could learn to be a radio operator myself and communicate with people all over the world. He explained that there was something called amateur radio—ham radio—and anyone could get a license to operate his own radio station.

The term "ham" had originated in the early 1900s as a way commercial radio operators referred to amateurs. It was meant as a pejorative term, but amateur operators took it on as a badge of honor, and the label stuck.

The idea that I could operate my own radio station seemed too good to be true. Fred and Leroy had been fascinated just to *receive* a signal on their little crystal radio. Now my father was telling me that I could operate my own radio station to both send *and* receive messages. I immediately plunged into the ham radio world. I started using my father's Morse code practice machine. I studied his theory books to learn the technical side of amateur radio. When I felt ready, I arranged to take the Novice Class amateur radio exam.

The Federal Communications Commission (FCC) was responsible for amateur radio licensing and regulation. The FCC administered the exams for most amateur radio licenses, but allowed third parties to administer the Novice Class exam. I arranged to take the exam at a local radio supply store, Shand Electronics. My father had sometimes taken me there when he was shopping for radio or TV parts. Shand was staffed by white men who were roughly my father's age and who had never seemed particularly friendly. My nascent racial awareness had me wondering if they would be supportive, but my concerns were unfounded. When the Shand guys learned I wanted to take the ham radio exam, they treated me like a valued apprentice.

The Novice Class exam consisted of two parts, a Morse code proficiency test and a technical examination. I passed both on my first try, and a few weeks later I received my FCC license in the mail. At thirteen years of age, I became a licensed ham radio operator.

I was itching to get on the air. First, though, I had to get some equipment and set up my radio station. I saw a listing for a used ham station in the *Flint Journal* classified ads. I called and made arrangements to see the equipment, and my mom took me to see the seller. He was a kid in his late teens, and his house was not far from Hurley Hospital where my mother worked. Although the kid was white, he didn't seem surprised that a black kid was interested in amateur radio. In fact, like the Shand guys, he seemed happy that I was joining the ham fraternity and that he could play some part in my introduction to it.

The kid seemed totally absorbed by his station equipment, and he painstakingly explained each component to me. He had built the Heathkit transmitter himself and had saved all the manuals and schematic diagrams. He said he was selling his gear because he had recently earned his General Class ham license and was going to buy a new, more powerful station. He seemed reluctant to part with his old station, though—especially his classic U.S. Army-issue J-38 telegraph key. It was mounted on a heavy brass base so it wouldn't move around during use, and to me it looked like a work of art. Even before the kid demonstrated his station, I was sold. I bought the gear and was in business. I was the only black ham radio operator on Sixth Street. I was the only black ham I knew.

The first order of business after setting up my second-hand radio gear was to erect an antenna so I could begin communicating with the world. My first antenna was called a folded dipole. It was simply a wire that ran from a tree across the driveway to the house, and then to another tree farther up the driveway. I connected one end of a feeder wire to the approximate middle of the antenna, and connected the other end to my transmitter and receiver. I was ready to go on the air.

My hands were clammy as I tuned the receiver to a quiet part of the radio band. I had listened to hams communicating for several months, so I knew the procedure, but there was nothing like actually communicating with someone. After tuning the transmitter to maximize power output, I began to nervously tap out "CQ, CQ," followed by my station's call letters, on the telegraph key. "CQ" is universal shorthand for "calling all stations," and it's what hams send when they are searching for someone to communicate with.

I listened intently after each round of CQs, but there was no response. Through noise and static, I heard other operators chattering back and forth in Morse code, but none of the chatter was intended for me. I went to another frequency and tried again. After what seemed like an hour of trying, I heard a faint signal echoing my call letters. I had made contact with another human being over the radio! It is hard to relate the joy I felt. Suffice to say that it was a moment I will never forget. We exchanged a few short messages about our respective locations, the weather, and our station equipment, and then we logged off. I was in ham heaven. I couldn't imagine being happier. I had reached the end of a long journey that began with seeing my father's weird green Morse code practice machine. I was thrilled, but didn't want to do any more communicating just yet—I wanted to savor the moment. I walked across my bedroom and flopped onto the bed. I drifted off to sleep while gazing admiringly at my second-hand ham radio station.

In the ensuing months, I communicated with lots of different hams via Morse code. The "conversations" usually covered the same topics—signal strength, location, weather, station equipment, antenna. Sometimes, though, I would get into extended "talks" with my fellow hams. I would tell them my age, what I was studying in school, and what I thought about current events. Those were some of my favorite conversations, because they provided a window into the feelings of my distant acquaintances. I wasn't just interested in technical discussions about signal strength and equipment specifications. I wanted to use ham radio to get to *know* people. I quickly discovered that, in some ways, you could get to know people better over the radio, because the conversations weren't colored by personal characteristics like race or gender or ethnicity. There was nothing in the dots and dashes of Morse code to distinguish a white grandmother from a black teenager. The anonymity was fascinating and liberating.

One rainy afternoon, I got into a Morse code conversation that lasted more than an hour. My contact and I "talked" about ourselves, our families, our towns, and the state of the world. Afterward, I received a letter saying that he had nominated me for the Rag Chewer's Club, a group that encouraged long over-the-air conversations. Rag Chewers believed that too many hams were only interested in making brief contacts with as many other hams as they could, rather than communicating in depth with their fellow hobbyists. Given my penchant for in-depth communication, I certainly understood the Rag Chewers' point of view. I was honored to be nominated, and a few weeks later I received my club certificate in the mail. I framed it and proudly displayed it on the wall above my station equipment. I felt like a member of a unique, close-knit fraternity.

During the next few years, I continued to advance in amateur radio. I got my General Class license, which allowed me to graduate from Morse code to voice communication. I bought a new transmitter and receiver from Shand Electronics. I bought a fancy Hy-Gain vertical antenna and erected it in the backyard. I talked with hams all over the world, and sometimes other hams told me my signal came in clear as a bell. For some reason, my signal was particularly strong in Managua, Nicaragua, and I enjoyed a long conversation with an American expatriate there.

I had great ham experiences, but none matched the excitement of setting up my first station and establishing that first contact. Nothing made me feel so much a part of the ham community as receiving the Rag Chewer's Club membership. I was welcomed into a community of highly technical people simply on the basis of my proficiency and participation. Most of my fellow hams didn't know my skin color. Among those who did, the fact that I was a ham seemed to loom larger than my race. At a time when blacks were struggling just to be able to eat at lunch counters in the U.S., I was a welcome participant in a worldwide, radio-based cyber-community.

Even though my father introduced me to ham radio, even though I used his curious green machine to learn Morse code, even though I used his radio theory books to study for the license exam, my father never saw me communicating with other hams. He didn't drive me to Detroit to take the General Class exam—my mother did that. Most curiously, my father never got any kind of radio license for himself. I never asked why. I suppose I was too wrapped up in my own pursuits to care. I always saw my father studying his radio books, so I figured he was up to something that would someday come to fruition. I also figured that, with his Signal Corps experience and vast knowledge, my father may have felt that amateur radio was beneath him. He had books about getting a commercial radio license, and I think that, if anything, was his desire. Despite my speculation, the relationship between my father and me (indeed, my father and most people) was such that we never had an in-depth conversation about his past, his desires, or his plans.

5. Kid Business

I DON'T REMEMBER WHEN OR HOW I DISCOVERED MONEY, but I learned to *acquire* money when I was just ten years old. That knowledge came from a variety of sources. Comic books advertised business opportunities like selling *Grit* newspaper. Neighbors offered money in return for doing household chores. Older kids handed down time-honored ways of making money, like stuffing ads into the Sunday edition of the *Flint Journal* or selling ice cream from a pushcart loaded with dry ice.

I engaged in several kid businesses. I wanted my own money to buy candy and soda pop, to save for my next bicycle, and to buy the latest gadgets. I mowed lawns in the summer and shoveled sidewalks in the winter. I sold Wallace Brown greeting cards and Burpee seeds and homemade potholders to family and neighbors. I hired a couple of neighborhood kids and started a bike repair shop in our garage. The money was nice, but I also found that I enjoyed commerce. I got a kick out of the idea that someone would actually pay me cash for goods or services. Seeing my enthusiasm for business, the Shaw boys, neighbors on the other side of Sixth Street, accused me of being "money hungry." That wasn't necessarily the case, but I admit the business bug bit me hard. I liked the feeling of power, the independence, and the ability to do things on my own.

One of the most popular kid businesses was collecting soft-drink bottles—we called them pop bottles. The stores charged a two-cent deposit for a small bottle and five cents for a large one. They refunded the deposit for every returned bottle, and didn't care who returned it—if you showed up with it, you got the refund. It was a great business opportunity for kids on The Block. We would knock on every door in the neighborhood and ask, "Got any pop bottles you don't want?" Most people didn't want the hassle of returning bottles to the store and were happy to have someone else do it.

One could easily collect a hundred bottles in a few hours, yielding a few bucks for candy, chips, and soda. My friends and I would often collect bottles in the afternoon, rendezvous at Waller's Drug Store to return them and collect our money, buy all sorts of goodies, and then go to my house for a backyard party. We would feast on Cokes and chips and Twinkies and laugh and talk about the day's exploits.

Bottle collecting was a great neighborhood business until my friend Johnny Reaves and his father got involved. Johnny lived just around the corner on Avon Street, between Fifth and Sixth. I simultaneously liked and loathed Johnny. I liked him because he was smart and was usually fun to hang around with, and because he had a nice house and friendly parents. I loathed Johnny because he was a "golden child." He had beautiful nut-brown skin and thick, curly black hair—what we called "good hair," because it wasn't thick and nappy like mine. All the neighborhood girls seemed to love Johnny, and when they saw the two of us together they would rush to him and pretty much ignore me.

Johnny seemed to live on a level above the rest of us kids. His father was handsome and articulate, his mother poised and beautiful. Johnny's folks seemed almost regal, and Johnny was like a young prince. His mother was exotic looking—Johnny said she was part American Indian. She had a reddish-brown complexion and long, straight black hair.

Politically, Johnny's folks were of a different sort. During the 1960 presidential election campaign, Johnny's father favored Richard Nixon when every other black parent in the neighborhood was for John F. Kennedy. Johnny himself paraded to and from Walker School professing his support for Nixon. That really irritated me, but upset as I was about Johnny's perfect life and his black Republican father and his movie-star-like mother, nothing upset me more than his bottle-collecting acumen.

Johnny wasn't satisfied to be an average player in the bottle business. He wanted *all* the bottles, and hence all the money, and didn't care if that meant the rest of us got nothing. With his father's help, Johnny set out to dominate the business. He did it with marketing.

Johnny's father printed up some business cards for him. The cards said:

Johnny Reaves
Professional Bottle Collection
"Get rid of the smells and the bugs—
let me take your bottles."
Call 555-5124 for prompt service.

The rest of us kids had never seen a business card before. We had never thought about advertising our services or talking about the benefits we were providing. We were essentially just begging. "Please give us your empty bottles so we can buy some candy" was the essence of our sales pitch. Johnny and his father turned that whole picture around. They presented Johnny as providing an important benefit. Johnny wasn't just begging for your bottles. No, Johnny was taking those smelly, bug-infested bottles off your hands, and with them he was removing the threat of pestilence. Johnny was doing you a big favor.

Johnny's pitch didn't end with his business card. His father taught him a little speech to use at each house he visited. The speech was craftily designed to close the deal. Johnny told each potential customer why they should call him instead of the other neighborhood vagabonds—us.

After a few weeks, Johnny's strategy began to take hold. I would arrive at a house and ask for pop bottles, and the owner would tell me they had already arranged for someone to pick them up. That "someone" was invariably Johnny. He had the whole frigging neighborhood under contract! I was able to collect bottles from a few houses here and there, but for the most part, Johnny had sewn up the business. For me and my other friends, bottle collecting was no longer a lucrative activity.

My buddies and I didn't know it, but we were entrepreneurs. If we *had* known it, we wouldn't have considered it anything special. We wanted to be like our parents, who had jobs—regular work with good, steady pay. They didn't have to worry about a shark like Johnny Reaves ripping away their livelihood. That's why when a real job opportunity presented itself, I jumped on it. The opportunity came in the form of a *Flint Journal* paper route.

My first route had only about twenty-five customers. I think the *Journal* considered it a training route, but I treated it seriously. My delivery bag had *Flint Journal* emblazoned in large letters, and I carried it proudly. I delivered my papers every day, and my customers paid me on time. I felt important as I punched each customer's subscription card to indicate that the bill was paid. One day, the folks at the *Journal* called with a proposition: A carrier in our neighborhood was giving up a huge ninety-two-customer route, and they wanted to know if I was interested in taking it over. Of course I was! The new route was nearly four times the size of my first one, and would nearly quadruple my earnings. I didn't mind that it would take a lot longer to deliver ninety-two papers, or that much of the route was in a not-so-nice area west of Lapeer Road. I regarded the new route as an opportunity I could not afford to pass up.

I left my old route just before Christmas, and was disappointed that so few customers gave me a Christmas bonus. I didn't mind, though, because I assumed I would get a Christmas bonus from many more customers on the new route. As it happened, the timing of my route change was such that the fellow who followed me on my old route got a holiday bonus from most of my old customers, while the guy who preceded me on the new route got the bonus from most of my new customers.

Despite that disappointment, I was excited to have a larger route and determined to do a good job. Soon after starting, I visited all the houses on my route to collect payment, and that was an eye-opening experience. I didn't know there were people in our area who didn't live in nice, neat houses like mine. I didn't know there were people who would do everything they could to avoid paying for something they had already agreed to buy. Some of the houses and apartments

I went into were an absolute mess. In a few cases, when the resident opened the door, a sickening smell hit me in the face. I hated going to those houses the most. I heard countless excuses as to why people couldn't pay for their paper. Everyone ended up paying, though, because they wanted their *Journal*, and if they didn't pay, delivery would stop. That paper was important to every customer on the route. If they couldn't find their newspaper, or if I delivered it late, they would call my house and complain. I was committed to delivering the paper on time to every customer, even the slow payers and the people with messy and smelly homes.

In the winter of 1967, that commitment was tested by an event I call the Big Storm. Every day that winter, a *Flint Journal* delivery truck dropped my newspaper stack in front of my house. Every day, that is, except January 27, 1967.

The Big Storm was the storm of the century, and no TV news hype announced its arrival. It just came. Snow fell so thick and fast that you could barely see your hand in front of your face. On the day of the Big Storm, I never gave a thought to canceling my newspaper deliveries. Neither did the *Flint Journal*. The paper called to say they would be delivering my newspapers not to the usual spot in front of my house but instead to the nearest intersection, Sixth Street and Avon. This was a minor inconvenience for me, but I understood because the snow was already pretty heavy and cars were having trouble traveling Sixth Street.

I took my red wagon to the corner and waited for my newspaper stack. The wagon was a prized possession. It had a sturdy steel undercarriage and big, businesslike wheels. It had a varnished pine carriage and removable red slats on the sides and back. It was the perfect vehicle for delivering the *Journal*, because I could get all ninety-two papers in there at once—even the Sunday edition. I stood at the corner and awaited my newspaper stack. It never came.

After a while, I went back to the house and called the folks at the *Journal*. "I don't know what happened to your stack," the dispatcher said. "Go back to the corner and we'll send the truck around again." After downing a cup of hot chocolate, I dutifully took my wagon back to the corner and waited for my stack. It never came. Now the snow was really getting quite heavy, and I was having trouble pulling my wagon through it. I thought *this is ridiculous!* I went back to the house and called the *Journal* again. For a third time, they said they would send my papers to the corner of Sixth and Avon. For a third time, I trudged back down there and waited. My stack never came.

I fought the snow all the way back to my house and called the *Journal* again. I was really peeved. "We have to try something else," I told the dispatcher. This just isn't working, and the snow is getting heavier and heavier." The dispatcher was ready with a solution. "Go to the other corner, at Sixth and Lapeer, and we'll drop your stack there." At that point, I was ready to try anything. Lapeer was a busier street than Avon, so it was probably plowed and passable. The downside was that Lapeer was farther from my house. I dragged my wagon through the drifting snow all the way up to Lapeer, but the struggle was worth it. There at the corner, wrapped in wildly fluttering orange cover sheets, was my stack of newspapers. Hallelujah!

I piled the papers into my wagon and set off on my route. By now, the snow was more than waist high, and I pretty much had to slide the wagon on top of it. The snow was blowing so hard I couldn't see more than a few feet. Snow and ice were caked inside my boots and my feet were freezing. It was tough, tough going, but I was determined to deliver those papers. In the end, I delivered them all, even though it took hours longer than usual. I stumbled home about 9:00 p.m., completely covered with snow and totally exhausted. My extremities were frozen and numb, and I was starving, but I was satisfied. If I could do my job on a day like this, I could do it on any day of any year.

As the calendar turned toward spring, snow from the Big Storm began to thaw. One day on my way to school, I noticed a piece of orange paper sticking through the melting snow at Sixth and Avon. Closer inspection revealed ink on newsprint. I got on my knees and started digging through the snow like an anxious archaeologist. There in the wet snow was a full stack of newspapers! I kept digging. I discovered another stack, and then another. The *Flint Journal* people had delivered my papers to Sixth and Avon, just as they had said... three times. The blowing, drifting snow had fallen so fast that it had covered each stack before I reached the corner.

Seeing those old, water-soaked stacks, I felt proud. On that winter day in 1967, the day of the worst snowstorm in Flint history, I learned something about myself. I wasn't thinking about my disappointing Christmas bonuses or my slow-paying customers. I would have been justified in staying home, but the adversity brought on by the Big Storm activated a deep inner impulse. My attitude was *I'm going to deliver these doggone papers, whatever it takes.* And I did.

I SURVIVED THE BIG STORM.
MY WAGON DID, TOO.

6. 1968

JULY 1967 CAME LESS THAN TWO MONTHS AFTER my fourteenth birthday. That month, Detroit was rocked by a devastating five-day riot that left forty-three people dead and thousands of buildings destroyed. Discontent in the black community had intensified to rage, then exploded in violence and ruin. In August, a few weeks after Detroit burned, my friend Kirk dropped by our house bursting with excitement. He extended an invitation that involved me in that summer of discontent and introduced me to the seismic changes the entire country was experiencing.

"Hey David," Kirk said. "There's gonna be a sleep-in at city hall tonight. A lot of people are gonna be there, and a whole lot of girls! All you need is a sleeping bag. You wanna go?"

I remembered seeing on the Channel 12 news that the Flint city commission had voted down something called an "open housing ordinance." Floyd McCree, Flint's first black mayor, was embarrassed by the vote and had threatened to resign. Civil rights activists organized a sleep-in on the city hall lawn to protest defeat of the ordinance and advocate its reconsideration.

I had never participated in a demonstration of any kind, and hadn't considered participating in the sleep-in. I loved my house and my neighborhood and had never thought about living anywhere else. I had never realized that blacks were restricted from living in certain areas of Flint, and that I was living in a segregated society. It wasn't something I pondered, it just was.

My church, Quinn Chapel A.M.E. (African Methodist Episcopal), was an all-black congregation. Our family doctor, Dr. Clarence Kimbrough, was black. Our dentist, Dr. Rex Weaver, was black. The drugstore over on Eighth Street was owned and run by a black pharmacist, Fred Waller. The weekly dances Kirk and I sometimes attended on the North Side at Father Blasko Hall were all-black affairs. My friends and I mowed lawns and went trick-or-treating in a white area on the other side of Burroughs Park, an area we called The Rich Neighborhood, because we realized that's where the money was, but we had no friends there. Charles Stewart Mott, one of the nation's wealthiest men, lived on an estate less than a mile from my house, but I wouldn't have recognized Mr. Mott if he walked up and shook my hand. Blacks and whites inhabited the same city and lived in adjacent neighborhoods but experienced little social contact. It was as if some invisible force separated the races, allowing us to coexist in close proximity without mixing, like oil and water.

MY MOTHER, GRANDMOTHER, AND ME AT QUINN CHAPEL A.M.E. CHURCH.

I definitely wanted to attend the sleep-in. The cause seemed worthy, and it sounded like fun. Kirk and I had done a lot of camping, mostly in my backyard, so we already owned sleeping bags. The idea of spending an entire night with a crowd of people on the city hall lawn was exciting, and the opportunity to socialize with girls at an all-night event was just too enticing to pass up.

I knew my father wouldn't want me to participate in any kind of demonstration, so I asked my mother: "Mama, there's going to be a sleep-in at city hall tonight for open housing. Kirk is going, and a lot of other people are going, too. Can I go?"

My mother was nothing if not civic minded. She acceded to my request.

"Just be careful, and come straight home when it's over," she said.

With that, Kirk and I grabbed our sleeping bags and ran the three blocks down Sixth Street to city hall. We arrived just before dark, and a large crowd had already assembled. We found an open patch of lawn and plopped our sleeping bags onto the ground.

As dusk approached, the program began. Several of the organizers gave rousing speeches. I had never been part of a crowd like this. The excitement and camaraderie were uplifting. Kirk and I met two sisters from the northwest part of town and convinced them to share our little camp. Together we cheered the speeches and chanted slogans. When the program was over, we sat and talked with our new friends until the wee hours. Finally, we climbed into our sleeping bags and drifted toward slumber, confident that we were witnessing the birth of a new world.

Next morning, we woke to chilly air and dew-covered grass. The electricity of the previous evening was gone, and participants were trickling away. It was time to go home and eat breakfast and sleep in a comfortable bed. Kirk and I gathered our sleeping bags, said goodbye to our prospective girlfriends, and trudged back up Sixth Street. We both wondered what civil rights adventures the coming days would bring.

The sleep-in proved successful. A few months later, the city commission passed the open housing ordinance. Black community leaders were ecstatic, but a backlash soon developed. A white citizens' group circulated a petition and forced a voter referendum, which took place in February 1968. At first, it looked as if the ordinance had failed, but a closer look by election officials revealed a "math error." The ordinance ended up passing by just forty-three votes out of some 40,000 cast. Once again, black citizens of Flint tasted victory. The front page of the *Flint Journal* carried a picture of a beaming Mayor McCree pointing to the election result.

I felt empowered by the sleep-in experience. The success of our protest told me that things really were changing. Mayor McCree had staked his political career on a critical issue and won. Black people could buy a house in any Flint neighborhood. Dr. Martin Luther King was leading the movement to affirm our rights on a national stage. I imagined that racial segregation would one day be a thing of the past.

Then came April 4, 1968.

When I first heard the TV news bulletin saying Dr. King had been shot dead in Memphis, my reaction wasn't disbelief. All I could think was *Oh no, not again!* When John F. Kennedy was killed in 1963 I was just ten years old, but that bullet made me realize that anyone could be killed at any time. Dr. King's murder left me shocked and numb and terribly sad. Mimama expressed the feeling best when, watching the TV coverage, she cried out, "Oh, Lord, why did they have to *kill* him?"

Then came June 5, 1968.

Bobby Kennedy was murdered while campaigning in California. I was still coming to terms with Dr. King's murder, trying to figure out what it meant for the civil rights movement, for black people, for me. After Bobby's murder, I could clearly see that the good people were being eliminated. First JFK, then MLK, then Bobby. Sinister forces seemed to be controlling the world. The exhilaration and hope I had felt after the open housing victory disintegrated, leaving only despair.

Then came October 1968.

It was a cold, gray day, and I was perched nervously beside my grandmother's bed at Hurley Hospital. An oxygen mask sat awkwardly over her nose and mouth, and with each breath she took I heard a faint "kuh... kuh... kuh." My mother had told me there was something seriously wrong with Mimama's heart. She said Mimama might die.

Mimama was nearly seventy-four. We were as close as two people could be. She had lived with our family since 1950, the year Grandpa Bill (William Wise Hayden Sr.) died. In 1953, soon after I was born, my mother told Mimama that if she would help take care of my siblings and me, she would have a home for life. Mimama took her up on the offer.

My mother worked third shift at Hurley Hospital, so she was rarely home when I awoke each morning. From the time I was a toddler, and well into grade school, Mimama was the person I sought at the start of the day. Rising from sleep, I would shout, "Mimama... Mimama" until she responded. Sometimes she would enter my room and sing, "Rise and shine, little David." Other times, I would hear her voice drifting upstairs from the kitchen: "Time for breakfast, David! I made you salmon croquettes and grits." She never had to call me twice. The breakfast aromas had me dashing downstairs within minutes. My day didn't start until I had my time with Mimama. She was my rock, my protector, my other mother.

Now Mimama lay teetering on the edge of life. I felt lost and alone, worried that my world was about to change.

I felt a hand gently squeezing my shoulder. It was Reverend Mitcham, the minister at our church. I was so lost in despair that I had forgotten he was in the room. From behind my chair, I heard him say, "Are you all right, son?"

"Yes, sir."

"I'm going to pray for your grandmother now. Will you join me?"

"Yes, sir."

Reverend Mitcham's gravelly voice was soft but firm. He was not one of those preachers given to sweeping oratory and theatrics. Reverend Mitcham was kind and steady, an earnest, devout minister of the gospel. He and his wife, Maymie, were among Mimama's closest friends.

"Dear Lord, if it be your will, bless Sister Hayden in this, her hour of need. End her suffering, Dear Lord, and if it be your will, return her to health and vitality so she may continue to be your kind and faithful servant. And Dear Lord, if it be your will to take Sister Hayden from us, surround her with your loving kindness and receive her into Heaven. These things we pray in Jesus' name and for his sake, Amen."

After Reverend Mitcham's prayer, I was even more afraid. I wasn't ready to lose Mimama. I didn't even want to acknowledge the possibility.

"Would you like me to give you a ride home, son?" Reverend Mitcham's words, though gently spoken, were a jolt of reality. There was a world outside that hospital room, and I had to return to it. Still,

I wasn't ready to leave Mimama's side. I was worried sick that I might never see her again.

"No, sir. I'm going to stay awhile. My mother will come and get me, or I'll walk."

"All right then, son. Keep praying for your grandmother. God bless you."

Out of the corner of my eye, I saw Reverend Mitcham step softly out of the room. Mimama and I were alone.

I took Mimama's hand and studied her face. Her skin was almost white, her face creased with wrinkles. Her gray hair was swept back from her forehead. Her eyes were closed, her expression peaceful and resigned.

I began to sob, and then I started to bargain with God. I prayed aloud:

"God, please make Mimama better. If you do, I will devote my life to doing good things for other people. God, please. She is a good person. She is so kind. I need her, God."

I didn't know if God had heard me.

Mimama didn't respond to my prayer. Her eyes remained closed, and the strange sounds continued to emanate from her throat: "kuh... kuh... kuh."

I sat there awhile longer, crying and praying. Finally, I decided to return to the world, and rose from the bedside chair to go home. I bent over the bedrail and kissed Mimama on the forehead.

"I love you, Mimama. I'll see you tomorrow."

No response.

I left the hospital and headed home on foot. Atwood Stadium was nearby, and I recalled prancing home during happier times with Fred after Flint Central football games. Today was different—I could barely put one foot in front of the other. I trudged all the way down Fifth Avenue to Saginaw Street. I turned right on Saginaw and headed downtown. Just north of downtown, I noticed the storefront that contained the local Hubert Humphrey presidential campaign office. A huge poster with Humphrey's picture was displayed in the window, and suddenly another large dose of despair was heaped upon my sadness. The Humphrey poster reminded me of the murders of Dr. Martin Luther King Jr. and Robert F. Kennedy. Those tragedies still hurt me to the core. I was overcome with grief. Dr. King was dead. Bobby was dead. Mimama was dying. And now this unexciting "substitute teacher" was running for president. It all seemed so unfair.

A few weeks later, Richard Nixon was elected president. I felt Nixon was far worse than Humphrey, because he and his supporters seemed uncommitted to the civil rights struggle. The "silent majority" Nixon often evoked seemed to codify people who wished to turn back the clock and stem the tide of black progress. I felt sad and defeated. The hope and empowerment I'd felt beginning with the open housing sleep-in were nearly gone.

I reached the lowest point in my young life, but then a miracle happened—Mimama recovered. She came home, and in a few weeks was her normal self. For the first time, I felt as if the hand of God had touched my life. Mimama's recovery gave me new life, new faith, new courage.

Reflecting on the open housing struggle, the murders of Dr. King and Robert F. Kennedy, and the riots in Detroit and elsewhere, I realized that progress wouldn't come easy, and that it would take more than protest to change things. I realized that, just as there were forces favoring expanded opportunities for blacks, there were opposing forces, too. I wondered: if some people were willing to kill JFK or MLK or Bobby to halt progress, what would that mean for my aspirations? I knew I was likely to forge a career in technology rather than social activism or politics, so I didn't know what form my personal struggle might take. I did know I was prepared to fight as hard as necessary to succeed. Mimama had taught me how. She had overcome much more adversity than I would likely ever face, and despite her many trials, she had maintained a warm, loving disposition.

Not long after those tumultuous days, my parents received a visit from a Flint city official. He said our neighborhood was part of the city's urban renewal plan, and that we would have to move. He indicated that two new freeways were going to be built, one north-south and the other east-west, and that a major interchange would be situated nearby. That visit signaled the beginning of my hometown's transformation. The ensuing years would bring rapid-fire changes. Our family would move to a new home on Flint's south side, in an upscale black neighborhood referred to as Sugar Hill. My parents would purchase the home, at 2038 Barks Street, for $25,000. Two new freeways would crisscross the city, making it easier for people to commute to and from the suburbs. A large shopping mall, Genesee Valley, would open in Flint Township, just outside the city limits.

At the time, I saw the changes as progress. Eventually, though, one side of Sixth Street—our side—was destroyed. Nothing was ever built to replace it, and the nearby freeways formed the scars of mortal wounds. What looked like progress was actually the seeds of destruction—of The Block and of Flint.

7. One of a Kind

MARCH 16, 1969, WAS A TRULY MOMENTOUS DAY. To understand just how momentous it was, we must travel back a decade and more, to the day in 1958 when I discovered romance.

It was the height of winter in Flint. Snow that had been shoveled from the sidewalk and plowed from the street formed small mountains between the sidewalk and Lapeer Road. On our way home from the old Walker School, we kids conquered each snow mountain, running up one side and down the other. I was in kindergarten, and Fred was in sixth grade. It was the only year we attended the same school. Fred caught me between snow-mountains and pulled me onto the sidewalk. He said, "I want you to meet this girl. Her name is Tiny. Tiny is in the first grade."

I didn't know why my brother was introducing me to this girl, but I was happy to meet her. She was cute—she had dark brown skin and delicate features and shiny black hair bundled under her snow hat. I grinned shyly and said, "Hi, Tiny."

Tiny said, "Hi, David." She had a big smile on her face, and her large brown eyes danced in the snow-reflected light. A brief conversation ensued.

"Where do you live?" I asked.

"On Seventh Street," Tiny said. She was still wearing that big smile.

"Want to walk home together?"

"Sure."

I grabbed Tiny's hand and we ran up and down the snow mountains together. When we reached Seventh Street, we turned and headed toward Tiny's house. When we got there, she said, "Thanks for walking home with me."

I said, "I'll see you tomorrow, Tiny."

Her real name was Claudia, but everyone knew her by the nickname neighborhood kids had given her. Tiny and I remained friends until she graduated from Walker School, five years later. We often walked home together, and I sometimes visited her house. There was nothing physical about our relationship, except that we occasionally held hands. Tiny was a girl, and she was my friend, so in my young and innocent mind she was my girlfriend.

After Tiny, there was Adora McCoy who lived in Mount Clemens, a Detroit suburb. Adora was a friend of my cousin, Aurora, my Aunt Mary's daughter. Aurora's parents were divorced, and her dad, my Uncle Odest, was living in Detroit. Aurora stayed with her dad during summer vacation, and that was when she and Adora met. Uncle Odest was dating Adora's mom. We called Aurora Rody, and Adora's folks called her Dody. Given their matching names, it is little wonder that Rody and Dody became fast summertime friends.

The summer after my tenth birthday, I visited Aurora at her dad's home in Detroit, and she introduced me to Adora. Dody was a cute girl. She had light skin (which we kids referred to as "high yellow") and long, straight, dark brown hair. She spoke with a pronounced Southern drawl, and she started flirting with me the moment we met. Dody had nice, full lips, and before long we were sneaking to different rooms in Uncle Odest's house, kissing and hugging and pressing our bodies together.

Dody and I continued our relationship the next two summers. Each time Aurora visited Uncle Odest, I would catch a ride to Detroit to see Dody. During the second summer, my mom invited Aurora to visit us in Flint, and Dody came along. My mom knew Dody was my girlfriend, but she figured it would be harmless to have her stay with us for a few days. After all, Dody and I were only twelve. What harm could we do?

When Rody and Dody arrived in Flint, they immediately became part of the Sixth Street gang. Aurora had visited before, so she was already approved and accepted by my Sixth Street pals. Dody, on the other hand, was new, and her light skin and long, straight hair caused some jealousy. My good friend Shirley Hampton was especially envious of Dody. Shirley and I had been pals since first grade, and she wasn't eager for any out-of-town girl to come between us.

One afternoon, Rody and Dody joined me and the rest of the Sixth Street gang for a bike ride. We wound up on a wide concrete walkway that led to the Whittier Jr. High School entrance. Next to the walkway, a steep, bumpy hill descended to a hard-packed sandy field. We paused to decide where to go next, and that's when Shirley urged Dody to ride her bike down the hill.

"We do this hill all the time, Dody," Shirley said. It wasn't true, but Shirley sounded convincing. "Go ahead and ride down. We'll be right behind you. You're not scared, are you?"

I didn't think Dody would do it. I was sure she didn't want to. I figured Dody would just say no and we would move on. Instead, she whipped her bike around and started down the hill. No one followed. Halfway down the hill, Dody fell and struck her head on the hard ground. All of us kids got off our bikes and ran down the hill to where she lay. Dody was out cold, and we were terrified. Fortunately, she came to after a few minutes. Aurora and I helped Dody stand up, and we supported her as she walked back up the hill. One of the other kids retrieved Dody's bike, and then we all rode back to Sixth Street in silence. I was mad at Shirley for goading Dody into riding down that hill. I was mad at myself for letting Dody do it, especially since I had no intention of doing it myself.

When we got back to my house, Dody was still woozy and had a splitting headache. My mother examined her, then drove her to Hurley Hospital. A doctor determined that Dody had a concussion and should stay overnight for observation. When the hospital released Dody, she and Aurora returned to Detroit. Dody's mother was upset about her injury, and for a long time Dody and I didn't talk. Our relationship was effectively over.

My next girlfriend was Janice. I met her in eighth-grade concert band at Whittier Junior High. I played trombone and she played clarinet. Janice wasn't exactly pretty, but she was interesting, outgoing, and sassy. She was tall and skinny and had a nut-brown complexion and short black hair. She wore thick-rimmed black glasses, so naturally our classmates called her four-eyes.

I was attracted to Janice because she was smart and spunky. We started flirting and passing notes in class, and before long I was visiting her home. Her father was a prominent black businessman in Flint. Her mother was a leading socialite and was active in civic organizations.

Janice's parents were pretty easygoing. She and I would spend hours kissing and hugging in her family room, and her folks never bothered us. Occasionally her little sister would come around and watch us kiss and make fun of us, but otherwise no one disturbed us. Her folks seemed to feel Janice was old enough to have a boyfriend and that kissing was okay.

After Janice and I had been dating for a year or so, things started to get weird. She became cold and distant. She didn't want me to visit. When I called her home, no one answered the phone. When I finally did reach her, she said she couldn't talk. I was confused because Janice and I had been so close. Later, I noticed she was hanging out with two boys from her south-side neighborhood. She seemed okay when she was with them, so I assumed she had made new friends and wasn't interested in me anymore.

Several weeks later, I called Janice again. This time, she seemed to miss talking to me. She said her father had received death threats, and that she hadn't been allowed to talk on the phone. She said her father had gone into hiding several times, and on a few occasions the family had gone with him. Despite our reconnection, the relationship between Janice and me changed. There were no more long romantic talks on the phone, no more kissing in her family den, no more visits to her home.

After my relationship with Janice ended, I was pretty much through with girls. Girls were too much trouble, and they were confusing. I was in the tenth grade, approaching age sixteen. I was thinking about school and basketball and band and the science fair, and I had no time for romance.

One afternoon after school, I got a call from my friend Denice Davis. Denice lived on Belmont Street, near the southern edge of our neighborhood. She had an unusual request:

"David, I'm having a sweet-sixteen party in a few weeks, and one of my guests doesn't have a date. I was wondering if you would be her escort. She's a nice girl, and she comes from a good family. Her father is my father's boss."

I was immediately apprehensive. I envisioned a light-skinned, pimply, plump girl with large, thick glasses—someone who was smart but not very attractive. I was inclined to say no, but Denice persisted: "She's a cute girl. I think you'll like her. Besides, she lives in Detroit, so if you don't like her you don't have to see her again. It's not a date—you're just escorting her to the party."

What the heck. Denice was a good friend. It was only one night out of my life. I decided to do Denice a favor and escort her father's boss's daughter to the party. Denice would owe me one.

"Okay" I said. "I'll do it."

"That's great—thank you, David. You won't be sorry. Her name is Gay—Gay Carlton. By the way, Raymond is taking Gay's younger sister, Pamela, to the party, so you guys can double-date!"

Raymond was one year younger than Denice and me. I thought: *Perfect—Raymond and me and two geeky girls from Detroit. This should be a great night.*

I called Raymond to get his take on the situation. What he had to say didn't relieve my discomfort:

"Gay's family is really bourgeoisie," Raymond began. That was neighborhood slang for *they think they're better than us*. "Pam and Gay probably won't like us. I think we should just take those girls to the party, be nice and respectful, and take them back home."

"Uh-huh," I said.

"Besides, Gay's father has these really big fingers. They call him Fat Finger Al. You don't want him to be mad at you!"

Oh, great, I thought. *Not only are we going to have a terrible time with these girls, we're going to have to deal with their mean, fat-fingered father. Maybe I should call Denice and back out.* I didn't back out though, because I was starting to get curious about Gay Carlton.

Denice's sweet-sixteen party was to be held on March 16, 1969, at the fancy new Voyager Hotel in downtown Flint. Raymond and I were to meet Pam and Gay at the home of Denice's uncle, Darwin Davis.

On the evening of the party, my mom drove me to the Darwin Davis home. Mrs. Davis greeted me at the door and said, "Hello, David. Come on in—Gay is already here."

I started to get excited as we walked toward the living room. I couldn't shake the image of the pimply, nerdy girl that had entered my mind when Denice first called me, but I felt a twinge of hope as I rounded the corner and entered the living room.

Oh, my God!

Sitting on the sofa in front of me was the most beautiful girl I had ever seen. My knees immediately turned to jelly—it took all my strength and willpower to remain standing. I could hear Mrs. Davis talking, making introductions, but I was barely listening.

Gay stood up and extended her hand. We shook hands in a semi-formal way, and Gay smiled and said, "Hello David. It's nice to meet you."

Gay's voice was magical. It cut right through my brain and straight into my heart. The moment I saw her, the moment I heard her voice, I was done. I was experiencing totally new sensations. I felt as if I had just been shown the answer to one of life's great questions.

Even before we arrived at the party, I knew Gay Carlton was the girl for me. Her skin, her smile, her voice, her bearing—her every attribute seemed perfect. No girl had ever affected me that way. I felt totally relaxed and natural in her presence.

Gay and I spent a lot of time talking, and we found that we had many things in common. We both loved music—she played piano and I played trombone. We were both serious students. We were both interested in world affairs. We were both mature beyond our nearly sixteen years.

In conversation, Gay and I fit together like long-lost friends. We talked easily and listened to each other intently. Gay seemed interested in everything I had to say. Her eyes twinkled as I described my science fair project, and she asked really good questions about it. That was unusual—most other girls would have written me off as a hopeless nerd.

We danced and danced on that magical night. We danced with each other and no one else. We sensed that the other people at the party could see what was happening between us. We noticed, but we didn't care. We were in our own world.

After the hotel festivities, a lot of the kids retreated to an "after-party" at Denice's house. Young couples slow-danced and engaged in quiet conversation. Gay and I sat on Denice's sofa and talked. Occasionally we got up and danced. Each time we did, we held each other a little tighter, and after each dance, our hands lingered a little longer. I really liked Gay, and I knew she liked me. At one point, I pulled Raymond aside and told him, "I think this is the girl for me."

I glanced at the clock in Denice's dining room and noticed it was nearly 11:30—the time I was supposed to be home. The evening had ended much too soon. I told Gay I had to go home and asked her to walk me to the front porch of Denice's house. As we stood there on the dimly lit porch, I said, "I had a great time tonight. It was really nice to meet you." Gay smiled warmly and said, "I enjoyed meeting you, too, Mr. Tarver." Then she extended her hand the same way she had when we first met. Calling me "Mr. Tarver" was her cute way of showing how far we had come in just a few hours. I took her hand and caressed it in mine, and then I said dramatically, "I want to see you again—soon!" Gay said, "I feel exactly the same way."

I gently eased my hand from hers while looking intently into her big brown eyes. I didn't want the night to end. I said, "I have to go home now, but I'm going to try to come back. I want to spend some more time with you tonight."

Gay's smile receded just a bit. She looked at me earnestly and said, "I understand if you have to go. I would love to see you again tonight, but if you can't come back, it's okay. We'll see each other again soon."

Gay's statement only intensified my desire to return. I wanted badly to kiss her, but I didn't want to appear forward. I felt we had established the potential for a really special relationship, and I didn't want to ruin it. I didn't try to kiss her. I just said, "I'll see you soon."

With that, I turned and ran down Avon Street toward home: Ninth Street, Eighth Street, Seventh Street, Sixth Street, up the street and through the front door. I was home.

The house was quiet. I wanted to go back to Denice's, but I knew I had to get my parents' permission. I knocked softly on their bedroom door at the foot of the stairs. No answer. I knocked again—still no answer. I went upstairs to my room to change from my suit into something more casual. Suddenly my father appeared in the doorway.

"What are you doing?" he said.

"I just got back from the party. I was thinking of going back over to the Davises' because a lot of the kids from the party are still there. The girl I took to the party is still there too, and I want to talk to her some more."

My father got a stern look on his face and said, "Just stay home."

I couldn't understand his reaction. "Why?" I said. Everybody else is still there. We're just at the Davises' house. I'll be back in about an hour."

My father wasn't budging. "It's twelve o'clock already," he said. "Just get ready and go to bed."

I was in no mood to submit. This had been the best night of my life, and the girl I might spend the rest of my life with was just a few blocks away. I could think of only one thing—sealing our first date with a kiss—and I was not going to be denied by anything or anyone.

"I'm going to go," I said. I'll be back in an hour."

My father was growing angry at my defiance. "You gon' stay right here," he said, and with that he came at me and tried to push me onto my bed. I dodged him and ran toward the stairs. When I got there, I bounded down three steps at a time. I reached the lower landing, did a quick pivot, and jumped over the last three steps. As my feet hit the floor, I heard a loud "clang" behind me. I looked back to see my steel drum major's baton crashing into the wall at the bottom of the stairs. My father had thrown it at me!

I didn't wait to see what he might do next. I dashed through the dining room, into the foyer, and out the front door. I ran all the way back to Belmont Street, back to the Davises'. I didn't care what my father thought. At that moment, I didn't care if I never went home again.

The party mood was still quiet and mellow. I found Gay sitting alone on the living room sofa, and I went over and sat next to her. Gay's face lit up—she seemed really happy that I had returned. I held her hand and said, "I had to come back and see you again." Gay gently stroked my hand and said, "I'm glad you came back."

We spent another half hour or so talking. I noticed that the other kids were beginning to filter out and go home, and then I remembered the encounter with my father. I said to Gay, "I'd better be going. I had a great time tonight. I'm glad Denice asked me to take you to her party. I want to see you again, real soon. Is that okay with you?"

Gay smiled, looked straight into my eyes, and said, "I'd love to see you again, David."

That was all I needed to hear. I was on cloud nine. I stood slowly to go, and Gay came along with me. I held her hand as I guided her onto the front porch. When we got there, we were alone. I turned to Gay and we looked tenderly into each other's eyes. She seemed warm and receptive, and I didn't feel at all uncomfortable as I pulled her toward me and planted a kiss on her lips. The first kiss was just a brief one. We paused, and then we kissed again. The second kiss was much longer, but still not long enough. I was caressing Gay, and holding her felt natural and wonderful. As our lips slowly parted, our eyes met again in a warm and loving gaze. Then Gay's eyes twinkled playfully, and she smiled and said, "Goodnight, Mr. Tarver." I said, "Goodnight, Miss Carlton."

I turned slowly and walked down the porch stairs and up to Avon Street. When I reached the corner, Gay was still standing on the porch. I turned and waved goodbye, and then slowly floated home.

I didn't see or hear anyone as I entered our house. My drum major's baton was lying, cracked and bent, on the landing. I hustled up the stairs, undressed, and slipped into bed. I was too excited to sleep. All I could think about was Gay Carlton. I replayed our conversations over and over. I replayed our kiss. Our night together had been perfect, and I couldn't wait to see her again.

The next thing I knew, it was morning. I woke to the sound of a song on the radio. It was one I had never heard before:

> *I don't remember what day it was*
> *And I didn't notice what time it was*
> *All I know is that I fell in love with you*
> *And if all my dreams come true,*
> *I'll be spending time with you...*

I lay on my bed in a daze, not sure if I was awake or dreaming. I felt as if I were in a magical world, and that the radio had produced a new song just for Gay Carlton and me.

After a while I got dressed and went downstairs to breakfast. My grandmother had cooked grits and salmon croquettes and eggs and bacon, and my mother and father were sitting at the breakfast table. Strangely, no one said a word about what transpired the night before. My father didn't seem angry anymore. It was truly a magical morning—the kind of morning every human being should experience at least once.

8. Science Fair Winner

MY LOSS IN THE FOURTH-GRADE SCIENCE FAIR left me bitter and embarrassed. I tried again in sixth grade, but I still didn't believe it was important to use the scientific method. I built a "Big Ear," a device that could detect and amplify distant sounds. It was impressive, but once again I failed to place in the top fifty. Add "really discouraged" to "bitter and embarrassed."

Each of the next two years I grew anxious as the science fair entry deadline came and went, but I didn't enter. Finally, in ninth grade—even as my friends were losing interest in favor of sports or girls or other teen fascinations—I decided to go with the program. If the fair officials wanted the scientific method and "Do plants grow better in white light than in blue light?" that's what I would give them. I entered the Flint Area Science Fair a third time, determined to show everyone I could win.

The science fair always took place in the spring. That meant I had to think up my 1968 entry and begin working on it in the fall semester of my ninth-grade year at Whittier Junior High School. My project idea came from a biology lesson about single-cell organisms—paramecia and amoebae. I found those microscopic creatures fascinating, and I thought they would be interesting subjects.

My entry was titled: "How does electricity affect one-celled animals?" My simple hypothesis was that electricity was bad for them. My experiment was to apply an electric shock to the little critters. The result: I electrocuted a few billion paramecia and amoebae. They died right away. The judges were pleased. I won First Division honors that year, fourteenth place overall. I was confident I had found the winning formula.

The next year, I got a little more creative. Instead of zapping the little one-cell buggers with electricity, I used radio waves. The result: I murdered billions more microscopic protozoa. It was harder to kill the little creatures with radio waves than with electricity, but they died just the same. I won First Division honors again, finishing in fourteenth place for the second straight year.

I could see that it was easy to win First Division honors in the Flint Area Science Fair. You just had to follow the scientific method, give your project an esoteric-sounding title, and make sure the judges could understand it. During my junior year in high school, I entered again. This time, I set my sights on something bigger. I wanted to win the whole thing—first place overall in the 1970 Flint Area Science Fair.

With an electronics revolution sweeping the globe, I decided to change my focus from protozoa to transistors. More and more products were incorporating transistors. The 1969 Apollo moon landing wouldn't have been possible without them. My project sought to determine how heat would affect transistor operation. It was entitled "How Does a High Environment Heat Affect the Operation of a Transistor and Its Associated RF Oscillator Circuit?" The title was elaborate, but the idea was simple. It was similar to the old "How does light affect plants?" experiment, only in this case it was, "How does heat affect transistors?" Nonetheless, the project was a ton of work. I worked day and night on it, even designed and built my own chamber to heat the transistors.

My project was classic science fair stuff. I posed a question: How does a high environment heat affect the performance of a transistor and its associated oscillator circuit? I formed a hypothesis: A high environment heat will adversely affect the operation of a transistor and its associated oscillator circuit. I designed and carried out an experiment to prove or disprove the hypothesis: Build the transistor-based oscillator circuit, subject it to heat, and see what happens.

On the appointed spring day, I loaded my project into the back of my father's station wagon and drove down to the IMA Auditorium to set everything up. Then I returned home to await the call from science fair officials. That call came the next night. I was a top finalist, and I was going to be interviewed at the IMA. I was beside myself with excitement.

The next day, a shiny new Buick Electra 225 came by the house to pick me up. A placard on each side of the car shouted in large letters, "Science Fair Winner." I sat tall in the back seat and gazed out the window. My neighbors had seen the car come to my house, and I saw several of them waving and pointing as we pulled away. I was ecstatic. I was finally realizing the dream I had carried with me since fourth grade.

At the IMA I was escorted to my booth on the exhibit floor. Several serious-looking white men in suits awaited me. My escort announced, "This is David Tarver from Flint Central High School. David, these gentlemen are the judges, and they have a few questions for you. First, please take a few minutes to describe your project."

I greeted the judges and started to summarize my project. I was nervous at first, but I loosened up quickly. I made sure to describe how I had employed the scientific method, and emphasized the fact that I'd built my own apparatus for heating the transistors.

The judges seemed impressed. Some asked good questions, but others asked questions that betrayed a lack of electronics knowledge. A few questions were downright absurd, but I did my best to answer carefully and respectfully.

At the end of the interview, I was confident I had a good shot at winning. My project was unique. I had addressed a modern, interesting technology, and I had adhered to the scientific method.

The next day, my mom and I arrived at the IMA for the awards program. We took our seats and I waited nervously for the winners to be announced. The Senior Division winners were presented last, and the announcer began with twentieth place. When tenth place was announced and it wasn't me, I started to feel giddy.

As the numbers got lower and lower, I grew more and more excited. Then I heard the announcer say, "Fourth place, W. David Tarver."

My immediate reaction was disappointment. I had hoped to win first place. Then the announcer added that I was the recipient of a special award from the U.S. Navy. As the night wore on, I started to feel better. Fourth place was a lot better than I had ever done before, and I was, after all, a top finalist.

The award from the U.S. Navy included a cruise on a navy ship. The cruise was to depart from and return to a port on the East Coast. I wasn't excited by the navy's offer, perhaps because I had never traveled outside Michigan by myself. I later declined the cruise.

I was disappointed, but had no doubt I would try again. I had one year left, my senior year, and I was determined to win first place.

Knowing I would have to do something spectacular, I decided to combine the best feature of my losing elementary school science fair projects—really cool technology—with the best feature of my winning junior high and high school projects—interesting and relevant science. Of course, my final project would scrupulously follow the scientific method.

I decided to stick with transistors. I had already tested them in heat, so I couldn't do that again. I decided to pursue a very chic topic: transistor performance in the presence of nuclear radiation. Surely that subject would knock the socks off those judges. The relevance of my project was obvious. Transistors were critical to the space program's success, and space vehicles routinely encountered radiation. Transistors were also key components in terrestrial communication systems. Would such systems continue to perform during a nuclear attack? This was significant stuff in 1971. I was sure the judges would consider my project both technically advanced and timely.

The science in my project was compelling, but I had to come up with some dazzling technology to impress the judges. I decided to build a sophisticated system to measure and display transistor performance, but I couldn't afford to build the system on my own. The components cost more than $300, and that was more than either of my parents' weekly income. I went to my ham radio mentors at Shand Electronics and asked them to lend me the money. I described my project and suggested they would get great publicity if someone they sponsored won the science fair. To my surprise, they gave me the loan.

That final science fair project took over my life. It was the most ambitious thing I had ever attempted. I stayed up late many nights building and testing and redesigning my elaborate transistor measurement system. Many nights I was so sleepy I didn't know what I was doing, but I kept going. One night I was working at 2:00 a.m. when a loud POP made me fall off my workbench stool. I lay on the floor for several seconds, heart racing. Slowly, I pulled myself up and looked at the circuit I was working on. I had installed an electrolytic capacitor backwards, and it had exploded in my face. At that moment, things looked pretty bleak.

Most days, I was too tired to concentrate in school. Most nights, I didn't get to bed until after midnight. When I did finally fall asleep, I had a recurring nightmare in which my physics teacher, Mr. Shaw, told me I had failed his class and would not graduate. I had been so obsessed with my science fair project that I'd missed weeks of classes. I always woke up at that point in the nightmare, sweating and shaking.

The months I spent building my transistor measurement system put me behind schedule. The science fair was fast approaching, and I had not yet begun the radiation experiments. They were the whole point of my project, so I would surely lose if I didn't complete them. With time winding down, I started a frantic search for a radiation source. It wasn't something one found in the *Yellow Pages*. Mr. Shaw—the teacher at the center of my nightmares—suggested I check with the University of Michigan, which was just fifty miles away in Ann Arbor. I didn't know anyone at U-M, and had never even been there, but after several calls I did find a professor whose lab included a cobalt radiation source. He agreed to let me use the source for a few hours. Several days later I carefully placed my equipment onto the soft back seat of my mom's Ford Galaxy 500 and beat it down to Ann Arbor.

Using the U-M radiation source, I was able to complete my experiments in a single afternoon. I finished my project on time, confident that no one would be able to compete with it. The science was new and relevant, the technology was state of the art, and I had designed and built everything myself.

On the day designated for setting up projects at the IMA, I was in the garage getting everything ready to go. My father was supposed to help me take my project to the IMA in his station wagon, but he was in the family room watching TV. I yelled to him that it was time to go, but he didn't respond. Then I yelled, "If you don't want to take me, I'll just take myself." That was the wrong thing to say. My father refused to take me and refused to let me use his station wagon. I couldn't believe it. He had to know that all my work would go down the drain if I didn't set my project up before the deadline, but he didn't seem concerned. I couldn't understand why he was being so disagreeable. I was crying and cursing my father under my breath when my mother suggested I call a neighbor, Mr. Spotsville, and ask him to take me to the IMA. With time running out, Mr. Spotsville agreed. I stopped speaking to my father.

The next night, I got a call from a science fair official informing me that I was a finalist. As in the previous year, a shiny new Buick came to pick me up at home and take me to the IMA for my interview with the judges. As in the previous year, the neighbors pointed and waved as I went by.

The interview went well. The judges were impressed with my topic, and were even more impressed with my fancy measurement system. Some seemed incredulous when I told them I had designed and built the whole thing myself, but my answers to their questions gave them no reason to think otherwise. After the interview I returned home fully believing I would win first place in the 1971 Flint Area Science Fair.

My mother and father must have had a big fight about my father's refusal to drive me to the IMA, because when I got home I learned he would be joining us at the awards ceremony. I really didn't care if he came or not. All I cared about was winning first place.

The next night, my mother, my father, and I attended the awards program. When the time came to present the Senior Division results, I was rapt. I listened intently as the winners were announced: "Fifth Place... Fourth Place... Third Place... Second Place, W. David Tarver." My heart fell to the floor. I watched a Northwestern High School student named Jan Stannard receive first-place honors. I had never seen or heard of Stannard, a skinny white kid with a pimpled face and thick black glasses. He looked like a bona fide "junior scientist," but I had no idea why his project was judged better than mine.

As if my second-place finish weren't bad enough, science fair officials had another surprise. They created a new division for previous winners who entered a project "similar to their past prize-winning effort," as the *Flint Journal* later reported. That division had only one entrant—Randy Brown, a suburban kid who had won first place the previous year. In prior years, the first- and second-place Senior Division winners represented Flint at the national science fair, but this year science fair officials decided Flint's representatives would be Randy Brown and Jan Stannard. I was livid. I felt that the brand-new rule had been created for one purpose—to shut out *black* David Tarver and send an all-white delegation to the national science fair.

Adding insult to injury, I had to pose for photos with the other winners. I felt worse than I had after the first science fair when I lost to Frank Stank and his tooth-decay project. I felt worse because I felt robbed.

After the awards ceremony I went down to stand by my project and greet attendees. My father and mother came by the booth. My mother gave me a big hug. "I'm proud of you," she said. We were both fighting back tears. Then I looked at my father. His face wore a look of sadness and resignation. In that moment, I felt we connected. I sensed the frustration and disappointment he must have experienced his whole life. I knew that, despite our fight and whatever feelings of resentment he harbored, deep down he had wanted me to win. He was the one who had introduced me to electronics, and I was demonstrating love for the same vocation he loved. The only words he could muster were, "Good job, boy." For me, that was enough.

Later, I wrote a letter to the *Flint Spokesman*, the local black newspaper, in which I railed against science fair officials. I was angry at the peculiar way they had shut me out of a trip to the national science fair. I was convinced my project was the best, and that I was excluded simply because of my skin color.

My science fair career was over, and though I was twice a finalist and four times a top-twenty finisher, I never won the whole thing. Though devastated by the result, I learned two important lessons from my years of participation. The first was that, to achieve success,

I would have to do meaningful things, not just "neat" things. My losing elementary school projects employed clever technology, but they completely missed the point. They were about technology, not science, and they didn't answer any questions or solve any problems. The second lesson was that I was good enough to compete with the best kids in the whole Flint area. Long after my peers in the neighborhood had stopped participating, I was right there competing with the suburban white kids, the kids with the professional parents and the money and the contacts. I was in the game, right there at the top, and I should have won.

Or should I? My science fair experience introduced me to a central ambiguity in race relations, a question at the heart of my expanding racial awareness. Was I deprived of first place because the judges were racially biased, or was my project simply not the best? Did the fair officials' decision to create a new award category and deprive me of a trip to the national science fair represent institutional racism, or did they simply want to send the best possible delegation to the event? If the judges or fair officials were biased against me, was it because of my race, or because I was a cocky kid who had earlier turned down a prized cruise on a navy ship?

The very next year, Harold Caldwell, a black kid from Northwestern High School, took first place in the Flint Area Science Fair.

FLINT JOURNAL FRONT PAGE, APRIL 19, 1971.

9. Most Likely To...

JUNE 1971. THE SCHOOL YEAR WAS ENDING, and all of us soon-to-be-graduates gathered in the Central High School auditorium for the senior awards assembly. I had mixed feelings. It was sure to be a long afternoon full of boring faculty speeches, but I felt my high school achievements merited some kind of award, and was anxious to see if I would receive one. Sitting among my classmates, I began to daydream and reflect on my eclectic high school career.

As a kid I had always been in awe of Central High—Whittier Junior High, too, for that matter. Central and Whittier shared a campus and a long and storied tradition. The letter emblazoned on every Flint Central varsity jacket was "F," a holdover from Central's days as Flint High School, the first and only high school in town. The Flint Central mascot was the Indian, and at halftime of every home football game, an "Indian" warrior (during my years at Central, a black kid in an Indian costume) would perform a war dance in the end zone.

In keeping with the Indian motif, the Whittier Junior High sports teams were known as the Braves. There was no dancing mascot at Whittier, though. Whittier Junior High was named for John Greenleaf Whittier. As far as I know, the poet had no connection to Flint, but at the time all Flint junior high schools were named for dead poets or dead presidents.

The Central/Whittier campus was a magnificent place. Both schools were constructed of red brick, with ivy climbing the walls. A grassy yard divided by cement walkways lay in front of the buildings.

Large oaks and maples contributed shade and added to the majestic ambiance. In those days, when someone spoke of an Ivy League school, the image that came to my mind was the Flint Central campus.

I began my career at Central as a tenth-grader in the fall of that terrible, tragic year, 1968. I was a diligent student, but I wasn't focused solely on the classroom. In the fall of each year, I participated in the marching band. In the winter, I played on the basketball team. In the spring I turned my attention to the Flint Area Science Fair. As far as I knew, I was the only student who had such a diverse mix of extracurricular activities, but few other students seemed to notice or care.

During my sophomore year, I played trombone in the band, and was the starting center on the sophomore basketball team. That spring saw my second straight 14th-place overall showing at the science fair. I had a good year, but it was nothing special.

In junior year, I was chosen to be assistant drum major of the marching band, a post that automatically led to being named drum major the next year. The selection happened in an interesting way. On the practice field one afternoon, our band director, Mr. Bruce Robart, asked the eighty-eight band members to form a single line. With the current drum major standing by his side, he asked every student who was interested in becoming assistant drum major to step forward. At that moment, I took a tentative step forward, and the other eighty-seven band members stepped back, leaving me a good distance in front of the line. They all knew I badly wanted the job. Mr. Robart said, "Mr. Tarver, looks like you're it!"

That same year, I earned a spot on the Flint Central varsity basketball team. I owed that notable achievement to my neighbor and friend Mike "Dookey-Hookey" Thompson. The summer after my sophomore year, Mike decided to help me make varsity. He said that to make the team, I needed to get much tougher. Nearly every day, we played one-on-one games on the driveway court behind my house. Every time I drove to the basket, Mike would hit me or push me and say, "You got to get tough, Turtle!" I was bigger than Mike, but his pushing and cajoling were hard to take. In the end, though, he made me a tougher player, and that fall I was fearless as I competed for a varsity spot. I not only made the team but six games into the season I became a starter. I made the starting squad because the coach saw me as a tough player—a hard-nosed defender, a determined rebounder, and a relentless hustler. In fact, my teammates gave me the nickname "Tarnacious."

(L-R) REGGIE BARNETT, WALTER MOORE, AND I APPLY THAT TENACIOUS FLINT CENTRAL DEFENSE.

I owed it all to those "Get tough, Turtle!" sessions with Dookey-Hookey. He made the varsity team too, but spent most of the season on the bench.

At the beginning of junior year, I got involved in school politics for the first time. Four of us black students, inspired by the civil rights movement, decided to run for class leadership. My friend Denice Davis ran for president. I ran for vice president and served as campaign manager. My friend and neighbor Shirley Hampton ran for treasurer. Another neighborhood girl, Jenny Dones, ran for secretary. To our surprise, we all won. In a school that was 30 percent black and 70 percent white, we managed to elect an all-black slate of class officers. Nearly every black student enthusiastically supported us, and we garnered enough white support to put us over the top.

We didn't have much time to savor our victory. Soon after the election, Shirley was summoned to the office of the principal, who confronted Shirley with the fact that her family had recently moved out of the Central High School district. Shirley was forced to leave Central and attend our arch rival school, Flint Northern. Denice and Jenny and I were upset. Not only were we losing one of our newly elected black officers, we were also losing our friend to another school—to *Northern*, no less! It was a tragic situation, and we immediately set out to determine who had ratted on Shirley. Our speculation quickly settled on Chris, a white girl who'd been Shirley's opponent for class treasurer.

Chris was a tall, attractive girl with light brown hair and pretty blue eyes. She was nice to just about everyone, and didn't seem to have a problem dealing with black students. I kind of liked Chris, and I hoped she wasn't the one who had told on Shirley. It didn't seem like something Chris would do. I appointed myself to question her on the matter, and one day after school I took her aside.

"Hey, Chris," I said, "Got a few minutes?"

Chris smiled and said, "Sure."

We walked toward Chris's house, a few blocks east of the school. I congratulated her on being named treasurer, and we made small talk about student council and school in general. Then we came to the subject of the election. Chris said, "I feel bad about what happened to Shirley. I didn't want to become treasurer like this. Denice and Jenny already don't like me. Now they'll hate me for sure."

I felt sorry for Chris. In our zeal to elect an all-black group of officers, we hadn't considered some of the goodhearted white kids in our class—kids like Chris. I knew that Denice and Jenny didn't like her, but I tried to downplay that.

"Denice and Jenny are just mad about what happened to Shirley. They think you might have had something to do with it—that you might be the one who turned Shirley in," I said cautiously.

Chris began to cry. "I would never do something like that," she protested. "I like Shirley, and I think it's terrible what happened. If you want me to, I'll resign!"

I felt myself starting to go soft. Chris seemed like a really nice girl. She had a heart, and she seemed sincere.

As we approached Chris's house, I grew apprehensive. This was not my neighborhood, and I didn't think her parents should see me walking her home.

"Chris," I said, "don't worry about it. I'll tell Denice and Jenny you had nothing to do with it. Things may be rough for a while, but you'll be okay. I look forward to working with you."

Chris said, "Oh, thank you, David." She turned to walk into her house and said, "Would you like to come in? You want something to drink?"

I declined. Just going into Chris's neighborhood was enough adventure for one day. I said goodbye and walked home to Sixth Street.

In the ensuing weeks, Chris and I got to know each other better, and I enjoyed being with her. I started visiting her home after school, and we spent hours studying and talking. We were just friends, but I liked her a lot, and I began to think about what it would be like if she were my girlfriend.

Someone at school started a rumor that I was dating Chris. Suddenly, some of the black kids shunned me and some of my basketball teammates gave me a hard time in the locker room. One of my teammates, Bobby, told the other black players: "Don't pass the ball to Tarver—he likes white girls." I knew that was stupid, but I still felt terrible. The folks who were getting on my case weren't my friends, so I didn't care much about what they thought. Nonetheless, their attitude made things pretty uncomfortable for me.

Meanwhile, at Chris's house, I was approaching a moment of truth. I was seeing her more and more after school, and we were spending more time talking and enjoying each other and less time studying. Finally one evening, sitting together on the sofa, we kissed.

Chris's parents seemed nice, and they didn't seem bothered that a black kid named David Tarver was spending a lot of time in the den with their daughter. One night, Chris's mother entered the den and presented a surprising offer. She said, "How would you kids like to go to a concert at Whiting Auditorium next week? I think you might enjoy it."

I was stunned. Chris's mother was actually encouraging me to take her daughter out. Until then, our relationship had existed solely in the den of Chris's house. I hadn't thought about what it would mean to go public—I just knew I was getting grief from the guys on the basketball team and some of the other kids at school. Chris was getting grief, too. A lot of the black girls at school hated her guts. I didn't know what to say, so I just muttered, "That sounds like fun, Mrs. Crawley."

For several days I thought hard about going to the concert with Chris. Something about it just didn't seem right. I already had a girlfriend, Gay Carlton, and although she and her folks had moved to the Chicago area, I didn't want to break up with her. Moreover, I realized I wasn't ready to have a white girlfriend—not at that time, not at Flint Central High School. I liked Chris a lot, but it wasn't that serious. I visited her house less and less. Eventually we saw each other only at school. We never went to the concert.

My junior year at Central ended on a high note—that fourth-place finish in the science fair. It wasn't quite the note I had hoped to hit, but it was still quite an achievement.

I entered my senior year with high expectations. I looked forward to leading the marching band and starring on the basketball team. Also, I looked forward to winning, finally, that elusive first-place trophy in the Flint Area Science Fair.

Marching band started out okay, but as the football season unfolded, conflicts arose. The band director didn't like the fact that I sometimes left band practice early to attend basketball practice. The basketball coach didn't like the fact that I arrived at basketball practice late, after band practice. I couldn't win. I was a fair drum major, but my heart wasn't in it. I was happy when football season ended so I could concentrate fully on basketball.

Our first basketball game was against Pontiac Central High School, one of the top teams in the state. The star player for Pontiac was Campanella (Campy) Russell. He was so good that people were already touting him as an NBA prospect. I had made my mark on our team when, during the previous season, I did an outstanding defensive job on Campy and we defeated Pontiac. My performance in that game earned me a starting position on the varsity team. I was ready to match up with Campy again. I thought if I played well, folks around the state might begin to think of *me* as a top player.

We played the game in the tiny, noisy Pontiac Central gym. The first time Campy got the ball, I overplayed his right hand. Campy simply drove around me with his left for an easy lay-up. The next time down the floor, I played him straight up, but Campy sealed me off with his left hand and drove around me with his right—another easy basket. The next time Campy got the ball, I stepped back so he couldn't drive around me. He simply pulled up and hit a jump shot—nothing but net. Campy grinned and shook his head as he trotted back down the floor. His teammates were laughing. Mine were furious. Campy scored eight points in the first two minutes of the game. That's when I heard the five words I dreaded most from our coach, Cliff Turner: "Get Tarver out of there!" Just like that, I was out of the game, and my backup, Eugene Wilborn, was in. I never started another game. Those two minutes effectively ended my career in organized basketball. I finished the season, but spent most of the time on the bench. Our team didn't have a very good year.

With my basketball career down the drain, I focused more attention on my final science fair project. I spent many hours on that effort, designing and building my apparatus, then traveling to Ann Arbor to use the U-M radiation source. For a few months, I neglected basketball and school and just about everything else. I've already indicated how that turned out: I finished in second place overall, but was denied the national science fair trip.

The school year ended on a high note. Every year, Central staged a talent show called *Kaleidoscope,* open to all students. Three buddies and I formed a Temptations wannabe group called the Sensations or Vibrations or some such name, and I was the lead singer. It was my first time singing onstage. We performed *Just My Imagination* and walked off with the only truly enthusiastic audience reaction. We won first prize.

A few weeks later, it was time for the senior prom. I had hoped my girlfriend, Gay, would travel from Chicago and attend the prom with me. To my immense satisfaction, Gay said yes. More important, her parents said yes, and they agreed to bring her to Flint. Gay and I attended the prom, and it was a magical night.

Suddenly, I was jolted from daydreams about Gay Carlton and recollections of my high school career. I realized I was sitting in the Senior Awards Assembly, and that our principal, Mr. Crowder, was about to announce the recipients. I wasn't sure what awards were going to be handed out, but I was curious to see who would be chosen "Most Likely to Succeed." I watched intently as the awards were given to my classmates. My friend Reggie Barnett received the top honor, the prestigious Red and Black Award, given each year to the top senior athlete, and he certainly deserved it. Reggie was an outstanding athlete in football, basketball, and track, and was an all-A student. Some of the other award decisions were dubious, and some I just downright couldn't understand. I entered the assembly ambivalent about the awards, but I left feeling angry and cheated. I felt that no one at Flint Central had taken on the variety of activities I had. I'd shown that I could lead the marching band, serve as a class officer, play varsity basketball, create winning science fair projects, sing lead in a band, and make the honor roll, all while taking tough college-prep courses. I couldn't understand why I didn't receive a single award.

A few days after the assembly, one of the school counselors, Mrs. Brown, beckoned me into her office. I had seen Mrs. Brown around the school on many occasions, and we had talked about my college plans. Mrs. Brown was an attractive young black woman who was popular with the students. She had always struck me as earnest and sincere, but she wasn't my counselor, so I was curious about why she wanted to talk to me.

Mrs. Brown asked me to sit down, and then she gently closed the door and sat behind her desk. She wore a pained expression. "David," she began, "I just wanted to talk to you for a few minutes about the awards assembly that was held the other day."

I had pretty much dismissed the assembly from my thoughts, except for the dull pain of rejection I still felt, but Mrs. Brown's words ignited my curiosity.

She continued: "I just want you to know that your name came up for every one of those awards, and for some reason the staff didn't see fit to give you any of them. I don't know what some of these people around here have against you, but it's just not fair. I wanted you to know because I think you're a fine young man, and I don't want this to affect how you see yourself."

Mrs. Brown's words lifted a big weight from my shoulders. I knew the Central faculty was responsible for selecting the award recipients, and I suspected some on the largely white staff resented me, for reasons I didn't completely understand. Perhaps it was because my varied interests and associations didn't fit their preconceived notions about black students. Notwithstanding the staff's decision, Mrs. Brown was telling me that I was an outstanding person with great potential. She reinforced in me the belief that I should focus on my goals and not be concerned about receiving accolades. I had started to feel that way after coming up short at the science fair, but Mrs. Brown's assurances really made it clear. She gave me an invaluable gift—a huge dose of self-confidence. She helped me see myself as a talented and versatile eclectic, not an awards-assembly loser.

When she finished, all I could say was, "Thank you, Mrs. Brown." I was overwhelmed that she was so concerned about my feelings and would take the time to reassure me.

Graduation came a few days later. We received our diplomas at the stunning Whiting Auditorium, and then it was time to party. Reggie and I were on top of the world. He was headed to Notre Dame on a football scholarship, and I was headed across town to General Motors Institute to study engineering. All of that was in the future, though. That night, we got into my little white Austin America sedan, rolled all the windows down, and blasted Sly and the Family Stone as we cruised slowly up Saginaw Street toward the north side.

> *"Sing a simple song.* Try a little do re mi fa so la ti do!
>
> Do re mi fa so la ti do!"
>
> *"Don't call me nigger, whitey.* Don't call me whitey,
>
> nigger!"
>
> *"Everybody is a star,* remove the rain and chase the
>
> dust away. Everybody wants to shine, who'll
>
> come out on a cloudy day?"
>
> *"Stand,* they will try to make you crawl, and they know
>
> what you're sayin' makes sense and all."

We screamed every song at the top of our lungs, and believed in our hearts that we were both "Most Likely To."

10. "GMI is Free, Boy!"

WHEN I GRADUATED FROM CENTRAL HIGH SCHOOL IN 1971, I knew exactly what my next moves would be: I would go to college and get a degree in electrical engineering, and then launch my own engineering business. I knew at an early age that I wanted to be an engineer because of all the days I spent watching my father fiddle with electronic stuff in the basement and all the nights I spent working on science fair projects. The desire to start my own company took shape when I was in junior high school, probably rooted in the many "kid business" ventures I engaged in on Sixth Street.

Choosing a career was easy. Choosing a college was tough. I applied to only three: Harvard University, the University of Michigan, and a little school in Flint called General Motors Institute.

My decision to apply to Harvard resulted more from passion than careful analysis. Harvard didn't confer engineering degrees, but it had a well-regarded physics program and a cooperative arrangement with nearby Massachusetts Institute of Technology, the best engineering school in the world. My girlfriend, Gay Carlton, was applying to MIT. If we both ended up in Cambridge, we would be living in the same place for the first time. That and Harvard's prestige were the sources of my passion. Unfortunately, I was not accepted at Harvard. I guess I didn't look like Ivy League material.

The University of Michigan was the big state school with the great reputation and the huge campus. U-M offered all the amenities I associated with a big university: lots of sports, lots of girls, lots of fun, and lots of academic options. When I visited the campus, however, I was intimidated by its enormity. I thought I would get lost in the shuffle, and just didn't feel comfortable. Though my application was accepted, I decided not to enroll.

General Motors Institute (now Kettering University) was an option I didn't consider until my last few months of high school. GMI, located on the west side of Flint and wholly owned by General Motors, was a top-notch engineering and industrial administration school. GMI had educated many top GM executives and had a great reputation. Flint Area Science Fair officials who were affiliated with the school introduced me to GMI.

Certain aspects of staying in Flint for college appealed to me. I could attend without giving up the comfort and familiarity of my hometown. GMI was a cooperative education (co-op) school—students alternated six-week academic periods (called academic sections) on campus with six-week work periods (called work sections) at a sponsoring GM division. At any point during the year, half the students were attending classes and the other half were at work. The salary a student earned allowed him to fund his own tuition, books, fees, and room and board, and still have money left over to buy a car. When my father found out about the GMI opportunity, his response was adamant and succinct. He said simply, "GMI is free, boy!" He couldn't understand why, under the circumstances, I would even consider going to Michigan or Harvard.

Case closed. In the fall of 1971 I began my college career as an engineering freshman at General Motors Institute. Never mind that Gay Carlton was attending college more than 700 miles away at MIT. Never mind that I had no intention of spending my life working as an engineer for General Motors. Never mind that I was giving up the opportunity to experience campus life at a "real" university. I had concluded that, all things considered, GMI was where I needed to be.

When I arrived at GMI, I found an environment quite different from Flint Central High School. Even though GMI was just on the other side of town, it seemed to exist in a different world. Physically, GMI consisted of only three buildings: an "academic building," a "campus center," and a dormitory. A large parking deck, filled with students' mostly late-model cars, sat adjacent to the dormitory. The physical environment, though small, was adequate.

The GMI social environment was something else. For one thing, hardly any blacks attended the school. In the wake of the 1964 Civil Rights Act and the social unrest of the 1960s, GMI was beginning to admit small numbers of blacks and women. The year before my class arrived, forty of the 3,000 students were black. There were thirty-two blacks in my freshman class, including four black women. Lots of white men and a handful of white women comprised the remainder.

The mostly white, nerdy GMI environment was a culture shock to most black students. Most of us had excelled in high school, and most came from supportive homes and communities. We may have been the rare blacks in our college prep classes, but outside the classroom there had been ample opportunity to socialize with our black friends. Being at GMI was like being exiled to a foreign and sometimes hostile world. The white students, on the other hand, were in a place that seemed designed for them. Sure, the white guys would complain that there were few girls at the school, but we blacks had few people we felt comfortable socializing with, male *or* female. The vast majority of white students came from similar backgrounds and shared similar aspirations. For white males at GMI, life was good. It was their world—we were interlopers.

Oddly enough, the foreign nature of the GMI environment yielded some advantages. For one, since there was so little social life, few things distracted students from studying. Second, the work was so intense and difficult that students had little time to worry about socializing. Third, and perhaps best, the academic sections were only six weeks long, after which time students returned to work at a GM division—and to some semblance of a healthy social life.

None of my high-school friends joined me at GMI. My best friend, Reggie Barnett, was at Notre Dame. Others went to U-M or Michigan State or other traditional colleges or universities. Initially, I found myself with no one to relate to, no one to relax and "kick back" with. Even though all freshmen shared a dorm, I felt isolated. The white kids generally hung out with each other. The several prosperous white fraternities on campus weren't exactly looking for black members, and most black students weren't interested in joining. A few black students hung out with the whites, but they were the type who liked rock music and white girls—the whole white "scene." I wasn't comfortable with that. Eventually I met and became friends with a few black students—two in particular: George Blocker from Detroit and Raymond Wood from Indianapolis. Their room was down the hall from mine, and we spent many nights talking and listening to music and passing judgment on the state of the world. We felt like exiles in a strange land, but we were determined, motivated exiles.

A few black students decided to start their own fraternity. It wasn't affiliated with any national organization—just a small group of guys getting together to support each other. They called their little independent frat Phi Eta Psi, and a rented house on Detroit Street served as headquarters. I visited one day to hang out, and on that particular occasion I was beaming. I had just purchased, for $400, the first-ever handheld scientific calculator, the Hewlett Packard 35. Until that time, the only handheld computing device students used was a slide rule. We had Wang desktop scientific calculators at school, but it was inconvenient to trudge over to the academic building to use one. I had in my hot little hand a device that was going to render the slide rule obsolete and free me from standing in line to use the Wang calculators. I couldn't wait to show my new tool to the Phi Eta Psi brothers.

When I arrived at the house, I pulled out the HP35 and began showing it around. Everyone was duly impressed, except one guy: Dorian Tyree. Dorian seemed pissed, and no doubt my gloating countenance had something to do with it. He insisted that my new calculator was nothing but a toy, and that he could do calculations much faster on his slide rule. He challenged me to a duel, right on the spot. We proceeded to calculate the square root of this number and the logarithm of that number, and each time I was much quicker and more accurate on the calculator than Dorian was on his primitive slide rule. That made him furious, and when our contest was over he picked me up and threw me over a hallway banister. I didn't know Dorian was so strong, and I never realized he was so fearful about the changing of the technological guard. I limped back to my dorm room with a big rip in my GMI jersey and my new, victorious HP35 in my pocket. Dorian the Luddite eventually got over his anger, but I never saw him use a calculator.

After a time, white students grudgingly accepted me and the other black newcomers. In general, we treated each other civilly. There were no big blowups, no fights, no campus demonstrations. There were smaller incidents, though. For example, my dorm neighbor Dan, a white student, liked pop rock music, especially Elton John and Don (*Bye-Bye Miss American Pie*) McLean. I favored R&B artists like Stevie Wonder and the Temptations and Marvin Gaye. On many occasions, Dan would crank up the volume on his stereo to the point where I couldn't hear my own music. I retaliated by cranking up mine. The situation escalated until, on some nights, no one on the entire corridor could hear himself think. We almost got tossed from the school as a result of our silliness, but eventually we learned to coexist. I even developed an appreciation for Elton John.

Academically, I didn't find GMI especially difficult. In general, if I studied hard and prepared for exams, I did well. Most classmates were serious students, so it was easy to focus and get down to business. One evening before a big calculus exam, a group of us sat in the lounge watching TV and talking. As the hour got late, my white classmates began to talk about the exam. "I'm not studying for that stupid exam anymore," one said. Another piped in, "That stuff is really easy. I don't plan to study anymore either." One by one, they filtered off to their rooms. I started to feel insecure and inferior because, though I knew the material pretty well, I wasn't as confident as my classmates seemed. I needed to study more, and that's exactly what I intended to do.

When I got back to my room around 9:00 p.m., I pulled out my calculus book and got to work. As midnight came and went, I began to tire, but I persevered. By 2:00 a.m. I was exhausted, and felt stupid to be studying while my confident classmates were asleep in their beds. *This is dumb*, I said to myself. *I'm going to bed. If the other guys feel like they're ready, I should be ready too.* I stood up and went to the window to draw the shade.

My window overlooked a courtyard. I noticed that most room lights around the courtyard were still lit. That was when I realized my classmates had been bluffing. They were still studying, just as I was! I went back to my desk and studied for a couple more hours. I made sure I knew all the material the exam might cover. The next afternoon I took the exam and got an A. Some of my confident white classmates didn't fare as well.

I was glad I hadn't let my classmates' bravado fool me into not studying, and I was especially glad I was able to see what was going on in their rooms the night before the exam. I realized their pre-exam trash talk was born of insecurity, not confidence. I resolved that in the future I would ignore my classmates' posturing, do the best I could, study as much as needed, and let the chips fall where they might. From that point on, the chips fell favorably for me at GMI.

11. AC and Me

A SPARK PLUG FACTORY MIGHT SEEM AN UNLIKELY LAUNCHING PAD for an electrical engineering career, but that's exactly where mine began. Every GMI student was sponsored by a General Motors division, and the AC Spark Plug division sponsored me. Dr. Karl Schwartzwalder, an AC executive and Flint Area Science Fair supporter, had encouraged me to attend GMI and arranged the AC sponsorship. Dr. Schwartzwalder was a world-renowned ceramic engineering expert, and because the core of every spark plug was a ceramic cylinder, he was a very important person at AC.

AC Spark Plug was only a few miles from the Sixth Street home that "urban renewal" had bounced my family from more than a year earlier. I had passed AC's main offices on Dort Highway a hundred times, but had never been inside. I didn't know what went on there besides the manufacture of spark plugs. I didn't realize AC made a long list of automobile components: oil and air filters, instrument panels, and sophisticated devices such as cruise control and anti-lock braking systems. I also didn't know the AC complex included an Engineering Research Center (ERC), located just one long city block from the spark plug factory.

I reported to work at AC in the summer of 1971. My contact in personnel was a genteel, dapper man named Clifford Gangnath, who was responsible for, among other things, coordinating GMI student activities at AC

The first order of business was an orientation meeting. I met AC's other incoming co-op students, and together we toured the complex. Then we were each given our initial work assignment, and after receiving mine I was delighted. I could have been assigned to a low-tech, mundane area like manufacturing or production control or accounting, but I was assigned to the one place at AC that did a reasonable amount of electrical engineering—the Engineering Research Center. I felt as if Dr. Schwartzwalder must have pulled a few strings to secure that plum placement.

ERC was *the* cool place to be. The sign outside the building was inscribed with a large representation of an atom, which in those days was a universal symbol that meant "High Tech Stuff Is Happening Here." ERC housed managers and executives, engineers, and technicians. The vast majority were white. I saw only an occasional speck of brown or black.

I was assigned to ERC's Ceramic Circuits Laboratory. I knew, of course, what circuits were, but had no idea what *ceramic* circuits were. I soon learned that the Ceramic Circuits Lab was a high-tech venture overseen by Dr. Schwartzwalder himself. The lab produced prototypes of advanced automotive electronics devices, which were fabricated on small ceramic wafers called substrates.

My direct supervisor was a genial, approachable man named Jim Spaniola. He and the other engineers and technicians seemed to enjoy working together, and the work environment was relaxed. My job was to assemble tiny electronic circuits on the ceramic substrates. I did that by looking through a microscope and connecting components such as transistors, capacitors, and resistors using miniscule gold wires not much thicker than a human hair.

Only a few months out of high school, I was doing important, leading-edge work. My coworkers accepted me and treated me with respect. I don't know how much my connection to Dr. Schwartzwalder had to do with that, but my colleagues' behavior seemed genuine. I enjoyed my job.

Every Friday morning, all the GMI co-op students at AC got together for a Tech Club meeting, at which we discussed work assignments and heard news from school. Occasionally, an AC executive gave a presentation on some aspect of AC's business, such as an oil filter manufacturing problem or a new efficiency scheme dreamed up by Operations Engineering. For students, Tech Club was a chance to get away from the daily grind and yuk it up with each other. After each meeting, we would leave the plant for an extended brunch at Angelo's Coney Island Restaurant. It was mainly at those brunches that I got to know my fellow Tech Club members.

All my fellow AC co-op students were white males. Most came from working-class families, and some had parents who worked for GM. They talked about cars morning, noon, and night—what kind of car they had, how they were going to modify it, what kind of car their brother/father/friend had, what car they were planning to buy next, and on and on. I liked talking about cars too, but after a while it became monotonous. The other subject we talked about was school—what classes were easy or difficult, or what a prick professor so-and-so was. One student didn't fit that mold; in fact he stuck out like a ninth spark plug on a V8. When he talked about cars, it was usually to say cars were destroying the planet. When he talked about school, he spoke of how easy and mundane the courses were. That student was Mike Ebel.

Mike was an interesting, if misplaced, GMI student. Most students were clean-shaven, but Mike wore a beard. It wasn't unkempt—Mike never arrived at work looking disheveled—but his beard was just messy enough to tweak his superiors. Most students came to work in neat dress slacks, a white shirt and tie. Mike usually showed up in fresh jeans and a neat, open sport shirt—again, not messy, but just nonconforming enough to demonstrate his rebellious nature.

In our sessions at Angelo's after Tech Club meetings, Mike would go on and on about how stupid it was for the country to waste "fossil fuels" (petroleum) in cars. He thought it was a crime to use a "finite resource" like petroleum for gasoline and plastic dashboards when that same petroleum was the main ingredient required for most medicines. Mike railed against cars, against General Motors, against the Vietnam War, against religion. He claimed to be something I'd never heard of—an agnostic. He quoted Maslow and Vonnegut. He listened to Bob Dylan. The more I listened to Mike, the more I wondered what the hell he was doing at General Motors Institute, but one day it hit me—Mike was probably at GMI because his family could *afford* GMI. Perhaps someone in *his* family had insisted, "GMI is free, boy!"

Despite his apparent discontent, Mike was a high-profile student at GMI and a high-profile co-op worker at AC, and he wasn't about to go anywhere. His AC managers put up with his rebelliousness and holier-than-thou attitude because Mike had the best grades at the school and because his work was unassailable. Management probably felt that if somehow Mike would just grow up, as so many previous rebellious GMI students had, they would have a diamond on their hands.

I learned a lot from Mike Ebel. I didn't agree with everything he said, but he definitely altered my thinking in several areas. I had never intended to stay at General Motors, but Mike's diatribes against the company further convinced me that GM was no place for me. More and more I believed my role as an engineer was to help solve the world's problems, not help create them. Mike helped me see that General Motors was not going to change anytime soon, and was driven by forces far more powerful than a few idealistic GMI students. It seemed more likely that GM would change me rather than the other way around, so I started to think about getting out.

Despite my growing Ebel-inspired disaffection with General Motors, I continued to enjoy my AC work assignments. I began to regard the co-op experience as something that would prepare me to reach my ultimate goal—starting my own company—and I approached each new assignment with that in mind. Eventually I was given jobs testing oil pressure sensors, working as a buyer in purchasing, and measuring worker efficiency in production control. I regarded each as a potentially valuable experience for an aspiring entrepreneur.

One of my favorite AC assignments was a stint in the Advanced Development Lab, which was charged with creating new systems that might one day be used in GM cars. I was excited about the assignment, because Advanced Development was the best department for an electrical engineer. The engineers there got to play with all kinds of electronic toys and dream up new and exciting products. During my first tour of the lab, I met an experienced black engineer who was developing a radar-based crash-avoidance system. I had never heard of anything like that, and I was doubly impressed because it was rare to see *any* black engineer at AC, let alone one who was managing such an impressive project.

My mentor in the Advanced Development Lab was Bill Kerscher. Bill was a recent GMI graduate, and he looked like the typical GM engineer: a skinny, white, pocket-protector-and-thick-glasses-wearing nerd. Bill and I were from different backgrounds, but we shared a love of cool gadgets and a passion for engineering. We got along well, not least because he drove a new Corvette and occasionally gave me a ride. Mostly, we got along because he understood how badly I wanted to be an engineer and gave me good assignments to feed my desire. My job in the department was to build and test electronic circuit prototypes that Bill designed, and I enjoyed and appreciated the advanced work.

One afternoon, representatives of a new technology company called Intel Corporation visited our department. They were touting two recent inventions. The first was a semiconductor memory device called a ROM—Read Only Memory—that could hold a whopping 256 *bits* of information. The second was a little semiconductor device called a microprocessor, which they referred to as "a computer on a chip." The Intel folks were anxious to get their new devices designed into things such as engine controls and anti-lock braking systems. They left samples for us to evaluate, and Bill Kerscher gave me the assignment of building and testing a prototype that incorporated the Intel chips. I spent the next few weeks painstakingly assembling and programming it.

I was unimpressed. The 256-bit memory chip was barely large enough to store someone's name and address, and programming it was a real chore. I had to sit at a big Teletype® terminal and type in each bit: "P" for a "1" and "N" for a "0." It was tedious and not the least bit interesting. I concluded that the Intel devices were novel but difficult to use and impractical. I preferred building circuits using more traditional components. Intel's stuff was cute, I thought, but would never fly. I had no clue that Intel would one day become one of the world's most successful companies.

I enjoyed working in the high-tech Advanced Development Lab, but one of my most enlightening assignments occurred in a decidedly low-tech department. That assignment, one of my last, was in production control, and I didn't relish it. Production control was intriguing to industrial engineers, but for an aspiring electrical engineer it was the pits. Nonetheless, when I got the assignment I didn't complain, because I thought it might help prepare me to start my own company.

When I reported to production control and got the details, I dreaded the assignment even more. My job was to sit next to an instrument panel assembly line and record each error the workers made. Worse, I was assigned to count the errors because management was planning to speed up the production line and wanted to be sure that action wouldn't result in more assembly errors. I realized immediately that I would not be popular with the workers. I was certain the assignment would be a miserable, unrewarding experience.

Things didn't get any better when I arrived at the production line. All the assembly workers were middle-aged white women, and when they saw me, their faces contorted as if I were a creature from Mars. They were uniformly shocked, and then they were curious. I could imagine what they were thinking. *Who is this black boy? What is he doing on* our *production line?*

The assembly foreman introduced me to the workers.

"Ladies, this is Dave Tarver, and he's a GMI student assigned to Production Control. He's going to be sitting out here on the line for a few weeks to observe how you assemble instrument panels. He won't get in your way, so don't give him a hard time."

One of the older "assembly ladies" seemed to be the ringleader. When the foreman was finished, she spoke up.

"What is he going to be observing? Is something wrong, or are you just making work for another GMI student?"

The foreman tightened a bit, as if he didn't want to tell the ladies the whole story. "He's just going to be taking some notes on the assembly process and checking for errors, that's all. No big deal."

I saw several ladies begin fidgeting on their stools as soon as they heard the words "checking for errors."

One of the ladies spoke up: "Checking for errors! Is that what this is about? We don't need no young GMI know-it-all checking our work. We don't make no errors!"

At that point, I wanted to leave production control and return to building circuits in the Advanced Development Lab. I was convinced this assignment was a waste of time and that working next to these women was bad news. It was too late, though. My assignment was set—the assembly ladies were going to be my workplace companions for the next few weeks.

"Look," the foreman said. "Just cooperate and don't make trouble. If you have any problems with him being here, take it up with your shop steward. But I'm telling you, this is just a standard production control job and it's no big deal."

I noticed that the foreman didn't say anything about speeding up the production line. I guessed he was saving that surprise for later.

The first day on the job was quiet—very quiet. I sat next to the conveyor belt and inspected each instrument panel as it went by. If I saw an error, such as a missing fastener or a misplaced light bulb, I marked its location on my clipboard pad. I counted fewer than ten errors during the entire shift. That first day, the ladies said absolutely nothing to me, and I said nothing to them. When they spoke to each other, they did so in hushed tones. Their demeanor was serious and unwelcoming. The tension was palpable. At the end of the day I couldn't wait to take my clipboard and go home.

The next day, things went pretty much the same way, except the ladies seemed freer about talking with each other. They stopped talking quietly and started to communicate in what seemed like their normal way.

"You goin' bowlin' with Jimmy again this weekend?"

"Yeah, I 'spect so, if he don't have to pull no overtime."

Things seemed to be returning to normal on the line, and the ladies were trying their best to ignore me.

After a few more days, some of the ladies started talking to me. Kathy, the ringleader, spoke up first:

"So, Mr. GMI, you havin' fun sittin' there all day with your little clipboard?"

The ice breaking surprised me. I had resigned myself to doing my job and ignoring the workers, but I had to respond.

"It's not a bad assignment. Besides, I get to show production control all the errors you ladies are making."

To my surprise, Kathy smiled at my attempt to tweak her.

"Don't you tease me, Mr. GMI. I can see you ain't makin' too many marks on your little clipboard, and I know we don't make no errors. That is, unless we *want* to make some."

She flashed a sly little smile, and several ladies within earshot smiled too.

"Actually, it's true," I said. "I'm not seeing many errors. You ladies must really know what you're doing."

It was a lame attempt to curry favor, but it seemed to work. Everybody seemed to relax, and the rest of the day passed smoothly. The ladies resumed talking to each other, but occasionally they threw a comment my way. At one point, Kathy said, "I'm going to the bathroom when the line stops, if it's okay with Mr. GMI." Everybody laughed at that one, including me.

That afternoon, after break time, the ladies started peppering me with questions:

"Where you from, Mr. GMI?"

"Flint."

"What high school you go to?"

"Central."

"Oh, so you're a big-time Flint Central Indian, huh? I went to Northern. I'm a Viking! You know, the ones that beat your butt in football last Thanksgiving!"

"Uh-huh."

How long you gonna be sittin' out here with us, Mr. GMI?"

"A few weeks, I guess."

After a few more questions, I felt as if I had earned the right to ask them some questions, and I did.

"Where are you all from?"

Most said they were from Flint or the surrounding area, but a few hailed from Arkansas, and one was from Ohio.

"How do you like working on this line?"

Kathy replied for everyone: "This is good-payin' work, and it's easy. We don't mind this at all, do we ladies?"

Some shook their heads in agreement.

Kathy continued, "Good hard work never hurt nobody, ain't that right, Mr. GMI?"

I felt like breaking the ice a little more. I said, "You don't have to call me 'Mr. GMI,' you know. You can just call me David."

"Okay, Dave" Kathy said. "And you can call us by our first names, too. Just call me Kathy."

We were making progress, and just like that, the shift was over. It seemed that the time had flown by, as if we had just returned from the mid-afternoon break a few minutes earlier.

The next day, the ladies were in a feisty mood. We were no more than ten minutes into the shift when Kathy turned to me and asked, "So, Dave, what do you think about busing?"

She caught me off guard. That was the last subject I expected to come up. A lot of stories about school busing had recently been in the newspaper, the result of a judge's order to desegregate the schools in Pontiac. White activists were loudly protesting the busing plans, and the stories were front-page news.

"Well," I began, "I think sometimes it makes sense. We still have too many segregated schools, and I don't think that's right."

To my great surprise, Kim, a lady who sat two positions over from Kathy and who had said next to nothing all week, spoke up vehemently. "I don't want my kids havin' to leave the neighborhood and go clear across town to school just so somebody can say the school is integrated. It just ain't fair!" Kim's face was red, and she was shaking with emotion. She looked as if she might cry.

I could see what was happening. The previous day, the ladies and I had experienced a little breakthrough. They were beginning to see me as a human being. Maybe they were starting to like me a little bit. Then they went home and thought about it. Maybe they talked to their husbands about the black GMI student sitting next to their assembly line. They saw the stories about school busing in the newspapers and on TV. They wanted to know what side I was on. Was I an "us" or a "them?" I could see what was happening, but there was no way I was going to sell out. I wanted them to know how I really felt.

"Kim," I said, "I know it may be inconvenient for your kids to go to a school outside your neighborhood, but lots of black kids have done that for years. My mother and father went to a segregated school in Georgia, and they had to walk past a white school to get there. I think things will work out better for everybody's kids if they can learn to live together. That's what busing is for—to teach the kids to live together."

Kim didn't say anything, but Debbie, who was sitting across the line from Kim, did.

"It ain't gonna work. You gonna get all the white people mad, and then where will you be? You *cain't* make people live together, you just *cain't!* It ain't right."

We sat there and argued about busing all day. Between arguments, the ladies assembled instrument panels and I counted errors. As we argued and they assembled and I counted, production control began speeding up the assembly line. I started counting more errors. In the beginning I had counted only five or six errors a day, but now I was seeing fifteen or twenty. I didn't know if the cause was our arguing or the sped-up line or both.

Before I knew it, the shift was over. All the arguing had made the day go by in a flash. I looked forward to coming in the next day and arguing some more. I think the ladies did too, because the next day they were ready with some new topics. We started with busing, but then we got on the subject of affirmative action. They were really trying to get under my skin! The next day, I decided to get under theirs, so I brought up the subject of interracial dating and marriage, and we argued all day about that. The subsequent days were a blur. Assemble... argue... count... assemble... argue... count. I was enjoying myself immensely, and then the assignment was over. Just like that, six weeks had passed and it was time to return to classes at GMI.

It was my last day on the instrument panel assembly line, and I really hated to go. Despite our arguing, I had developed affection for the assembly ladies, and I was beginning to understand them. I think the affection and the understanding were mutual. After all our arguments, I could see that the ladies were just human beings with concerns and fears and hopes. They could see that I was just an aspiring young black man full of hopes and dreams. We didn't necessarily resolve our differences, but I was convinced that the next time a young black man entered their world, they would receive him with more open minds. At the end of the shift, all of the ladies came up to me in turn to say goodbye and wish me well.

The last one in line was Kathy, the ringleader, and in her hands she was holding a gift for me. It was a fully assembled Chevy Camaro instrument panel—and it contained no errors.

12. A Fork in the Road

GMI WAS AN EXCELLENT SCHOOL, but it just didn't seem like the place for me. I couldn't envision spending my career as a cog in the enormous General Motors machine. I wanted to pursue bioengineering or telecommunications, not automobiles. Besides, Gay Carlton was a student at MIT, and I longed to be with her.

During a freshman break at GMI, I traveled to Cambridge to see Gay. We ate hoagies at a little sub shop on the Boston side of the Charles River Bridge. We took a bus trip to Cape Cod and rolled in the dunes, kissing and getting sand all over our clothes and in our mouths. We held hands and talked as we walked around in Harvard Square and Cambridge. It was a blissful experience, and I yearned for more.

Gay and I had always maintained a long-distance relationship. We saw each other infrequently, usually for only two or three days at a time. During that first visit to Cambridge, I thought seriously about transferring to MIT. The idea of being with Gay all the time, while studying engineering at the world's best technical institute, was extremely enticing. I decided to give it a shot.

I applied to MIT and was accepted in the spring of 1972. My excellent, if brief, GMI record served me well, because I would not have been accepted on my high-school grades alone, even with my science fair successes and my activities in band and basketball. But I had a solid A average after one semester at GMI, and I had Gay lobbying for me at MIT.

Despite my enthusiasm for the transfer, I found MIT intimidating. First, it was one of the most expensive schools in the country. I guess the expense was appropriate given MIT's stature, but that didn't make the high tuition any less daunting. MIT offered a financial aid package that consisted of grants and loans and part-time work, but my folks would still face a substantial financial burden, and I would be repaying student loans for years after graduation.

Then there was the apartment situation. I faced the prospect of paying through the nose for a dingy little apartment in Boston or Cambridge and sharing it with a roommate I didn't know. Most apartment ads indicated NO HEAT. Being a naïve kid from Flint, I assumed that actually meant the apartment was not heated. I envisioned myself huddled in the corner of a cold, barren apartment next to a small space heater and a hanging incandescent bulb, trying desperately to finish my calculus homework. I discovered much later that "no heat" merely meant that the cost of heating fuel was not included in the rent.

Despite my apprehension, I continued preparations to attend MIT, and in June 1972 I drove to Cambridge to get acclimated and purchase books. Gay was home in Chicago for the summer. My mother and Aunt Mary accompanied me on the trip. We piled into my 1971 Camaro and proceeded along the northern route from Michigan to Boston—into Ontario at Sarnia, back into the U.S. at Buffalo, then over the I-90 turnpike to Boston.

The air was quite chilly on that June night, so I closed the Chevy's windows and turned on the heat. By the time we reached the New York Thruway I was experiencing severe intestinal distress—gas. Maybe it was nerves, or perhaps something I had eaten. In any case, my condition made the trip quite uncomfortable not just for me but for my mother and Aunt Mary. We had to choose between opening the car windows and shivering, or keeping the windows closed and breathing the fumes of my intermittent noxious flatus. We chose the latter.

The next morning we arrived in Boston. I dropped my mother and aunt at my cousin Thelma's house in the Roxbury section of Boston and immediately headed to Cambridge to see the folks at MIT. Gay had arranged for me to meet with an assistant dean she knew, so

I went straight to his office. The assistant dean, a serious, middle-aged black man, gave me lots of information about MIT and its engineering program. We also talked about financial aid and housing. Discussing those topics with the affable administrator made them seem much less daunting. He explained that many transfer students were frightened by the cost of attending MIT, but after settling on campus they always seemed to find a job and line up financial support. I felt better, but I was still uneasy. It felt as if I were standing at the edge of a cliff, and people were shouting, "Go ahead and jump—everyone else does! By the way, here's a parachute." The parachute might make you feel better, but you wonder if it will work. You wonder where you'll land.

Uncertainties aside, I left the dean's office and went to purchase books. I wanted to get a head start on my fall classes by studying over the summer. I purchased the calculus and physics textbooks and immediately found them complicated and intimidating. The calculus text, written by an MIT professor, was twice as thick as the one we used at GMI. The physics text was highly mathematical, and there was no namby-pamby introductory chapter—the incomprehensible equations began on page 3!

I could see that MIT was not going to be easy.

Walking around campus, I couldn't help but notice the nerds. I had seen them during my earlier visit to Cambridge, but now they were really apparent. MIT students seemed focused and serious. I didn't see much laughing or goofing off. I saw very few black students. I didn't see many women. Absent Gay Carlton, MIT seemed cold and uninviting.

Toward the end of my walk, I approached the main academic building. Its wide stairs and huge marble columns were a powerful and imposing symbol of academic excellence. I walked up the stairs and wandered around until I found an empty classroom. I felt tired and lost, so I sat at a student desk and stared at the blackboard, which was filled with some of the most complicated math I had ever seen. I didn't even recognize a lot of the symbols and notation. I felt intimidated and began to doubt my decision to attend MIT. Could I succeed? Could I compete with the elite students there? Was I transferring to MIT because I felt it was best for my career, or because I was afraid of losing Gay?

Something didn't feel right. Was it fair to subject my family to the financial burden of MIT when GMI was free? With the distractions of financial aid, and having to work for spending money, and living in an apartment with "no heat," would I be able to focus on my studies? Was I apprehensive because this wasn't the right move, or because I was afraid of the challenge? I didn't know what to think or do. I looked up at the blackboard again. I cried.

Despite that breakdown, I returned to Flint determined to get my affairs in order and make the move to Cambridge. The first task was to quit my job at AC Spark Plug, and with it my enrollment at GMI. I made an appointment to see Mr. Gangnath, the AC personnel manager who oversaw GMI students.

I had always liked Mr. Gangnath. He had been nice to me from the start, and seemed to take a personal interest in my success. News of my imminent departure was sure to disappoint him, but I couldn't predict his reaction.

On my way to Mr. Gangnath's office in the AC administration building, I passed through the accounting department and paused at Oscar Barnett's desk to chat. Oscar was my friend Reggie's older brother. Talking with Oscar helped calm me down and put me in the right frame of mind for meeting with Mr. Gangnath.

Mr. Gangnath's secretary ushered me into his office. He was reviewing some paperwork, and when I entered he looked up and gestured to me to sit in the chair facing his desk. I sat and prepared to utter the words that would make my career change official. I had decided to cast my departure in positive terms, since everyone knew it was an honor to be accepted at MIT.

"Mr. Gangnath, I have an offer to attend MIT and I've decided to accept it."

Mr. Gangnath raised his eyebrows ever so slightly, but he didn't say anything.

"I've really enjoyed working here at AC and going to GMI," I continued, " but I think MIT is a once-in-a-lifetime opportunity that I need to take advantage of."

Mr. Gangnath smiled slightly. "That does sound like a great opportunity, David," he said. "It sounds like you've made up your mind, so I won't try to change it. I'm disappointed you'll be leaving us, but I know you'll do well wherever you go."

I was relieved. Mr. Gangnath understood how important this opportunity was. He didn't seem to have any hard feelings. That was a big load off my mind.

We sat there for a while and talked logistics: *When would I actually leave?* At the end of the current work section. *When would I start at MIT?* In the fall, after Labor Day.

Finally Mr. Gangnath said, "You know you will always have a home here at AC, David. Come back and see us anytime."

I left Mr. Gangnath's office feeling both sadness and anticipation. I was leaving a place I knew well, a place where I was well regarded, a place where I was experiencing success—but I was facing an outstanding opportunity.

I had many goodbyes to say. The first was to Bill Kerscher, my mentor in the advanced development department. I admired Bill for three reasons: he had a great job, he drove a brand-new Corvette, and he wore that distinctive beaver-inscribed class ring denoting his grad-school attendance at MIT.

I told Bill about my plans, and he was happy for me. I knew he would be. Then I told my fellow GMI co-op students. Most of them were happy for me, too. Given their working-class backgrounds, most weren't attending GMI because it was the school of their dreams. They were attending because it was a sure path to GM management and because it was "free."

My remaining time at AC passed quickly, and before I knew it, it was time to depart for Boston. On a Friday night in late August, I packed all my belongings into the Camaro. I said emotional goodbyes to my father and mother and Mimama. Then I got into the car to begin the twelve-hour drive to Cambridge.

On the way out of town I stopped at Allen Thompson's house. Allen and his family had been our next-door neighbors on Sixth Street, and Allen was still a good friend. I said my good-byes to Mr. and Mrs. Thompson and to Allen. He joked that Gay Carlton must really "have my nose" if I was willing to go all the way to Cambridge to be with her. I explained that I wasn't going there just to be with Gay—that MIT was a great school. Maybe I sounded a bit defensive.

After leaving the Thompsons', I turned onto Center Road and headed north to I-69 East. As I pulled onto the freeway, feelings of doubt began to well up. Speeding east toward Port Huron, the questions echoed inside my head. *Am I doing the right thing? Am I going to MIT just to be with Gay?* Allen's words were echoing, too: *She got your nose, man!* I thought about the impossibly complex equations I had seen on that blackboard, and thought about the hardship going to MIT would cause my family. *GMI is free, boy!* I thought about the Cambridge apartment with "no heat." No more warm room at my parents' house. No more waking up to Mimama's grits and salmon croquettes. *What if I don't do well at MIT? What if Gay and I break up? What if I end up alone on campus with a bunch of white nerds?*

Through the windshield I saw the I-69 pavement disappearing beneath the Camaro's hood. My head felt as if it weighed a ton. Then, moments later, the weight suddenly vanished. The feelings of doubt and confusion ceased. My head was clear. I saw the M-15 exit ahead and veered onto it. I turned left, crossed over the freeway, turned left onto the entrance ramp for I-69 West, and headed back to Flint.

I hadn't felt right since the day I sat alone and cried in that MIT classroom. It wasn't only fear I was reacting to—it was also a feeling that transferring to MIT wasn't the right move for me. True, MIT was a great school, but in many respects it was like GMI. It wasn't a university where I could get to know lots of different kinds of people. It wasn't a place where I could interact freely with people from lots of different disciplines. It wasn't a place where I could spend Saturday afternoons in a huge football stadium. No, MIT was just another technical institute—a bigger, fancier one in a bigger, fancier area. I wanted a qualitatively different college experience, but that desire had been obscured by my intense desire to be with Gay. When I turned the car around, all of that became clear to me, and my body's reaction told me I was making the right decision.

When I got home, everyone was happy. My mother gave me a huge hug. She said she hadn't wanted me to go, but had wanted me to decide for myself. Mimama just cried with joy. Even my father gave me a smile and a pat on the back. I was sure he was relieved at not having to pay for an MIT education, but I was also sure he was proud that his son had been accepted there.

I was back where I belonged, but I sensed that my decision would mean the end of Gay and me. We hung on for a few months, and then she met someone else. I was devastated. Losing Gay left a void I could not easily fill.

I stayed at GMI for another year before transferring to the University of Michigan. It took tremendous focus and determination to keep moving forward, but my professional goals and dreams sustained me. Though I hurt a lot, I didn't regret the path I chose.

13. Moog Magic

WITH MIT IN MY REARVIEW MIRROR, I began my second year at GMI. I resumed living on Barks Street, in the home my folks had moved to after "urban renewal" destroyed our old neighborhood, The Block. On a luscious late-summer evening, I was sitting in my bedroom reading and daydreaming. Most of my GMI friends had gone home for the weekend. Through the open window, I could hear cars rushing by on Lippincott Boulevard, and I could hear kids playing over on Seymour Avenue. The delicious smell of the night air and the sounds of life lured me from the house. Longing for something to do, I drove over to campus to see what was happening.

Only a few students were at the Campus Center—a couple nuzzling and kissing on a corner sofa, a few white frat boys playing cards around a long table. Pretty dull. I decided to get a bite to eat and go back home. On the way out I stopped to look at the bulletin board near the door. Amid the usual notices—people seeking roommates, cars for sale, and intramural sports schedules—I noticed a plain-looking flier. It said simply:

MOOG Concert
Gymnasium 7:00 PM

It was 1972, and my taste in music was pretty much restricted to Motown R&B. I didn't know who or what Moog was. I puzzled over the flier for a moment, then concluded that it must be the name of some music group—probably a rock band. I envisioned a group of loud, stringy-haired rockers, probably Beatles wannabes. That kind of concert wouldn't satisfy my entertainment appetite, and it wasn't the way I wanted to spend a nice summer evening. I had no intention of attending, but to satisfy my curiosity, I stopped by the gymnasium to see if my hunch about Moog was correct.

The gym revealed a much different scene than I'd expected. Perhaps a hundred students—all white—sat scattered around the gymnasium floor. Three large benches sat on a stage that had been erected at one end of the gym. Each supported a large wooden cabinet with a black metal faceplate. The faceplates were covered with switches and knobs and electrical jacks and patch cords. The contraptions reminded me of an old-fashioned telephone switchboard. A piano-style keyboard in a wooden enclosure sat in front of each cabinet. A small black panel at the left side of each keyboard contained a few more switches and knobs. The whole setup was impressive, but I wondered what kind of concert this would be. And I still had no idea what Moog was.

I wandered around the gym for a few minutes, then picked an empty spot on the floor and sat down. I turned to a nearby student and asked, "What's going on here?"

"A Moog concert" was his terse response.

"What's that?" I continued.

"Electronic music," he replied.

I thought about leaving. The guy's tone didn't make me feel particularly welcome. The idea of electronic music wasn't exactly enticing, either. My only experience with that was a device called a theremin that Fred once constructed. The theremin emitted an eerie sci-fi tone, and you controlled its pitch by waving your hands near it. It sounded weird and seemed useless, and I wasn't up for a night of sounds like that—especially in a gym full of GMI nerds. Still, the size and complexity of the cabinets onstage piqued my curiosity, so I decided to stay around to see what would happen next.

Three people walked onto the stage and sat down, one at each keyboard. The gym lights dimmed, and the musicians began to play. I use the words "musician" and "play" loosely, because the performers weren't making music in the traditional way. They alternated between fingering the keyboards and working the knobs, switches, and patch cords on the big black panels. It was a very unusual concert.

But the sounds! The sounds coming from that stage were at once beautiful and otherworldly. I sat captivated as the musicians performed several classical pieces. I recognized a few from my time with the Flint Junior Symphony Orchestra, but the electronic instruments made the music sound totally new and exciting. I got lost in the occasion—completely mesmerized by the sights and sounds. The experience touched me on several levels. The intersection of art and physics and engineering was compelling. I stayed for the entire concert, and at the end was dying to learn more about what I had just seen.

The next several days, I tore into the subject of electronic music. I learned that the devices I had seen on the GMI stage were music synthesizers, and that they were designed by Robert Moog. So that's what a Moog was—an electronic music synthesizer.

I found a book on electronic music in the AC Spark Plug library, and I read and reread it until I understood the basic technology. Then I went shopping for electronic music recordings. I scanned the record albums until I saw one that froze me in my tracks. The album jacket pictured the same Moog synthesizer I had seen in the GMI gym, and a guy dressed up like Johann Sebastian Bach was standing in front of it. The title of the album was *Switched-On Bach.*

I bought the album and rushed home to play it. I had heard Bach compositions such as the *Brandenburg Concertos* and *Jesu, Joy of Man's Desiring,* but the synthesizer gave them a whole new flavor. I was amazed that music so beautiful and stimulating could be produced electronically.

A few months later, my favorite musician, Stevie Wonder, released an album called *Talking Book.* On that album, Stevie made extensive use of music synthesizers, and the result was popular music taken to a new level. When I heard *Talking Book,* I became completely obsessed with electronic music, and made a decision: I would design and build my own electronic music synthesizer, and I would start a company to manufacture and sell my creation. I decided to call my company Tarv Acoustic Synthesis—TAS.

14. Goodbye, Pops

MY FATHER—WE CALLED HIM POPS—DIED OF PANCREATIC CANCER in September 1973, a few days before his fifty-sixth birthday; a few weeks after I began my studies at the U-M College of Engineering. Pops' cancer was diagnosed in April. The doctors first thought he had a stomach ulcer, then they suspected diabetes. Finally, they diagnosed the cancer, but offered hope that it was not "the bad kind." By the time Pops' cancer was identified as a particularly nasty strain, it was too late. My father received treatment at Hurley Hospital for a few months, but when it was obvious he would not recover, he was discharged so he could spend his final days at home.

MY MOTHER AND FATHER, CIRCA 1972.

Pops took up residence in my old bedroom. It was a convenient spot, adjacent to a bathroom and steps away from the family room and his favorite TV. A poster-sized photo of me, a remnant of my high school vanity, was plastered on the back of the bedroom door.

My father must have seen that picture during each of his final days. I wonder what he thought about it. On his last day, he got up from bed, maybe to go to the bathroom or maybe as a final effort to affirm his existence. Whatever the reason, he collapsed outside my bedroom door and died. My mother went downstairs sometime later to check on him and found him sprawled beneath the built-in bookshelf that contained our family photos and his beloved electronics books.

I was at school in Ann Arbor when my father died, and my main concern was the first exam in my electronic networks course, ECE 214. That was the first course in what I considered my main area of expertise, and I looked forward to it. The professor, Richard A. Volz, had a reputation among students as a no-nonsense guy, tough but fair. I was confident I would do well no matter who the professor was. I was anxious to establish my academic bona fides with Dr. Volz and my classmates, but my U-M concerns quickly faded when I got the call from Flint.

The call came from my brother-in-law, Evans. "Rich, son," he began, invoking the nickname he had given me years earlier, "I have some bad news. Your father passed away at home this morning."

"Oh," I said. I wasn't surprised, because I knew how seriously ill my father had been. Still, I had held out hope that he would recover. Perhaps it was wishful thinking, or perhaps denial. The news of his death left me numb. I didn't cry.

I hung up the phone and thought about when I'd last visited my father in the hospital. He was sitting up in bed watching TV, and I was sitting in a chair beside him. Pops was a shriveled version of his former self. He had lost a lot of weight, and his skin was drawn around his bones.

While we sat together, a preacher and two deacons stopped by. They weren't from our church and had never met my father. The preacher asked my father if he and the deacons could pray with him. My father nodded, and the preacher and his disciples positioned themselves next to the bed and began to pray.

"Lord, you are the Master. You are the way, the truth, and the light. You brought Daniel from the lion's den. You raised Lazarus from the dead. Now bring your healing power to our brother and lift him out of this bed. You alone have the power! Cast out the demons of sin. Cast out the sickness and infirmity, and let this brother be a living witness to your power! Cast the demons out Lord! Cast them out! Cast them out!"

With each "Cast them out!" the preacher pushed hard on my father's forehead, so hard it seemed the force might snap his neck. I wanted to jump up and tell the preacher to stop, but my father didn't seem to mind. He seemed at peace.

After the prayer, the visitors left as suddenly as they had come, and Pops and I were alone again. I felt guilty, because I was moving to Ann Arbor and wouldn't see much of him after that. I thought of his mantra "GMI is free, boy!" A year earlier, he had wanted me to remain at GMI rather than transfer to MIT. Though U-M was much less expensive than MIT and only an hour away, I didn't know how he felt about my decision to go there. I tentatively explored the issue with him.

"Pops, you know I'm supposed to start at Michigan soon, right?" I said.

"Yeah, I know. You go down there and work hard at that school. You finish that school."

The intensity of my father's response startled me. I suddenly realized that his medical situation was grave and that, in his own way, he was telling me to go on with my life and not to worry about him. A few days later, I moved to Ann Arbor and took up my studies at U-M.

I was standing in the middle of my living room pondering my father's death and thinking about driving up to Flint to be with the family. I couldn't leave Ann Arbor right away, though, because I had classes all week and didn't want to fall behind. I remembered the impending ECE 214 exam and realized it would probably take place around the same time as the funeral, but I didn't worry about it. I figured I would work it out somehow.

A couple of days later, I learned that the funeral was scheduled for the day after the exam. I wanted to spend some time with my family, and I didn't think I could concentrate on the exam with the funeral looming. I decided to ask Professor Volz if I could delay taking the exam, and I went to see him at his office.

A colorful poster on Professor Volz's door was captioned: "*Concha y Toro, Vino de Oro.*" I had no idea what that meant, but I was pretty sure it had nothing to do with electrical engineering. I found the poster amusing, and it eased my considerable anxiety as I knocked on the door.

I heard a gruff voice: "Come in." I entered and saw Dr. Volz perched behind a big wooden desk loaded with stacks of books and journals and papers. He gave me a stern, businesslike look and said, "How can I help you?"

"Dr. Volz," I began, "My father just died. His funeral is this Friday, the day after the 214 exam. I have to be up in Flint with my family this week, and I was wondering if you would let me take the exam next Monday, after the funeral."

Dr. Volz's face softened to a sympathetic gaze. "I'm sorry to hear about your father. Are you all right?"

I was surprised by his sympathy. A simple yes or no was all I was looking for. I felt uncomfortable having anything other than a formal relationship with this stern professor.

"Yes, I'm fine," I said. "I just need to be excused from taking the exam on Thursday. My thoughts aren't exactly focused on my classes right now."

"Yes, I understand," Professor Volz said. "But I can't excuse you from taking the exam. You can miss it, but if you do I'll have to base your grade on the remaining exams in the course."

I was floored. I had been sure Professor Volz would let me reschedule. I felt myself getting angry—angry at his refusal, angry at his sympathetic but unhelpful remarks.

"Fine, I understand," I said curtly. I got up and abruptly left the office. I didn't want a confrontation, but I wanted Professor Volz to know I was pissed. Later that week, I took the exam.

Immediately after the exam I drove up to Flint to be with my family. By the time I arrived, most of the out-of-town relatives were there, and everyone was busy socializing and reminiscing. My mother seemed to be doing okay, probably because so many people were there to support her. I thought it odd that the family could get together for such a sad occasion but still manage to laugh and party. Then I realized that they were just trying to find refuge from the sadness, and that most hadn't seen each other in a very long time.

The funeral took place the next day at Quinn Chapel A.M.E. Church. Most of the family rode there in the funeral-home limousines, but I walked. The church was only three blocks from our house. It was a nice day, and I wanted to be alone with my thoughts. The family sat in the customary pews up front. I sat next to my mother as we gazed upon my father's open casket. I got up for a few minutes to take one final look. He didn't look much like the person I knew. The cancer had left his face drawn and emaciated. Standing there, I tried to summon feelings of grief, but I couldn't. I felt sadness and pity, but I couldn't cry. I stood there for a few seconds more before returning to the family pew to receive the guests who were filing by.

After the service the family boarded the funeral-home limos for the trip to River Rest Cemetery. I entered the one reserved for immediate family and sat next to my mother. We were directly behind the hearse that bore my father's casket. As we pulled away from the church, my eyes fixed on the casket, and thoughts of my father poured into my head.

The first thought was that my father and I had not been close during most of my teen years, and that we would never be able to change that. At that moment, grief consumed me, and I began to sob—quietly at first, and then uncontrollably. My mother squeezed my hand, but I continued to sob. After several minutes, the sobbing subsided, and I continued to reflect on my father's life and on our relationship.

I thought about the simplicity of my father's life. He had never been able to secure a job in the electronics field despite his time in the U.S. Army Signal Corps, his busy basement shop, his tinkering and self-education, and his undeniable technical proficiency. Instead, he went to his job at the post office every day. Until his final illness, I never knew him to call in sick. He left every day in one of the dark green J.C. Penney work suits my mom bought for him. He carried a lunch pail and a Thermos. The lunch pail contained a couple of sandwiches—bologna or pimento loaf or some other lunchmeat. The Thermos was filled with hot coffee, cream, and sugar.

I thought about the fact that my father didn't do much socializing. He and my mom never went out on the town. We didn't take family vacations. To my mom's constant chagrin, we never went out to eat. Whenever my mother raised that subject, my father would say, "I like your cookin'." That always ended the discussion.

From the day in 1951 when my parents arrived in Flint until the day he died, Pops never again set foot in Georgia. His mother and several siblings lived in Atlanta, but he never saw them except on those rare occasions when they visited Michigan. Pops never talked about his reasons for staying away from Georgia.

Pops' main avocation during his two decades in Flint was golf. He learned the game as a young man while caddying in Albany. He was constantly working on his game, and he won a few local tournaments.

Pops also loved to fish. He often dragged Fred and me along to his favorite spots. On the night before a big fishing trip, Fred and I would go out into the yard with flashlights and catch night crawlers for bait. Early the next morning, usually before sunrise, we would hop into Pops' station wagon and head for one of the many nearby lakes. Sometimes the trips were fun, especially when the weather was nice and we caught lots of fish. Other times, a fishing trip could result in a day of pure misery—few fish, bad weather, and severe hunger. My father always seemed to enjoy our outings, even if the weather was bad and even if we caught no fish.

Pops was a man of few words. I never heard him say, "I love you, son," or anything remotely resembling that. He seemed incapable of discussing anything on an emotional level. I had no idea how he felt about me. I was his son, and he was my father, and feelings just didn't enter into it. That's why an incident that occurred during one of our fishing trips surprised me so much.

My father and I were angling from the banks of Kearsley Lake, east of Flint. I was sitting near the water next to a rather large white man while my father sat up higher on the embankment trying out a fancy new spinning reel. I wasn't having much luck, so I cast my line out to different spots to see if I could find the fish. The white man kept glaring at me, as if I were bothering him. Then he said, "Why don't you move to another spot? You don't know what you're doing, and you're scaring the fish away."

I was surprised by his nasty, condescending tone, and for a moment I was speechless. I said, rather sheepishly, "I'm just trying to catch some fish like you are."

That seemed to infuriate the fisherman, and he made a nasty comment under his breath. I wasn't sure, but I thought he called me a name. My father must have heard what he said, because the next thing I knew he came crashing down the rocky embankment. Pops stood between the white man and me, and in a threatening tone, he asked, "What did you say?"

The white fisherman seemed taken aback. I was, too. I had never heard my father talk to anyone that way, and now Pops was facing down this white stranger over something the man had said to me.

The fisherman stood and confronted my father. Their faces were only inches apart, and I was sure one of them would throw a punch. The white man's face was beet red. My father's face looked angry and resolute. I was terrified—I didn't know what was going to happen.

The stranger said, "Your son doesn't know how to fish."

"What did you call him?" my father insisted, refusing to be drawn into an argument over fishing etiquette.

"I didn't call him nuthin'."

My father said, "Let it be nuthin' then."

Pops continued to stand there glaring at the white man. After a few uneasy seconds, the man turned and began collecting his fishing equipment, and then moved to a new spot. My father returned to his perch and resumed fishing.

I stood there shaking with fear and shock. Fear because my father could have been mortally wounded in a fight with the white stranger. Shock because my dad was willing to fight for me. I didn't know he cared.

I wasn't the only one in the family who had a tenuous emotional connection to my father. My brother and sister also had a strained relationship with him. My mother had trouble relating to Pops, too. I don't remember the two of them ever just talking and laughing together. Their conversations seemed to center on household chores

(Mom: "Did you cut the grass yet?" Dad: "What time are we going to eat?") and family finances.

My mother earned more money at Hurley Hospital than my father made at the post office, and she managed the family budget. Every week, they sat down at the kitchen table to review expenses. My father would present his on the back of a number-ten envelope. His expenses included things like "haircut" and "drinks" (meaning soda pop). Those were his personal spending priorities. Items like "groceries" or "school clothes" were left to my mother to manage.

Sometimes my parents would argue, usually about money, and sometimes the arguments became heated. During one particularly nasty dispute, Fred and I grew concerned. Our parents were in their bedroom, and my brother and I were crouched behind the living room sofa. When the fight reached peak intensity, Fred came up with an idea to defuse the situation. He crawled over to the hi-fi system on the other side of the living room and put a Mahalia Jackson gospel album on the phonograph, and slowly increased the volume until our parents could hear it. Then he snuck back to our hiding place and waited. I don't know if it was the gospel music that did it, but after several minutes the arguing subsided.

I resented my father. I thought those arguments with my mother were his fault. Pops was hard to relate to; my mother was easy to get along with. Pops hardly ever came to events I was involved in at school; my mother was almost always there. Pops found it hard to relate to people socially; my mother was articulate and outgoing. After years and years of work, the highest position my father achieved was postal maintenance technician. My mother advanced from floor nurse to assistant director of nursing at Flint's largest hospital.

Sitting near the head of the funeral procession, watching Pops travel to his final resting place, I gained a new perspective on his life. I began to understand his discomfort with interpersonal relationships and his attempts at escape. I surmised that he must have felt humiliated by his career circumstances because he knew he was capable of much more. He didn't possess the confidence and communication skills needed to change his situation, so he grew frustrated. Electronic gadgets, golf, and fishing were diversions from an otherwise stifled existence.

As much as I was coming to understand Pops, I also realized that his way was no way to live. Emotionally estranged from his wife, unable to communicate with his children, fearful of society—my father must have been burning up inside.

Despite frustration, insecurity, and disappointment, at least my father was there. However difficult it was for him, however he came up short, he fulfilled the roles of husband and father in the best way he could. He put on his work clothes and picked up his lunchbox and Thermos and went to work every day. He worked at a job he didn't love, but he knew he had to do it to help provide for his family. Through it all, he had enough imagination to maintain interesting hobbies, and he shared those hobbies with his sons—not in the manner of "Look, son, I'm going to show you how to design electronic circuits," but by osmosis.

I thought about how badly my father had wanted to be an engineer, and how he brought home all those books about radio and television and vacuum tubes and transistors and lasers. He had established the environment that led to my interest in engineering, and I had never thanked him.

The most pitiful thing about my father was that he died before he could fully enjoy the fruits of his work. He didn't get to retire and spend all his time fishing and golfing and playing with gadgets. And he died before I could get beyond my teenage angst and really get to know him.

As we approached the cemetery, I vowed I would live differently than my father had. I vowed that I would control my destiny, that I would work hard and go wherever I needed to go to be in control, to keep life from painting me into a corner.

I whispered under my breath, "I'm going to make it, Pops, you'll see."

A few minutes later, we buried my father.

Back on the U-M campus after the funeral, I learned I had earned an A on the ECE 214 exam. I picked up the blue exam booklet, and when I saw my grade I thought, *kiss my ass, Professor Volz*. That A was the best possible tribute to Pops.

At the end of the semester, I got an A in the course. I earned the respect of Dr. Volz, and he earned mine. I came to understand that he was not biased against me—that true to his billing, he was tough but fair. I realized, too, that my achievement in that first important electronics course would not have been possible had my father not blazed the trail. Despite stifling bias, Pops would not be stifled. He couldn't pursue his interests at work, so he pursued them at home. Despite disrespect and racism, he maintained his dignity, and he provided for his children.

Fred Douglas Tarver was a talented, decent, and earnest man. After he died, I finally came to appreciate him.

Part 2: Ann Arbor

THE AUTHOR AT UNIVERSITY OF MICHIGAN, 1974

15. Michigan Man

POPS' DEATH NOTWITHSTANDING, I was ready to meet the challenges and opportunities posed by the University of Michigan. I had worked too hard and had endured too much emotional turmoil to turn back. And then there was the silent promise I'd made to my father—that I wouldn't let anything or anyone stop me; that I would control my destiny.

My life as a "Michigan Man" began weeks earlier with preparations for the move from Flint to Ann Arbor. GMI had ingrained in me the importance of work experience, so before moving I contacted the U-M College of Engineering to seek an on-campus job. I wanted the practical experience, and I needed the spending money. Unlike GMI, a U-M education wasn't free, and despite my mother's willingness to pay for my education, I didn't want the entire burden to fall on her.

My job-seeking call to the dean of engineering's office scored a referral to Professor William J. Williams. I phoned Professor Williams, who sounded like a reasonable, no-nonsense guy. I asked if he had any jobs in his laboratory, and right away he hired me as an undergraduate research assistant. I was surprised because Professor Williams knew little about me other than that I was transferring from GMI and was interested in his research field: biomedical engineering.

Just prior to Labor Day, I loaded up the Camaro and took off on the fifty-five-mile jaunt to Ann Arbor. It was a bright, sunny day. Stevie Wonder had just released the album *Innervisions*, and I had it blasting from my cassette player as I cruised south on U.S.-23. My first Ann Arbor stop was the G. G. Brown Laboratory on North Campus. That was where Dr. Williams's research lab was located, and I wanted to meet him right away.

I found an office marked BIOELECTRICAL SCIENCES, and saw a white, middle-aged man with curly dark brown hair seated behind a desk. I knocked softly on the open door, and when the man looked up, I said, "Excuse me, I'm looking for Professor Williams."

The man got up from his chair, extended his hand, and said, "I'm Professor Williams. And you must be David Tarver, the man from GMI." Professor Williams smiled as he said, "the man from GMI," and I couldn't help smiling back. He seemed like a nice guy, not a cold and detached "professor type." He didn't seem surprised by my skin color—his face didn't distort into that weird but all too familiar expression that said *I wasn't expecting you to be black*. His hairstyle and his disarming, confident manner reminded me of John F. Kennedy. I took an immediate liking to my new boss.

Dr. Williams told me about his research as we toured the Bioelectrical Sciences Lab. He said his work focused on understanding how the brain communicates with the rest of the body. I was fascinated as he explained that the brain and muscles communicate via electrical signals, and that by decoding those signals he hoped to create medical breakthroughs for people with spinal-cord injuries. As I followed Dr. Williams around the lab and listened to his mini-lecture, I thought *this is important work that can really make a difference in the world*.

Of course I had other reasons to be excited about my move to the University of Michigan. One reason was girls. I was happy to trade GMI's near-monastic environment for a campus populated by thousands and thousands of voluptuous girls—and not just engineering girls, but social science girls and education girls and biology girls and psychology girls. I couldn't wait to start meeting them.

Another cause for excitement was the sports scene. Michigan had a full-blown major-university sports program that stood in stark contrast to the few intramural activities at GMI. More than 100,000 people gathered in Michigan Stadium for each home football game. In the winter, the action shifted next door to Crisler Arena for Big

Ten basketball. Even the intramural sports programs at Michigan were much bigger and better than those at GMI. With more than 40,000 students on campus, compared with 3,000 at GMI, everything seemed more grandiose at Michigan.

I was filled with excitement and anticipation, and a bit of apprehension, as I eased onto central campus in my Camaro. I was apprehensive because Michigan was so big, and because the engineering school was full of smart kids, each from the top of his or her high school class. The apprehension dampened my spirits, despite the beauty and warmth of the sunny late-summer afternoon. I remained dispirited until I remembered Reggie's Rule.

Whenever my friend Reggie Barnett and I found ourselves the only blacks in a high school class, and I expressed concern, he told me not to worry. He summed the situation up this way: "Look, Dave, in any of these classes, you're only competing against twenty-five percent of the students. Twenty-five percent are incompetent, twenty-five percent are lazy, and twenty-five percent are just not as good as you. That means you have a B in the class even before you get started. If you work hard, you should be right up there at the top of the class."

That was Reggie's Twenty-five Percent Rule, and he must have taken it to heart, because he graduated from Flint Central with a 3.95 grade-point average while starring every year in football, basketball, and track. As I thought about Reggie's Rule, I smiled softly, and the euphoria brought on by the beautiful fall afternoon returned.

My first stop on campus was the housing office. I needed a place to stay, but didn't want to live in a dormitory. I had heard too many stories about the wild and crazy goings-on in the dorms. That environment didn't seem compatible with the intense studying my engineering curriculum would require. My problem was that, as a transfer student, I didn't know anyone I could share an apartment with, and I didn't think I could afford my own.

The clerk at the housing office listened patiently as I explained that I was a transfer student in need of an apartment. She looked as if she had heard my predicament a thousand times. She suggested an apartment complex called University Towers, commonly referred to as U Towers. She said the rooms there were reasonably priced, so I wouldn't need a roommate. That was exciting news. U Towers sat at the eastern edge of campus, only a few blocks from the engineering buildings. It was a dull, grayish high-rise, one of the tallest buildings in Ann Arbor. A few retail shops were housed on the ground floor, and the building was surrounded by a variety of shops and eateries. I was thrilled by the idea of living in a relatively new high-rise, and pleased at not needing a roommate.

The U Towers rental agent confirmed that units were available for $106 per month. I gladly accepted an apartment at that rate, sight unseen. The agent then said I would be sharing an apartment suite with two other tenants, but that each tenant was separately responsible for his own rent payments. I was taken aback. It seemed that I *would* have roommates, people I hadn't chosen or even met. I was uneasy, but the school year was about to begin, and I needed the apartment, so I accepted.

I unloaded some belongings from the car and carried them up to my first-ever apartment. I unlocked the door and peered inside. The front door opened into the kitchen. The living room sat just beyond, with a bedroom on either side. A bathroom was situated just off the living room. The agent had told me my room was on the right-hand side of the apartment, so I immediately went there to settle in. I was shocked to find two beds, and even more shocked to find an olive-skinned fellow with a scraggly beard sitting on one of them.

"Hello" the scraggly one said. "My name Mehmet." His English was heavily accented but understandable, and I appreciated his cordiality.

"Hello" I replied tentatively. "I'm David. I was told that this was an apartment for three people."

"Yes, my friend and I are sleeping in this bed. You can have that one." He pointed to the other bed. "Another couple stays in the other room. They are from India, I think."

I was burning with anger, not at Mehmet, but at U Towers, as I realized I would have four roommates, not two: Mehmet and his "friend," and a couple from India who would be occupying the other room. And I would be paying a third of the rent even though five people were living in the apartment. I felt as if I had been taken.

Still, it wasn't a bad apartment, and Mehmet seemed like a nice guy. I needed a place to stay, and I didn't know anyone else I could share an apartment with. I had to give the arrangement a try.

I decided I might as well get to know my new roommate. "Where are you from, Mehmet?" I asked.

"I and my friend are from Turkey. We are here at the university to study education. From where are you?"

"I'm from Flint, just up the road," I replied. "I'm in the engineering school."

Mehmet and I continued to make small talk for a few minutes. He told me he liked Ann Arbor but missed his home. I told him I was looking forward to living in Ann Arbor, and that I was excited to study engineering. We talked, but we didn't seem to have much in common. After a while I started unpacking boxes and organizing my side of the room. Mehmet sat on his bed and read a book for a while, and then he went out.

That night, things began to heat up in our little apartment. First, Mehmet and his friend arrived together. I was in the living room setting up my stereo equipment when they came in. Mehmet introduced me to his friend, but the second Turk spoke hardly any English. His name was difficult for me to pronounce and even more difficult to remember. After the introductions, the two Turks went immediately to the bedroom. A few minutes later, my other roommates, the Indian couple, made their entrance. The man wore a wrinkled tweed sport jacket and an open-collar white shirt. The wife wore traditional Indian garb. When they saw me standing in the living room, they stopped dead in their tracks. They seemed shocked to see me, and I recognized that kind of shock. The last thing they expected to find in their living room was a large American Negro.

The Indian man tried to conceal his surprise, and from the spot where he and his wife stood frozen in the kitchen, he made his introductions.

"Hello," he said, tentatively. "I am Gopal and this is my wife Dipa." It was a terse and choppy introduction, and Dipa's gaze never left the kitchen floor.

"Hello, Gopal," I said, trying to sound cheery. "My name is David, and I guess we're going to be sharing this apartment. I'm a student in electrical engineering."

I thought the knowledge that I was an engineering student would put Gopal's mind at ease. It didn't.

"Nice to meet you," Gopal said. The brevity of his response was matched by its insincere tone. Gopal and his wife retreated to their room and closed the door. I heard the click of the deadbolt lock. Gopal and Dipa were in for the night.

I finished setting up my stereo system and put on some vinyl—*360 Degrees of Billy Paul*—at a respectful volume level. I chuckled when the cut *Am I Black Enough For You* came on:

> "We're gonna move on up, one by one
>
> We ain't gonna stop until the work gets done
>
> Am I black enough for yah?
>
> Am I black enough for yah?"

The song put me in a defiant mood. I finished unpacking and storing my things, and then sat back on the sofa to contemplate the situation. The apartment was far from what I had envisioned. It was nice enough, and it was modern, but it wasn't glamorous or fun. The Turks and the Indians certainly weren't ideal roommates. The rent, which had looked attractive at first, now seemed like a rip-off.

On the other hand, I had a place to live. I had moved to Ann Arbor on my own and had found a job. My roommates, while not exactly fun folks, probably wouldn't bother me or intrude on my business. And I lived just two blocks from the East Engineering building ("East Engin"), where most of my classes would be. *It's not quite what I expected, but at least I'm here*, I thought. *Things can only get better.* I decided to focus on my classes and on having the best possible time in Ann Arbor.

Most engineering students took three or four difficult courses in a semester. That first semester, despite being new to U-M, I took five. I had carried heavy course loads at GMI, where intense study was the rule, but U-M academic life presented new challenges. For one thing, most of my course subjects were new to me. Sure, I was familiar with electronic circuits, but not with subjects like electromagnetic field theory and linear algebra. Then there was modern physics. I had earned an A in each classical physics course I took at GMI, but modern physics at U-M was a different animal. We studied Einstein's theory of relativity and other new-age topics. The ideas were abstract, and I had a hard time grasping them.

The whole approach to teaching and learning at Michigan was different. At GMI, genial and experienced instructors led the classes, and their main role was to teach. It was similar to high school, but with more advanced, more specialized subjects. At Michigan, professors lectured but seldom interacted with students. Their primary charge was to do research and publish books and get famous, to bring acclaim to themselves and the university. Dealing with students was often delegated to graduate teaching assistants who explained the lectures, answered questions, and administered and graded exams. Many teaching assistants were from Asia, so dealing with them sometimes presented considerable language and cultural barriers.

At GMI, if you paid attention in class, did your homework, and studied hard, you would almost always do well on exams. At Michigan, you could study as hard as you wanted, but you might not even *recognize* the questions on an exam. U-M professors wanted students to demonstrate that they could extrapolate from what they learned in the classroom and apply that knowledge to new and unfamiliar situations. GMI instructors were usually satisfied to have students regurgitate what they'd learned in class. That difference made for some challenging exams during my first semester in Ann Arbor. I felt fortunate just to be keeping up.

The 1960s civil rights protests, the Detroit riots, and the U-M Black Action Movement (BAM) protests spurred diversification of the U-M student population. In 1970, BAM sponsored student strikes and took over buildings, demanding that the university admit more blacks. I was part of the first wave as Michigan began admitting more than a token number of African Americans. Even so, being in the College of Engineering avant-garde made for a lonely existence. Other areas of the university, such as the School of Education and the College of Literature, Science and Arts, had far more black students. I rarely saw them, though, because the College of Engineering was a world unto itself.

One day, feeling isolated and alone, I was walking toward the dean's office in West Engineering when I heard a female voice behind me.

"Hey, there! You're new here, aren't you?"

I turned around and was amazed to see an attractive black woman running up the hallway toward me. I wasn't sure if she was a student or an employee. She was casually dressed and wore her hair in a neat Afro.

"Yes, I just transferred here from GMI, up in Flint." I tried my best to sound cool and confident, seeking to make a good impression.

"Well, I'm Anne Monterio, and I run MEPO. Have you heard about it?"

"Uh, no, I haven't," I said. I didn't want to seem uninformed, but I had to tell the truth.

"Well, Mr. GMI, what's your name?"

"Oh, I'm David Tarver, and I'm in ECE." ECE was the acronym for Electrical and Computer Engineering—no one ever used the full name.

"Oh, a double-E," Ms. Monterio said. "I'm impressed." Double-E was college slang for electrical engineer, and was a cooler expression than "ECE."

"Well, Mr. Tarver, you need to come by my office and get acquainted! That's how we do things around here."

Anne didn't strike me as pushy. Despite her unfamiliar Brooklyn accent, she seemed warm and welcoming, in sharp contrast to the drab hallways of West Engineering. I walked with Anne to her office, and that's when she elaborated about MEPO, the Minority Engineering Programs Office. Anne explained that the office was set up by the engineering dean to help blacks and other minorities deal with life in the College of Engineering. She said if I had any problems, if I needed academic help, if I had trouble with a professor, I should come to see her. Anne had a completely disarming way about her, and I immediately felt at ease.

I was glad to meet Anne and find out about MEPO, but I didn't think I would need any academic help. I had always thought of myself as a top student, and given Reggie's Twenty-five Percent Rule, I planned to be in the upper echelon of any class I took. MEPO was a welcome sight, though, a place where black engineering students could congregate, commiserate, and exchange information. MEPO seemed like a little black island in the midst of a vast engineering sea.

And what a sea it was! In most classes I was the only black student, and the environment was often hostile. Repeatedly, when I raised my hand to ask a question, I heard snickers or saw exasperated, impatient expressions on my classmates' faces. It was as if they felt my question proved I wasn't qualified to be in the class. At first, I was intimidated and reluctant to ask questions, but I soon noticed that when the professor answered my question, other students scribbled in their notebooks. I realized that they had wanted to ask the same question but had been afraid to. I developed a thick skin. I decided that if I didn't know something, I was going to ask, and to hell with what others thought. I knew I wasn't stupid, and even if I was, I wasn't going to get smart by not asking questions.

Classroom life wasn't all bad. After a few weeks, I bonded with a couple of foreign students, a guy named Carlos from Brazil, and a woman named Adilah from Kuwait. Both were shunned by classmates, just as I was, and both seemed to be serious students and interesting people.

Carlos was a studying maniac. When he left class, he would immediately go somewhere to study, either his room or the library or one of the small study compartments in the East Engineering building. Carlos was a bright guy, and was always prepared for class, but he had no practical engineering experience. Since I had so much experience as a hobbyist and as a GMI co-op student, I was able to add practical perspective to the things we learned in class. Carlos and I were great study partners. He would come up with the right answers to the homework problems, and I would tell him how each answer applied in the real world.

Adilah was another story. From the first day I saw her in class, I felt sorry for her. None of our classmates ever spoke to Adilah. On those few occasions when she asked a question in class, other students treated her even worse than they treated me. Adilah dressed conservatively and seemed extremely shy, adding to her isolation. One day after class, I decided I would befriend Adilah and get to know her better. I knew she also lived in U Towers, because I often saw her entering the building. On the way home one day, I struck up a conversation.

"Hello," I said.

"Hello."

"That was a pretty tough class today, wouldn't you say?"

"Yes. I think I will need some help before the next exam."

"Well, I'd be glad to study with you if you want."

"Yes, that would be nice."

After that, Adilah and I studied together a few times in the East Engineering building. We often walked back to U Towers together after class, and we started to talk about things other than school. One day, Adilah invited me to her apartment for dinner. I was quite surprised, because I thought she was much too shy to do anything like that.

On the day of our dinner, I arrived at Adilah's apartment not knowing what to expect. I didn't feel a romantic attraction to her, and I hoped the sentiment was mutual. I suspected and hoped Adilah was just expressing appreciation to her study buddy.

Dinner was a huge disappointment. No, it was more than disappointing—it was the worst meal I ever had. I didn't know if it was the lamb—I had never eaten lamb—or if it was the unusual spices, or if Adilah just couldn't cook. Whatever the reason, the food tasted horrible, and the dinner was torture because I had to be polite and pretend I was enjoying it, though I wanted to spit it out.

After dinner, we talked about our backgrounds. Adilah told me about her upbringing in Kuwait, and that her father wanted her to be an engineer. I told her about my life in Flint, and that I had loved electronics since grade school. The Yom Kippur war was in the news at the time, and I was curious about Adilah's view of the Arab/Israeli conflict. I figured that being from the Middle East, she might have a special perspective on the situation, so I asked.

"What do you think the chances are for peace in the Middle East?"

Adilah turned very serious and stern. Without hesitating, she looked straight into my eyes and said, "There will be no peace in the Middle East until the Palestinians have a home."

I didn't even know who the Palestinians were.

The relationship with my roommates rapidly unraveled. Mehmet and I continued to get along reasonably well, but his friend never spoke to me. I attributed that to the language barrier, but I couldn't be sure. Gopal and Dipa kept pretty much to themselves; Gopal spent almost the entire day in class, and Dipa holed up in their bedroom. Sometimes while I was away at class, Dipa would venture into the kitchen and cook a pungent Indian stew. On those few occasions when I returned to the apartment and found her in the kitchen, Dipa ran to her bedroom and locked the door. I found that humiliating.

The Turks never seemed to change their bedding, and between that and Dipa's aromatic cooking, my nose was never wanting for stimulation. After enduring those conditions for a few weeks, I retreated home to Flint to rest and sleep in my comfortable old bed. As I lay there thinking about the good old days, I felt an itch. I looked down at my forearm and saw a small scab, and I immediately did what any red-blooded young male would do—I picked at it. To my dismay, the scab began to fight back. Suddenly, it leaped off my arm and ran across the bed.

My heart started to beat rapidly. *What the heck was that?* I was frightened out of my wits. The anxiety began to subside as I realized the "scab" had been some kind of insect. I spent the rest of the weekend at home and didn't think much more about it.

I returned to Ann Arbor on Sunday night, rested and ready for another week at the apartment. When I arrived, the Turks and the Indians were absent. Before turning in for the night, I went to the bathroom. While standing in front of the toilet, I noticed a bug on my shorts that resembled the one I had seen in Flint. *So that's where it went!* Then I saw another, and another, and I started to get frightened all over again. Upon closer inspection, I noticed what must have been hundreds of the little creatures.

A few of the bugs were dead, so I pinched one between my fingers and inspected it more closely. It was one of the strangest bugs I had ever seen, because it looked like a little crab, complete with near-microscopic pincers. I couldn't bear to think that thousands of those little things were residing in my pubic region at that very moment. What were they? Would they kill me? If I went to sleep, would the Turks find me dead in the morning, half-eaten by tiny crustaceans?

I went to bed, but between the itching and the anxiety I couldn't sleep. First thing in the morning, I dashed over to the student health center. As soon as I entered the building, I saw a huge banner that depicted a giant version of the bugs that were infesting my body. I didn't know if I should be relieved or terrified.

My condition turned out to be well known and common: I had lice. The little creatures had occupied every patch of hair on my body, including my afro. The health-center nurse said they probably came from the bedding in the not-so-tidy room I shared with the Turks. She brought me some medicated shampoo and told me to shower with it a few times. I did, and within a couple of days the lice were gone.

Back at the apartment, I resolved that the Turks had to go. The Indians too. I was sick of my living conditions. I was sick of the Turks' smelly sheets. I was sick of Dipa's pungent stews, and I was sick of hearing the Indians' door slam and lock every time I entered the apartment. I had to take control of my environment, so I decided to drive them all from the apartment with music—very loud *black* music. I cranked up my stereo late at night. For extra emphasis, I played along on my trombone. The Turks and the Indians were too timid to complain. They just got up the next morning, bleary-eyed, and went about their business. My strategy worked. Within a few days, Gopal and Dipa, Mehmet, and Turk #2 all moved out.

I had the place to myself for several days, but then U Towers assigned two new tenants. My new roommates were brothers—brothers as in siblings, and "brothers" as in African Americans. The brothers were well dressed and clean, and they seemed like nice guys. We communicated easily, and they were Michigan natives, just like me. I was confident my roommate problems were solved.

One afternoon, I returned to the apartment to find the brothers sitting on the living room sofa, listening to *my* records on *my* stereo. A huge pile of marijuana was stacked on the coffee table in front of them, and they were calmly transferring it to little plastic bags. The older of the two brothers looked at me severely and said, "Don't tell nobody 'bout this shit, man."

I was too frightened to discuss the situation, so I just said okay and left.

It was my turn to move out. A few days later, I found a one-bedroom apartment far away from U Towers. The cost was much higher—$180 per month—and the location much less convenient, but I never wanted to see the brothers or the Turks or the Indians again, and I never wanted another roommate.

All things considered, I felt fortunate to have that first semester behind me. I survived the loss of my father. I survived the transition from GMI to U-M. I survived U Towers. Sitting in my nice one-bedroom apartment at the Village Green, near U-M's North Campus, I resolved to do better.

16. Cheated

MY SECOND SEMESTER AT MICHIGAN WAS MUCH LIKE THE FIRST—I survived but didn't distinguish myself. My courses didn't interest me much and didn't seem all that relevant. I was still learning the ropes, and felt as if I hadn't made the transition to full-fledged Michigan Man. At the end of the first year at Michigan, my junior year, I had a solid B average. Trouble was I never saw myself as a B student. I resolved to do better in my senior year, come hell or high water.

The summer after junior year, I stayed in Ann Arbor and got a job at a place called Environmental Research Institute of Michigan—ERIM. The name suggested an outfit that was devising ways to counter air and water pollution, but in fact most of ERIM's work dealt with remote sensing and surveillance, and was financed by the U.S. Department of Defense. My job at ERIM was to write computer programs to simulate the operation of airborne radar systems, specifically Synthetic Aperture Radar. The work was interesting, and it allowed me to hone my math and computer skills. Summer went by quickly, and I was ready to attack my senior year courses.

I signed up for some interesting, tough courses in the first semester. One was Electrical Biophysics, ECE 471. It combined engineering, biology, physics, and math to describe the human nervous system as an electrical communication network. A two-professor team taught it: one was my Bioelectrical Sciences Lab mentor and employer, Dr. William J. Williams; the other was Dr. Spencer Bement.

Williams and Bement were an odd team. Whereas Professor Williams was Kennedy-esque, cerebral, and original, Professor Bement impressed me as stern, uptight, and dull. In 1970s pop-psychology terms, I regarded him as "parental"—definitely not an "I'm Okay, You're Okay" type of guy. During the first few class sessions, I noted a distinct difference in the way the two professors acted toward me. Professor Williams was friendly and familiar. Though he was my employer at the G. G. Brown Lab, our classroom relationship was strictly professor/student. I didn't expect special treatment, and Professor Williams didn't signal that he was willing to provide any. Professor Bement was another story. His behavior made me feel that he resented my very presence in the classroom. While he sometimes spoke informally with other students, he was strictly formal with me. His answers to the questions I posed struck me as terse and condescending.

I considered Professor Bement's contributions to the course minimal. He seemed to leave the more esoteric aspects of the material to Dr. Williams, who seemed to be the intellectual driver of the course. For me, Bement was little more than a source of bad vibes.

Despite my apprehension about Professor Bement, I was bubbling with excitement about ECE 471. The course dealt with fascinating topics such as detection of taste and smell and light, and transmission of sensations from nerve endings to the brain. The course included laboratory exercises such as measuring the speed of nerve impulses and analyzing electrocardiogram (EKG) signals. This was exactly the kind of interesting, stimulating work I had hoped to find at Michigan.

I worked hard in ECE 471, and had an A going into the final exam. I thought if I could just ace the final I would be well on my way to establishing myself as a top student in the College of Engineering.

Final exams were the week before Christmas, and by then I was mentally exhausted, on the verge of crazy. I wasn't walking around campus muttering to myself like some other burnt-out students, but I was close. In spite of my precarious mental state, I was prepared for finals and confident I would get an A in ECE 471.

On the day of the final, I drove down to Central Campus and had a nice breakfast at the Brown Jug restaurant. Then I made my way to the exam room in East Engineering. I was early. I was prepared, relaxed, and ready. I didn't want charity, and I didn't want luck—I just wanted an A. I *expected* an A.

Professor Bement passed out the exam. I thought it odd that he was administering it, because that role was typically delegated to a teaching assistant. When the exam came to me, I opened my blue book, picked up a fresh #2 pencil, and started to have at it. The questions on the exam were the tough, open-ended type I'd come to expect at Michigan. When I finished, I was confident I had done well, but I wasn't sure. The exam was complex, and would be graded "on a curve," which meant my grade would depend on how the other students did.

Toward the end of finals week, a huge snowstorm hit Ann Arbor. By December 23, most students had left town for the Christmas break, but I stayed around to confirm my grades. This was my make-or-break semester, and I was determined to see it through to the bitter end. Professors Williams and Bement had said they would leave our graded exam booklets in the East Engineering study room, a tight little space where students crammed between classes. The professors had indicated that they would place each student's overall numeric score and course grade on his exam booklet.

That snowy morning, I cautiously made my way to the study room and immediately found the exam booklets stacked on a chair. I glanced at the cover of the blue book on top of the stack. It wasn't mine, but sure enough, there was a number and a grade on the cover. The number was 212, and the grade was A+. I surmised that the owner of that blue book had achieved a total numeric score of 212, and that score qualified him for an A+ grade. Lucky bastard! Alone in the deserted study room, I proceeded to thumb through the stack to find mine. I was nervous and excited as I examined the booklets, one by one. I certainly can't remember everyone's exact grade, but the following approximates what I discovered:

212 ... A+ ... not me
201 ... A ... not me
195 ... A... not me
190 ... A ... not me
185 ... A– ... not me
176 ... B+ ... me!

I was dejected. After all my hard work in ECE 471, I barely missed getting an A. In fact, it looked as if my grade was the highest B. I slipped my blue book out of the stack and turned to leave the room, but something made me stop. I went back to examine the other booklets. I found the place where I had removed my book from the stack and continued to examine the remaining ones.

174 ... A–
171 ... A–
165 ... B+
160 ... B

What? Oh shit! Apparently, two people with a lower overall score than mine received an A– as their final grade. I couldn't believe it. I went through the entire pile again from top to bottom. There was no mistaking it—my score was higher than two others that qualified for A's.

Shaking with anger, I ran to the Electrical Engineering office and asked the secretary if I could use the phone. I reached Professor Williams at his G. G. Brown office. I told him there seemed to be a mistake with my course grade, and Dr. Williams agreed to see me.

I ran outside and jumped into my Camaro. The snow was really coming down, and the streets were completely covered. I sped out to North Campus, skidding through snowdrifts at every turn. My heart was racing. There had to be some mistake. This wasn't fair. I was in no mood to be toyed with. I was half-crazed from the long semester and the seemingly endless hours of studying.

I arrived at G.G. Brown and bolted into Professor Williams' office. He wore a bemused look as he sat behind his desk and listened to my story. He must have noted my desperation, because my voice quivered with each word. He asked to take a look at my blue book. He looked at the cover and started flipping through the pages, then consulted a journal that contained a record of the course grades. After what seemed like an eternity, he spoke.

"Well, David, it looks like you might have an issue here. Professor Bement handled the grading of the exams and the assignment of the final grades, so I'll have to speak to him about this. I'll get back to you."

I knew it! I had gotten bad vibes from Professor Bement the moment I set foot in his classroom. I was sure Bement was trying to screw me, but why? Was he a racist? His body language, the way he treated me in class, and this apparent manipulation of my grade all said yes. Whatever his motivation, I was not going to let the injustice stand. I had worked too hard to let Bement spoil my achievement. I was determined to get my rightful grade no matter what.

I went home for Christmas seething with anger. All I could think about was getting my grade changed, and doubt about the outcome spoiled the holidays. The day after Christmas I was on the phone to Dr. Williams. Sure enough, he had followed up on my grade.

"Well, David," he said, "I spoke with Professor Bement. He tells me he based a portion of each student's grade on his own subjective criteria. That's why you had a higher numeric score than a couple of other students but got a lower overall grade."

"But that's not fair!" I blurted into the phone. I was hot and could feel myself trembling, and my voice was somewhere between shouting and crying. "I got a higher score than those other students, so I shouldn't get a lower grade than they did!"

Professor Williams said, "We've decided to change your grade. You'll be getting an A-minus in the course."

It took a moment for what he said to register. I said, "Thank you, Professor Williams." I was mentally and emotionally exhausted.

"You're welcome, David. I guess we won't be leaving the blue books in the study room anymore."

I found Professor Williams' last comment odd. It seemed to imply that Professor Bement's mistake wasn't denying me my rightful grade—it was allowing me to find out about it.

Professor Williams changed my grade to A–, which was an A in the official record. I had foiled what I regarded as Professor Bement's attempt to cheat me. Though his bias seemed blatant, I still didn't want to believe it. Prior to that episode, most people in positions of authority seemed supportive of, or at least indifferent to, my aspirations. The experience with Professor Bement was a real eye-opener.

That semester, I made the dean's list in the College of Engineering for the first time. I proved to myself and my professors that I could excel, and I believed that my success would be limited only by my desire.

I was proud of my academic achievement. I was even prouder that I stood up and didn't allow myself to be cheated.

17. No-Plea McAfee

MY ACADEMIC PERFORMANCE IN THE FALL SEMESTER OF 1974, the first semester of my senior year, was truly a breakthrough. It was a great performance, tempered only by the fact that I didn't get straight A's. As it turned out, two courses prevented me from achieving that lofty goal. One was ECE 380, Semiconductor Physics, and the other was ECE 312, Electronic Circuits.

Semiconductor Physics was taught by Professor Chai Yeh. According to his bio, Professor Yeh received his PhD in physics from Harvard University in 1936. As a Chinese national who earned a doctorate from the world's foremost university, at a time when segregation against non-whites was rampant in the U.S., he must have been truly brilliant. Professor Yeh was a short, slight man. He probably weighed no more than 120 pounds, including the uniform he wore to every class: plain black shoes, a neat, old gray suit, and white cotton gloves. Professor Yeh wore the gloves to protect his hands from the chalk that flew from the blackboard during his lectures.

I liked Professor Yeh. Trouble was I couldn't understand a word he said. Yeh spoke with a pronounced Chinese accent I had never encountered in Flint. I did my best to decipher his words, and religiously transcribed every equation he wrote on the board. At night, I reviewed my notes and read the relevant chapter in the textbook, but in the end my efforts only took me so far.

After six weeks in Professor Yeh's class, I made a breakthrough.

I realized that when I thought he was saying "arfa," he was actually saying "alpha," the first letter in the Greek alphabet, a common engineering symbol. All of a sudden, a notebook full of nonsensical class notes fell into place as I scrambled to replace the word "arfa" with the symbol "alpha." That realization helped a lot, but alas, there were too many other undecipherable "Yeh-isms" in my notes to yield a complete understanding of the course. In the end, I felt fortunate to receive a B in Semiconductor Physics. The only thing that saved me from an even lower grade was that the course was graded on the curve, and most of the other students couldn't understand Professor Yeh, either.

The other course that precluded a perfect semester was ECE 312, Electronic Circuits. At the beginning of the semester, I felt supremely confident about that course for two reasons. First, I had been working with electronic circuits since elementary school, and felt I could probably teach my professors about them. Second, the instructor for the course, Dr. Leo McAfee, was a black man—a brother! He was the only black professor in the U-M College of Engineering. Surely I would not fail to get an A in a course that dealt with my favorite subject, a course taught by a black man. I was brimming with confidence, certain that ECE 312 was going to be a piece of cake.

Curious about Professor McAfee, I carefully examined his bio. He had received his bachelor's degree in electrical engineering from Prairie View A&M University in Texas, a school I had never heard of. He received his PhD from the University of Michigan in 1971, a couple of years before I arrived. The U-M degree was impressive, but how had he come to be the only black engineering professor at the university? Was he really that good? Did he know his stuff about electronic circuits, or was I going to have to teach him a thing or two? I was cocky to the point of giddiness, and couldn't wait for the class to begin.

Professor McAfee was a dark-skinned, husky man who spoke with a Southern accent consistent with his Texas roots. He seemed confident but in no way arrogant. He seemed like a down-to-earth, nice guy. As he reviewed the course material, the quizzes and exams we would have to take, and his method of assigning grades, I could see that he was a competent, no-nonsense person. After the first class, I stopped to introduce myself.

"Professor McAfee," I said. "I'm looking forward to taking this course. I have a lot of experience with electronic circuits, so I think I'm going to like it. I got an A in 214, so I think I should do well in your class."

It was tacky of me to mention my A in the first electronics course, but I wanted Professor McAfee to know that I was no slouch, and that I expected an A in his course.

Professor McAfee looked at me and smiled. "This class is a lot harder than 214," he said, "but if you work hard you should be able to do well."

Had we made a "brother-to-brother" connection? I couldn't tell.

The first few weeks of the course went pretty much as expected. I was familiar with the material, and enjoyed demonstrating my electronics knowledge to Professor McAfee and my fellow students. Around the middle of the semester, the material started to get more difficult. I was familiar with the circuits, but the course delved into detailed analyses, complete with math-intensive computer models. Computerized circuit analysis was new to me, and before long I found myself struggling to keep up. Dr. McAfee seemed to revel in my struggles, as though he was putting me, the "uppity young brother," in my place.

I wasn't the only student he had to straighten out. From the first day of class, I noticed that several of the white students were doing their best to catch Professor McAfee in an error. They would question the accuracy of an equation he presented on the board, and when that failed to expose any inaccuracy, they would throw out all kinds of obscure "what-ifs" to test the range of his knowledge. They seemed more motivated by a desire to show Dr. McAfee up than by their interest in the course material. I found myself quietly rooting for the professor in those encounters, even as I tried to show off my own knowledge. Dr. McAfee was unruffled. He seemed pleased by our enthusiastic participation in his course, regardless of our motivation.

As the end of the semester drew near, I found myself spending more and more time in Professor McAfee's office. He was one of the few professors who observed regular office hours, during which he personally dealt with students' questions. I took advantage of his accessibility as often as I could. During our discussions, I sometimes brought up my other courses and mentioned that I was intent on getting A's in all my classes. I tried to get to know Dr. McAfee—where he came from, how he liked being a professor at Michigan, what his family was like. True, I was still hoping for that "brother-to-brother" connection, but I was also genuinely interested in this strange and rare beast, this black electrical engineering professor. Dr. McAfee readily answered all my questions, academic and non-academic, and soon I found myself in awe of him, and a bit jealous. He was a black man who had already achieved much of what I aspired to in the field of electrical engineering.

Through diligent but not extreme effort, I had a B+ going into the final exam. I felt confident I would be able to excel on the final and get an A in the course. On the day of the final, I arrived at the classroom rested and ready. Dr. McAfee had assigned a teaching assistant (TA) to administer the exam, and as soon as I received it, I opened my blue book and started. As I reviewed the questions, a quiet despair set in. The exam was a Michigan classic—one in which the questions seemed barely related to the course material. First I struggled to understand each question, then to determine which course principles I needed to apply to arrive at a solution. It was the exam from hell, and when the allotted time for completion rolled around, I was sure I had blown it. Then a miracle happened: The TA didn't collect the blue books! Was his delay intentional, or was it a fortuitous oversight? I didn't care; I just continued to work. Most of the students turned in their blue books and left, but several of us continued to plow through the exam. The TA never came. Finally, after another hour or so, having dealt with every question as best I could, I gave up and turned in my blue book. I was hopeful—the extra time was just what I needed to get things sorted out.

Shortly after the exam, the course grades for ECE 312 were posted. I received a B+. I was disappointed, but had no reason to suspect that Professor McAfee had been unfair. He knew how much I wanted an A in his class, and I assumed that if he hadn't given me an A, I must not have earned one. Still, I decided to ask him for an explanation of my grade, and to see if there was some way my little B+ could magically turn into an A–.

I arrived at Professor McAfee's office ready to plead for an ever-so-slight adjustment of my grade. Of course I wouldn't explicitly ask him to "help a brother out," but the implication would be clear. When I sat down next to his desk, I decided to start with a simple, rational approach.

"Dr. McAfee, I got a B+ in your course, and I was just wondering if you would review the grades for me so I can see how close I was to getting an A."

"No problem, Dave," Dr. McAfee began. He took out his grade book and opened it to the relevant page. He smiled as he showed me my final exam grade, and then he said, "You did pretty well—you actually got the highest B+, but there was a significant gap between your grade and the lowest A–. There is no way I could justify giving you an A in the course."

The highest B+! The highest B+! Professor McAfee's words kept ringing in my ears, and his smile only accentuated my pain. Was he playing a game with me? I couldn't say a word. I couldn't argue with his math or his logic. I was disappointed I hadn't worked harder in the course. I had taken for granted that I would get an A, and my overconfidence had come back to bite me.

As I left Dr. McAfee's office, I felt no anger toward him. In fact, I felt an emotion that brought me close to tears, an emotion born of admiration and respect for Dr. McAfee. He had given me the grade I'd earned, no more and no less. He wasn't afraid to show me how he had arrived at my grade, and wasn't at all defensive about presenting it to me. In fact, he smiled. It would have been easy for Dr. Mac to "help" the only black student in his class. Perhaps he had been tempted to do so. In the end, though, his integrity wouldn't let him.

Professor McAfee gave me the grade I deserved, and in the process I came to understand how he earned his nickname: "No-Plea McAfee."

18. Gotcha!

JANUARY 1975 MARKED THE BEGINNING of my final semester at Michigan. Most of the previous two years were a blur, obscured by long hours of study, exams, difficult professors, and the general intensity of campus life. Sitting in my apartment at Village Green, I felt satisfied and confident. In the just-completed semester, I had finally achieved an academic breakthrough. If I had just one more great semester, I would graduate *cum laude* from the College of Engineering. Some might have considered that quite an achievement for a black kid from Flint, but from the beginning, I had expected nothing less.

I needed to take only three courses that final semester to complete my bachelor's degree. That seemed like a pretty light load, because I was accustomed to taking four or five. I could have maintained a light load to focus on getting A's, but I felt confident, so I decided to be more aggressive and take two graduate courses along with my last three undergraduate courses. I planned to stay at Michigan to get a master's degree, and getting the first two graduate courses under my belt would allow me to finish sooner. I didn't want to stay in Ann Arbor any longer than necessary. I was anxious to finish college and start my professional career.

One of the undergraduate courses I chose proved that I was either truly crazy or unbelievably cocky. That was ECE 472, Biomedical Instrumentation Design, and the instructor was none other than Dr. Spencer Bement. I hadn't spoken to Bement since his attempt to give me a B in Electrical Biophysics was overruled by Professor Williams.

I regretted not reporting Professor Bement's behavior to the dean of engineering. I hadn't done it because I was able to get the A after all, and because I was afraid of retaliation by him or other engineering professors. Bement couldn't have been happy about being forced to change my grade, but I hoped we could put that episode behind us. I needed to take his course in case I decided to pursue a bioengineering graduate degree.

On the first day of class, I was optimistic. I didn't want to believe Professor Bement would repeat his earlier shenanigans. Besides, after the grading fiasco in Electrical Biophysics, he had to know I would be right there in his office if I didn't get the grade I deserved. I was convinced that ECE 472 would be easy. Sure, the course title sounded esoteric—Biomedical Instrumentation Design—but in fact the technology was pretty basic. I was confident I would receive nothing less than an A.

At the end of week one, Dr. Bement dropped a surprise quiz on the class. He administered it himself, and he stayed in the classroom while we completed it. I found the quiz unbelievably easy, but easy quizzes weren't unusual during the first week of a course. I was glad to start off with a quick A.

A few days later Bement returned the graded quizzes. When I looked at my paper, I couldn't believe my eyes. The numeric grade was eighty-two, and the letter grade was B. I was incredulous. There had to be a mistake. Professor Bement was beginning the day's lecture, but I wasn't listening. I was going over the exam to see why my score was eighty-two and not one hundred. I quickly came to the quiz problem Bement had marked wrong. It had to do with an operational amplifier (op-amp) circuit design. I had been working with op-amps since junior high school, and was certain I had answered the problem correctly. I went over the problem again and again, but I couldn't see why Bement had marked my answer wrong.

After class I went up to Professor Bement to ask about my grade. I pointed to the quiz problem and asked him to show me my error.

"Your answer was correct," he said, "but you didn't show enough of your work to demonstrate how you *arrived* at the answer. Besides, it looks a little suspicious because the guy next to you had the same answer."

I was so angry I could hardly speak. Who the hell was Bement to mark my answer wrong because he couldn't see how I got it? And was he accusing me of cheating? I was shaking with disbelief and anger, and I felt like fighting.

"I would *expect* both of us to have the same answer if we were both right, wouldn't you?" My tone was defiant. In those few moments any optimism I felt about Bement was gone, and I lost any residual respect I had for him. "I know this stuff inside out!" I continued. "I didn't write down all of the steps because I didn't have to. I had the right answer, so I can't understand how you would mark it wrong!"

Professor Bement's face was red, his body erect. His expression was somewhere between disdain and hatred. He didn't raise his voice, but I could tell he was agitated. I could imagine him thinking *how dare this uppity black kid challenge me? I'm going to put him in his place, and Williams won't be around to rescue him this time.* If those were his thoughts, his response didn't betray them. He simply sneered and said, "Next time, be sure to show your work, and be sure to work independently."

I snatched the paper back from him and stormed out of the classroom. I needed to talk to someone about this, to come up with a way to keep Bement from cheating me again. Whom could I talk to? Professor Williams? Dr. McAfee? After reflecting for a few minutes, I decided to take my concerns to Anne Monterio in the Minority Engineering Programs Office. Anne had been around the College of Engineering for a while, so she would know what to do. If I needed to see the dean of engineering about Bement, she could tell me how to go about it, since she was on the dean's staff.

After talking to Anne, I felt a little better. She advised me not to go to the dean. Anne said faculty supported each other, and that the dean would be reluctant to come to the aid of an individual student unless the professor's transgression was well documented. She told me to stay in the course, continue to work hard, and document my work to the hilt. She said I shouldn't give Bement any excuse to reduce my grade. It wasn't the advice I wanted, but I knew it was sound and sincere. I decided to heed it.

For the rest of that semester, I was a model student. I meticulously diagrammed my answers on every homework assignment, every quiz, and every exam. I would not be accused of failing to show my work adequately, or of cheating, again. I participated in classroom discussions, asked good questions during lectures, and generally tried to be a good, respectful student. None of that seemed to change Professor Bement's demeanor. His bearing was aloof, and his manner of responding to my questions was condescending. I was barely able to take it with a straight face, but I did.

Several weeks into the term, I had to complete a project for Bement's class, and the deadline was fast approaching. If I waited until the next class period to submit my project, it would be late, and Bement would penalize me heavily. He wasn't scheduled to be in his office, so I couldn't drop it off there. I called Bement to tell him that my project was finished and that I wanted to turn it in as soon as possible. He confirmed that he didn't have any office hours scheduled, but he didn't offer a solution. I wasn't about to let him off the hook. I remembered Anne's advice: "Don't give him any excuse, David!"

I said, "Professor Bement, I'd be glad to bring the assignment anywhere you like. I'll bring it to your house if you want."

Bement paused as if he was trying to figure a way out of that proposal. Finally he said, "Okay, you can bring it to my house, but have it there tonight. I won't be available at any other time."

"Fine," I said.

Professor Bement gave me directions and told me what time to show up. He lived on the outskirts of Ann Arbor, in a rural area less than fifteen minutes from my apartment. I was happy to deliver the project to him, and I thought that by doing so, maybe, just maybe, I would be able to break the ice that had formed between us.

Later that night, I drove out to Bement's place. I expected to roll up his driveway, get out, and go inside. I thought I might meet his wife and kids, if he had any. Maybe we would sit near his fireplace and talk. Maybe Bement would begin to see me in a different light. As soon as I pulled into his driveway, all such hopes were dashed. My headlights illuminated a long driveway snaking toward the car. Professor Bement was walking toward the car dressed in a heavy coat and knee-high boots. He was accompanied by two large Dobermans. Bement stopped next to the car, and I rolled down the window. He didn't say hello. He simply asked, "Do you have the assignment?"

I pushed the papers through the window into his outstretched hand. Bement took them, turned around, and marched back up the driveway with his dogs. There was no invitation to the house, no introduction to the family. I felt sure that the dogs were meant to send a message: *Don't mess with me!*

I finally accepted that Professor Bement and I would never be friends. I really didn't care about that, because I didn't like the man. I just wanted an A in his class so I could graduate from the University of Michigan with honors.

The ECE 472 final exam took place a few weeks later. Regardless how well I had done prior to the exam—and I had done very well—I knew I would have to ace the final to get an A. I knew Professor Bement would never give me a break, and I didn't want one.

The final was easy. Unlike most U-M exams, the questions didn't require much intellectual extrapolation. I only had to recall what we had covered in class. I had no problem finishing in the allotted time, even though I carefully documented my thought process along with each answer. At the end of the exam, I handed in my blue book, certain that I had earned an A in the course. I left the exam room feeling relieved and happy.

Grades were posted the week after the exam. Bement had shocked and disappointed me twice, but I didn't think he would do it again. I was wrong. The grade I received was a B. I had already calculated that if I received a B in ECE 472, I would miss graduating with honors by less than seven one-hundredths of a point.

I couldn't let the grade stand. I immediately called Bement's office. I had to see him to hear firsthand how he arrived at the decision to give me a B. I decided that if he wouldn't see me, I'd go directly to the dean.

I reached Professor Bement by phone at his North Campus office. To my surprise, he agreed without hesitation to see me. When I arrived at G. G. Brown, he was alone, seated behind his desk.

"Hello, Dave, come on in," Bement muttered, half under his

breath. I sat across the desk from him. "How can I help you?"

I started right in, and I didn't attempt to hide my anger. "Professor Bement, I see you gave me a B in 472, and I don't understand how that could be. I was doing very well going into the final, and I'm sure I did well on the exam. None of the questions presented a problem for me."

"Well," Bement intoned, "I guess you didn't do well enough relative to everyone else. That's why you got a B."

"If you don't mind, I'd like to see my exam book so I can see what I missed. I'd also like to see how my total score in the class compared to everyone else's."

"I'm sorry," Professor Bement said. He didn't look sorry; he looked smug. "I'm not revealing any total scores this semester, and I'm not returning the blue books. The course is over, and the grades are assigned. That's it."

"But that's not fair! I should be able to see how I did. I'm able to see that in every other class!"

"Well, that's not the way I'm running *this* class," Bement snarled. "You've done pretty well, and I understand you're graduating next week. You should feel satisfied about that."

So that was it. Bement thought I should be satisfied with a B. I couldn't believe what I was hearing, and couldn't believe I was sitting there taking all of his crap without screaming at him, or worse, jumping over the desk and punching him. I decided to appeal to him for understanding, for sympathy. He had never shown those qualities toward me, but I thought it was worth a try.

"Professor Bement," I pleaded, "getting a B in this course knocks me out of graduating with honors. I believe I deserve an A, and all I want is for you to show me why I didn't get one."

Bement looked at me sternly. "You didn't get an A because in my judgment you didn't deserve one, and that's it."

I didn't know what to do or what else I could say. Professor Bement's unfairness was overwhelming. I was completely broken.

I cried.

I could see what Bement was doing to me, and I knew why he was doing it, yet the dominant emotion I felt wasn't anger or hatred, it was hurt. Through my sobs, I said in a low voice, "Oh well, I guess graduating with honors doesn't mean all that much. It's just a number."

I was through pleading. I was rationalizing my B to myself. I was trying to find common ground with Bement. In a perverted way, I was trying to make sure he didn't feel too bad about giving me a B.

Bement responded, "Oh, no, it's very important. Graduating with honors means a lot."

Bement was rubbing my nose in my B, and that alone convinced me that he was intentionally cheating me. It seemed I had caught him the previous semester, and now he was getting his revenge.

In the ensuing days, I tried to forget my hurt feelings as I prepared for commencement, but something hardened within me. The hurt I felt slowly turned to resolve. I resolved that I wouldn't let Professor Bement or anyone else stop me. My ultimate goal wasn't to graduate from Michigan with honors, though in my mind I had achieved that. My ultimate goal was to build and run my own electronics business. My undergraduate studies at the University of Michigan completed, I was one huge step closer to achieving that.

19. Almost Obsolete

COMMENCEMENT WAS HELD AT CRISLER ARENA, the U-M basketball palace. It was May 1975, the near-midpoint of another tumultuous decade. Graduates from many disciplines, along with their families, completely filled the arena. Most were full of excitement and anticipation as they received their degrees and contemplated a post-collegiate existence.

Despite several opportunities, I hadn't interviewed for a job during my senior year. I told potential employers I would be staying at U-M for my master's degree, and that I hoped to interview with them when I finished. I again took a summer job with ERIM, doing mostly defense-related work, and did my best to enjoy the break.

In the fall I began the full-time quest for a master's degree. With two graduate courses already under my belt, I expected to complete the master's in just one year. I was eager to finish, find a job, and start laying the groundwork for my own business. My core graduate school courses involved math, communication systems, and signal processing. I was focused on the technologies of the future, technologies enabled and enhanced by the invention of the transistor, the integrated circuit, and the microcomputer.

Because entrepreneurship was my ultimate goal, I also enrolled in a U-M Business School course called Small Business Policy. I had never taken a business course, and on the first day of class I sat somewhat uneasily amid a roomful of MBA candidates. I was nonetheless confident because business, like electronics, was in my blood, and because I felt that nothing could be more difficult than an engineering course.

The Small Business Policy instructor was Professor Larue Tone Hosmer. His eccentricity seemed to extend beyond his name—his patrician bearing and manner of speaking struck me as Eastern Elite. On the first day of class, Dr. Hosmer spent a few minutes describing his background. He had received his MBA from Harvard, but his bio then took a surprising turn. Having noticed that most of his fellow Harvard MBA grads were heading to the New York City banking and investment industry, Hosmer did something different. He struck out for the South American rainforest and launched an exotic-wood export business. After making a large sum of money, he returned to the U.S., got his PhD, and began his teaching career.

After he related that story, Dr. Hosmer had my undivided attention. He was an excellent role model, having already accomplished what I only hoped to do. I wanted to start a business and make millions, then, as I had told friends, "study anthropology." Professor Hosmer's career seemed like a perfect template. I especially liked that he had chosen the path less traveled. I also admired that he'd walked away from a successful career at midlife and begun a new one. I imagined that few Harvard alums on Wall Street had careers as fulfilling as Hosmer's.

Small Business Policy turned out to be everything I had hoped and more. Each week, we analyzed an actual small business. We reviewed the businesses' organization, financial performance, operations, sales, and marketing plans. Our assignment was to understand the business and formulate a strategy for success. Dr. Hosmer then let us know what had actually happened to the business, and in some cases the business founder visited our class. Small Business Policy showed me the challenges faced by a variety of businesses and how the owners dealt with those challenges. I gained insight that might otherwise have taken years, and learned to avoid some common business pitfalls. By the end of the course, I was more convinced than ever that I absolutely had to start my own business.

I also took two graduate courses from "No Plea McAfee." Dr. Mac and I had become friends and tennis buddies, but I knew our friendship would not carry over into the classroom. The first class I took from him was an electronics course called Digital Logic Design. It was interesting, and I worked hard at it, determined to avoid getting any more B's. The other course I pursued with Professor McAfee was Directed Study, ECE 598. I chose the course topic myself: the design and construction of a digital electronic music synthesizer. Directed Study allowed me to get academic credit for pursuing my passion. I looked forward to leaving Ann Arbor with a master's degree in electrical engineering and a completed synthesizer.

I had spent countless hours during undergraduate school working on my synthesizer, but it was strictly a hobby. The project was much more fun and motivating than my class work. I assembled and tested the electronic circuits on the workbench in my Village Green apartment. All of my gear—an imposing array of electronic instruments and design prototypes—was perched a few feet from my bed.

Several classmates teased me about my bedroom electronics lab. My friend Nick said, "You're never going to get any girls in here with all that junk next to your bed!" The teasing bounced off me. The synthesizer was my passion, my first love, my all-consuming avocation. In fact, I still planned for it to form the core of my entrepreneurial adventure.

My zeal for the synthesizer project was so strong that I sometimes found classes a distraction. There were, however, occasions when I was able to apply something I'd learned in class to my synthesizer design, such as when I learned to design a universal asynchronous receiver transmitter (UART).

A UART was an electronic system that recognized when a key on a keyboard was pressed, then transmitted a digital code representing that key to another device such as a computer. My synthesizer keyboard looked like a regular piano keyboard, but it worked like a computer keyboard. Its sole function was to determine what key the musician pressed and transmit a code to the synthesizer, which would then play the appropriate note. It was a slick design, and I was convinced I was one of the first to devise such an innovative music keyboard. I had learned about UARTs in my undergraduate Digital Computer Hardware Design class and had immediately begun building one in my bedroom laboratory. The design required lots of logic chips and a "rat's nest" of interconnecting wires. After a considerable time, I got my unwieldy UART working, and was proud of my achievement.

I shared the details of my synthesizer project with few other students, but I did discuss it with Les McDermott, a fellow electrical engineering student who hailed from Detroit. Les was an albino African American. His light beige skin had a pinkish hue, and his hair was nappy and blonde. Les was legally blind, and wore eyeglasses with lenses that looked like small telescopes. He was tall and skinny, and his face was covered with pimples and reddish splotches. Les looked and acted like the quintessential engineering nerd.

I respected Les because, though he was barely scraping by academically, he had lots of practical electronics knowledge. He was also an audiophile par excellence, and maintained a state-of-the-art hi-fi system and an extensive record collection. Les's most endearing feature, though, was that he was absolutely crazy—right out there on the edge. I relished his uninhibited, freethinking spirit.

Les' girlfriend—he called her his concubine—was Vickie, a young white woman. They had a baby boy named Wesley. Les loved Vickie to death, and Vickie certainly loved Les. The two of them often invited me to their university-housing apartment for spaghetti and wine, and over dinner we covered a wide range of topics. We discussed the College of Engineering, where Les seemed to know every professor and every student. We discussed the state of the world—the Vietnam War, the civil rights struggle, the Nixon impeachment. Of course we discussed hi-fi audio technology and music. Regardless of the issue, Les always delivered a thoughtfully considered and vociferously stated opinion. I enjoyed being around him and Vickie, which was good, because they had sort of adopted me.

During one of our dinners, the conversation turned to the topic of my synthesizer project. I had just finished building the UART circuit, and couldn't wait to tell Les about it.

"Les, I've been working on building a UART for the past few weeks, and I finally got it to work! I'm really coming along on the synthesizer now."

I expected Les to raise his glass and give me one of his famous attaboys, but he just glared at me through his telescopic lenses. He seemed incredulous.

"You been working on what?" he asked. He had a blank stare on his face. He looked as if I had just told him I'd taken a crap in his bathroom sink.

"A UART circuit," I said.

I couldn't understand Les's response. Was he jealous? Was he surprised I could build a UART by myself?

"This thing took about twenty-five chips to build, and the wiring is kind of messy, but it works. That means my keyboard design is pretty much done, man. It's great!"

"Wait right here," Les said. He got up from the table and walked over to one of his cinder-block-and-plywood bookshelves. He grabbed what looked like a paperback book, came back to the table, pulled his chair over, and sat down next to me. Les opened the book and bent over until the little telescopes on his glasses nearly touched the pages. He started leafing through, stopping when he came to something that said General Instrument D AY-5-1013 Universal Asynchronous Receiver/Transmitter. Les then ceremoniously laid the book on the table in front of me. I couldn't believe what I saw.

Les didn't say a word. He just sat there and let it sink in. He wasn't laughing at me. He wasn't even smiling. He wore a serious, officious look, as though he were teaching me an important lesson. I resented that, because I was a much better student than Les. In fact, he'd already been at Michigan nearly five years and still didn't have his bachelor's degree. I had demonstrated the knowledge to design and build my own UART, and Les was showing me how to take the easy way out. Typical.

"You mean this one chip does everything my twenty-five-chip UART does?" I croaked. A sick feeling was gathering inside me. I felt lightheaded and disoriented.

"Yep. All that and a lot more," Les said apologetically.

"Can I take the book home with me?"

"No problem—plenty more where that came from." Les wasn't gloating, but I could tell he was full of himself. He knew he had taught me a lesson, based on his superior practical knowledge. Me, a grad student no less!

So ended the fun spaghetti-dinner evening. When I reached my apartment, I sat at my workbench and opened the book to the page that described the UART circuit. Sure enough, Les's one chip did a lot more than my twenty-five chips and rat's nest of wires. Still, I was proud of my achievement, and couldn't bring myself to concede that I should abandon my circuit for Les's fancy new chip. I lay on my bed, fully clothed. The only light in the room was the green glow from my workbench oscilloscope. Feeling terrible, I fell asleep.

I awoke to 6:00 a.m. sunlight streaming through my bedroom window. At first I didn't know where I was, much less what had transpired the night before. Slowly the events of the previous evening oozed into my consciousness. Was it a dream? I rolled over and spotted the words "General Instrument Data Catalog 1975." It was real. Then it hit me, sure as the sunbeams' blinding glow: I had to use the chip. No matter how proud I was of my design, I knew in my heart it couldn't compete with the General Instrument chip, which did more and cost less. It was a no-brainer. My UART circuit was obsolete, and if I didn't use this latest bit of technology, my entire synthesizer would be obsolete, too, before it ever saw the light of day. That prospect was unacceptable.

Realizing it would be unwise to spend precious time and money building something I could easily buy, I used the General Instrument UART chip. With it, I finished my digital electronic music synthesizer project, and earned an A+ in Directed Study from Professor McAfee.

I had accomplished the goals I set for myself at University of Michigan. I'd proved to myself that I could make it at Michigan, and could compete with the best engineering students. I'd proved I could handle the interminable hours of study, the abstract concepts, and the intellectual extrapolation required of a Michigan engineer. I'd proved I could overcome a doubting, antagonistic professor whose grading of my work looked like, talked like, and walked like racism.

By August 1976 I was more than ready to leave Ann Arbor and begin the next phase of my career. I had a lot more to prove.

20. Not Dr. Tarver

ON A WARM JUNE EVENING IN 1976, I sat on the living room floor of my apartment mired in confusion and doubt. The room was dark, the only sounds the occasional passing car and the muffled music and voices from other apartments. A dull ache started in my gut and rose toward my neck. My head felt unusually light, and my eyes felt as if they would burst forth tears at any moment. I should have been happy. I was about to receive my master's degree in electrical engineering, leave Ann Arbor, and begin my professional career; however, I was anything but happy, because I had an important decision to make, and uncertainty was tearing me apart.

In the preceding months, I had interviewed on campus with several corporate recruiters and had visited several potential employers. I was looking for a place where I could find stimulating, challenging work and gain the skills I would need to start my own business. The main things I wanted from an employer were money and on-the-job training. I planned to work only one or two years before starting my own company. In that respect, I was like many other ambitious college grads who dreamed of starting a business. Employers counted on the fact that interesting work, steady pay, marriage, and kids would make most would-be entrepreneurs abandon their dreams. I was determined to be different from most.

My job search included some of the crown jewels of American corporate capitalism and engineering technology: I visited Ford Motor

Company in nearby Dearborn; IBM Office Products in Austin, Texas; NASA Lewis Research Center (later named Glenn Research Center for astronaut and U.S. senator John Glenn) in Cleveland, Ohio; and Bell Telephone Laboratories in Holmdel, New Jersey.

Ford was an impressive company, and working there would have allowed me to stay in Michigan, but I couldn't see myself working in the auto industry again. That ship sailed when I left GMI. I was quite impressed with IBM, and even more impressed with Austin, but I felt IBM's regimented corporate culture wouldn't allow the flexibility or the time to pursue my business plans. I enjoyed visiting NASA, and was tempted by the prestige of working there, but experience in space research wasn't the kind I needed to start a manufacturing business. Each of those companies was attractive in its own way, and each extended a job offer, but the only place that truly fit my plan was Bell Telephone Laboratories.

Bell Labs scientists had pioneered countless breakthroughs in electronics and telecommunications, including the transistor—arguably the most important invention of the 20th century. Bell Labs scientists also developed much of the communication theory I learned at Michigan. The scientists and engineers there performed both fundamental research and product development. "The Labs" was the best of both worlds—a place where one could explore the fundamentals of physics and electronics, and where one could develop cutting-edge electronic products. Bell Labs seemed to have all the ingredients I needed to prepare for my own electronics business.

There was only one problem: Bell Labs was located in New Jersey. Prior to my interview visit, a *Newsweek* article about cancer clusters reported that Monmouth County had the country's highest cancer rate. The Bell Labs location I visited was in the heart of Monmouth County.

New Jersey had several other undesirable features. I encountered two on the drive from Newark Airport to my hotel. The first was the smell. Oil and gas refineries lined the New Jersey Turnpike southwest of the airport, and gave off an odor so noxious I had to hold my breath and hope the stench would pass. My nose ached and my eyes watered, even after my limo driver closed the car windows. After twenty minutes or so, the smell subsided, and I encountered another undesirable feature: toll roads. I watched in amazement as the driver stopped to pay a toll collector upon exiting the New Jersey Turnpike. A few minutes later I was dumbfounded as he pulled to a stop at a row of tollbooths that stretched across the Garden State Parkway. He tossed a quarter into a basket and off we went. Michigan had nothing like that. It seemed an affront to one's freedom and dignity to stop every few miles and throw coins into a basket.

New Jersey was not my kind of place.

My Bell Labs interviews were the most extensive I experienced. I met with three different departments, and was asked at the end of the process to list my first, second, and third choice.

The Digital Terminal Laboratory was the first department I visited. The engineers there were developing electronic switching systems—sophisticated devices that formed the nervous system of the modern telephone network. Before the interview began, my host, a supervisor in the department, took me on a tour of his lab. Just as we were getting started, we encountered a black technician who seemed engrossed in his work. The supervisor asked the technician to get us some coffee, whereupon the tech dutifully dropped what he was doing and went to fetch some. At that moment, I knew I would not be working in the Digital Terminal Laboratory.

My next stop was the Private Branch Exchange (PBX) Systems Engineering Department. The folks there were nice enough, but they weren't actually designing anything. Their function was to set the guidelines for the engineering departments that did the actual design. In short, they were paper pushers, and I wasn't interested in that sort of work.

I was beginning to think Bell Labs might not be the place for me after all. When I saw the name of the third department, I all but threw in the towel. It was the Special Services Maintenance Systems Department. I didn't know what its members did, but the name suggested they might be designing better ways to sweep floors and take out trash.

The lead interviewer was a thirty-something white guy, John Colton. His title was group supervisor, and he seemed pretty sharp.

As we walked the halls of his department, I was surprised to see blacks, Asians, and a few women mixed in with the ubiquitous white males. Colton's department seemed far more diverse than the others. His description of the department's work was quite different from what I'd imagined. The department was developing computer-based systems to maintain far-flung electronic communication networks. The word "maintenance" had thrown me off. They weren't designing ways to maintain buildings; they were devising systems to diagnose and maintain the health of the most advanced telecommunications networks. I was comfortable with John Colton and intrigued by his department. I decided my first choice—my only choice—was the Special Services Maintenance Systems Department.

Colton really wanted me in his group. He didn't seem concerned when I told him about my music synthesizer project and my desire to start a business. He said I would gain experience in his group that would aid my aspirations. Colton's open, supportive attitude impressed me. I was sold.

I returned to Ann Arbor convinced I would go to work for Bell Labs. Still, the New Jersey location nagged at me. I hated what I had seen of the state, and wasn't anxious to make New Jersey my home. Austin, Dearborn, and even Cleveland seemed like much nicer places. I couldn't shake the thought of moving to sunny Austin and making a nice, comfortable living. On the other hand, in New Jersey I would be working and learning at the leading edge of my field. I knew where I needed to go, but needed to talk it over with someone.

I called my brother-in-law, Evans, and related my misgivings about New Jersey. He listened carefully and then told me I was making the right decision. He assured me I could carve out the kind of life I wanted in New Jersey—that it would be a matter of finding my way around and adjusting to the differences. At any rate, he said, I didn't have to stay in New Jersey forever. I could get the knowledge and experience I needed and get out.

Evans was right. I had to go to New Jersey.

Several days later Bell Labs offered me the job. As I stood in my apartment holding the letter, I was overcome with pride. It seemed I was ending one long journey and beginning another. The offer felt like a huge accomplishment, because most engineering graduates would have killed for a job at Bell Labs.

I was ready to accept the offer, but an offer I hadn't expected quickly followed it. The University of Illinois wrote to offer me admission to its PhD program in electrical engineering, plus a fellowship and a teaching assistant position. I had spoken with Paul Parker, a black assistant dean at Illinois, a few months earlier, and had completed an application merely to keep my career options open, but when I received the letter, I realized for the first time that getting a PhD was a real option. It was exciting to think that in just a few years I could be Dr. Tarver. *Dr. Tarver!* A kid from Flint who couldn't even place in the top fifty elementary science fair contestants could actually reach the highest academic station in the electrical engineering profession.

It was almost too much to comprehend. In one hand, I held a job offer from the top industrial research organization in the history of the world—Bell Telephone Laboratories. In the other, I held an offer to study for a PhD at the highly regarded University of Illinois. I was proud, but also confused and a bit afraid. Which was the right path? Could I adjust to life in New Jersey? Could I succeed at Bell Labs? Was I really PhD material, and could I succeed at Illinois? Would I encounter a supportive, fair faculty there, or might it be a hostile environment where racists would attempt to block my path?

Holding the two letters, I sat down on the floor and propped my back against the patio-furniture loveseat that served as my sofa. I had to work through the two offers and make a decision. I knew I faced a fork in the road, one that would determine how my dreams would unfold. I was convinced that working at Bell Labs would best prepare me for starting my own business. I knew pursuing a PhD was an unnecessary diversion that could take years out of my life. Still, the prestige of the PhD was enormous, and achieving it would be a huge accomplishment. It would open every aspect of engineering to me—product development, research, teaching, management. The words "Doctor Tarver" would resonate far beyond the walls of the university, far beyond even the field of engineering. A PhD meant power and credibility, and a part of me wanted that. If I accepted the Illinois offer, I could become Dr. Tarver. I would merely have to defer, or perhaps sacrifice, my dream of starting a business.

Defer or sacrifice my dream. As my mind replayed those words, the fog inside my head lifted. I had never dreamed of studying for a PhD. I had never dreamed of writing research papers or teaching at a university. My dream since junior high school was to start my own electronics business. If that dream was to have any meaning, I had to pursue it. It was now or never.

With that realization, tears started streaming down my face. Sitting on the floor holding letters that invited me down two very different career paths, I chose one. I had invested too much to change course.

Part 3: New Jersey

"GIANT TRANSISTOR" WATER TOWER AT ENTRANCE TO BELL LABS, HOLMDEL, NEW JERSEY.

21. Mecca

TWO MONTHS AFTER CHOOSING TO BEGIN MY CAREER at Bell Labs, I was driving toward my first day of work. Despite the gravity of my career decision, I wasn't exactly thinking about my new job. The morning of August 9, 1976, cruising along Crawford's Corner Road in Holmdel, New Jersey, I was thinking about a photo I had seen in a Bell Labs recruiting brochure. The photo showed a trim young Asian woman playing tennis on a pristine court in a grassy field.

The caption identified the woman as Leesing Pang, a Bell Labs MTS: "Member of Technical Staff." Ms. Pang's short tennis skirt revealed taut and athletic legs, and her right arm was stretched in perfect form—wrist back, slight bend at the elbow—as she struck the tennis ball. Despite her obvious physical exertion, Ms. Pang's expression was calm and confident, and her delicate features sparkled with the joy of friendly competition. The photo was meant to capture the reader's imagination and to offer a glimpse into the nature of a Bell Labs MTS. The subliminal message was: Bell Labs people are attractive, intelligent, fit, hardworking, cosmopolitan, and international—the best of the best.

I had long been in awe of Bell Labs because of its unique place in the history of technology. The Bell Labs mission was noble and inspiring: to discover new facts about the physical world; to blaze new trails in fields such as mathematics and physics and chemistry; to devise new ways for people to communicate. I couldn't imagine a better place to begin a career as an electrical engineer.

As I approached my destination, the Bell Labs water tower came into view. The massive white structure, a local landmark, was constructed in the shape of Bell Labs' seminal invention—the transistor. The tower communicated several messages to the technically astute Bell Labs community. It commemorated the genius of Bell Labs people. It represented practical innovation, because Bell Labs planners turned what could have been a dreary water tower into an interesting architectural feature. It signaled the confident and fun-loving nature of Bell Labs people. I chuckled as I passed the tower and veered onto the Bell Labs property, knowing I was about to join a very special company.

Lush sugar maples lined the entrance drive and formed a partition between the road and the many acres of manicured lawn on either side. The gleaming Bell Labs structure was directly in front of me. An enormous rectangular prism, the building was only six stories high but very long, as if a minimalist skyscraper had been laid on its side. Bathed in morning sun, its glass skin reflected the surrounding bucolic landscape. Geysers erupted from a large pond in front of the building. The view was breathtaking and otherworldly, as if a gigantic alien spacecraft had settled amid a rolling cow pasture.

I parked in the shrubbery-enclosed visitor parking lot between the pond and the building, and walked up elegant slate steps to the entrance. I told the guard at the door I was reporting for my first day of work, and he pointed me straight ahead toward the reception area.

I paused for a moment to gather the scene. I had been at Bell Labs only once, during the interview visit, and the place still amazed me. The renowned architect Eero Saarinen, designer of the Saint Louis arch, had designed the Bell Labs Holmdel building. The outer shell was subdivided into four inner buildings, each situated in one quadrant of the outer structure. The space between the inner buildings formed a huge cross-shaped atrium, and catwalks lined the interior perimeter.

I looked up to my right and left at the futuristic catwalks, with their steel and glass balustrades, that ran along each level of the building. The reception area was directly in front of me. It was sunken and seemed carved into the stone floor, and was covered in modern industrial-grade carpet. Long bench seats covered with thick red leather cushions were carved into the perimeter of the reception area. A large stone desk rose from the center of the space. Beyond the reception area, I saw the massive center atrium surrounded by still more catwalks. The scene could have been from a science-fiction movie, but it wasn't. It was a scene from my new life.

A small sign near the front of the reception desk carried the Bell Labs logo and the name "Ann Walling." A middle-aged woman looked up as I approached, and in a pleasant Scottish accent she said, "Good morning, young man. You have the look of a new employee." Her friendly manner put me at ease. Ms. Walling signed me in and presented an ID badge, and a few minutes later I was seated in the personnel department being processed onto the payroll.

The induction process was more interesting than I had expected. Orientation stressed the many benefits of working at Bell Labs. These included not only full medical coverage and generous pension and savings plans, but also a few benefits that seemed especially unusual and generous. I learned about the Bell Labs Club, a collection of many company-sponsored employee activities including basketball, soccer, travel, gardening, and even model railroading. I also learned that Bell Labs offered in-house courses in numerous technical subjects, usually taught by the company's own experts during work hours. Then there was the Doctoral Support Program, in which Bell Labs paid qualifying employees to obtain a PhD, with time off for the required classroom work. The orientation session left me even more enthusiastic. I was eager to get to my job and find out what life at Bell Labs was really like.

A secretary escorted me to John Colton's office. My new supervisor was seated behind his desk poring over a memo. He saw me and said, "David! You're finally here!" He jumped up and walked around his desk to shake my hand, and then offered me a seat. He seemed genuinely happy to see me.

Colton's office was small and not at all fancy. The flooring was the gray tile that was everywhere in the building. The furniture included the standard-issue metal desk and spring-loaded executive chair. A couple of guest chairs sat next to the desk, and a five-foot table stacked high with monographs, memos, and drawings sat behind it. Colton's office told me he was a workhorse, not a show horse.

"How was the trip out from Michigan?" Colton asked.

"Oh, it was fine," I said. "I really enjoyed the drive."

"So, where are you staying?" Colton continued.

"Well, that's another story," I said. "Personnel put me up at the Molly Pitcher Inn in Red Bank. It's not bad, but it's not exactly the Hyatt."

Colton laughed. "Oh, yeah, the good-ol' Molly," he said. "Well, in a couple of weeks you'll be in your own place and the Molly will just be a bad memory."

I was surprised at how easy it was to talk to Colton, and how readily he identified with my situation. In just two minutes he seemed more colleague than boss. Most important, he didn't seem at all uncomfortable having a black engineer in his group.

"Before we get down to business, I've got to tell you about basketball," Colton said with a sly smile. "When you were here for your interview, you told me you were a good player, so now it's time to show what you've got. We play basketball every night after work at the Holmdel Village School. Some of the guys from our league team, the Knicks, will be there. If you're really a good player, the other teams are going to try to recruit you. You don't have to play for the Knicks, but just remember who signs your paycheck!" Colton laughed to make it clear he was joking.

"I'm definitely ready to play some ball," I said. "I look forward to getting some exercise, too. Anything is better than sitting around at the Molly. I'll bring my stuff with me tomorrow."

After that icebreaker, Colton got down to business. He said my assignment was to design new equipment for Remote Test System 5 (RTS-5). He started describing how my assignment fit with the organization's other projects, and the acronyms came flying out of his mouth faster than I could write them down: PC-1A, SMAS, DACS, CMS-3, and so on. Seeing the confused look on my face, Colton stopped abruptly and said, "Don't worry, we'll go over all of this again later." Then he changed the subject.

"How's the music synthesizer project coming along?" he asked.

I was pleasantly surprised that Colton remembered. I was also glad to switch to a technical subject I was familiar with.

"It's going pretty well," I said. "I just have to tie up some loose ends, and then get some musicians to try it out."

"Well, the synthesizer technology you described when you interviewed here is very similar to what we'll need on RTS-5," Colton said. "That's why I wanted you in my group. Most new hires don't have the experience you have. I think we can learn a lot from you, and I think you'll learn things on this project that will help you with your synthesizer."

I suspected Colton was just blowing smoke when he talked about learning a lot from me. It made me feel important, though, and it was flattering to think the folks at the venerable Bell Labs *could* learn something from me.

"Well, I'm excited about the project," I said. "And I'm anxious to get started."

Colton's look grew more serious. "This is a great project, Dave, and the organization is growing fast. I know you're an ambitious guy. I think you're going to do well, and I think you'll move up fast."

That was exactly what I wanted to hear.

Colton went on to deal with some "housekeeping" issues. "I'm sure they told you in personnel that starting time is eight thirty. I don't worry too much about what time my guys get here—just try not to get in too far past nine o'clock. I know you're going to have a lot of things to take care of as you get settled, so if you have to leave during the day, just let me know. You can always make up the time later. And I'm sure they told you in personnel we don't have sick days here. If you're sick and need to take a day off, just call the department secretary and let her know."

My mind flashed back to when I was a GMI co-op student working at AC Spark Plug. Each morning a horn signaled the start of the workday. The horn sounded again at lunchtime and again at quitting time. Bell Labs seemed much less regimented than that, much more relaxed.

Colton glanced at his watch and said, "Oh, it's lunchtime! Lunch is on us today."

He got up and walked into the hallway, and I followed closely.

"We're having lunch with Ed Spack and Bob Yeager today, but first I want to introduce you to some of the other guys in the group."

Colton and I walked along the corridor, stopping at several different offices. At each stop, Colton said, "Hey guys, I want you to meet Dave Tarver. He just joined the group today." Almost every person in Colton's group seemed to be from a different race or ethnic group. I was surprised and relieved to see another black group member—a fellow from Jamaica. I had never encountered such a diverse group at any place I had ever worked or interviewed.

Colton and I then joined the department head, Ed Spack, and the director, Bob Yeager, for lunch in an executive dining room. Ed and Bob were much older than John, and much less engaging. The food was excellent, and the atmosphere pleasant, but I felt almost no connection to the older execs. After a few minutes of small talk, they resorted to quizzing John about the status of the group's projects, and I just listened and ate. I realized that, for Ed and Bob, the meal was merely a perfunctory aspect of the new employee induction process.

After lunch, Colton and I parted company with Ed and Bob, and John escorted me to my new two-person office. It was a little bigger than his, and a doorway in the rear wall led to another office. One of the two desks was occupied by a fellow whose head was buried in a technical drawing. He swiveled in his chair and faced us. Colton said, "George, this is your new office mate, Dave Tarver. Dave, this is George Callahan."

George jumped up and snapped to attention, military style. He was of medium height and build. Despite his graying and thinning hair he was trim and energetic. George gripped my hand firmly and said, "Yes, sir, nice to meet you."

"Nice to meet you, George," I said.

"George is an old-timer here," Colton said. "He'll show you the ropes and help you get whatever you need to get started. I'll let you get settled. Feel free to wander around."

Colton left, and George and I spent a few minutes getting acquainted. He seemed quite curious about my background, and wanted to know where I had gone to school, what degree I had obtained, and what my Bell Labs job title was. I told George I was an MTS, and that as far as I knew I would be designing modules for RTS-5. I was surprised to learn that George wasn't an MTS—he was an STA (senior technical associate), a few steps down the pecking order. George's role was to provide technical support to the group. He had spent a career in the navy before joining the Bell System at Western Electric, then working his way up to Bell Labs. He seemed proud of his accomplishments and pleased with his position.

While George and I talked, an even older white man and a middle-aged white woman entered the office. Again, George spun his chair toward the doorway, and said, "Ron, Ruth—this is Dave. He's a new MTS in Colton's group. He just started today." George's tone was cheery, as if he was happy to have a new office mate.

Ron and Ruth issued a curt hello, then hustled to the back office and closed the door. I didn't know what to make of their ungracious response.

George must have seen the puzzled look on my face. He said, "Ron and Ruth don't work for Colton. They're in Gilmore's group."

I was relieved. I didn't know anything about "Gilmore's group," but was glad I wasn't in it.

George abruptly changed the subject. "If you're going to work here, you're going to need some supplies. Let's take a trip down to the stockroom."

I jumped at the opportunity to get out of the office and find out more about Bell Labs. "Okay, let's go," I said.

The stockroom, on the lower level of the building, resembled an electronics and office supply store. George said, "You can get whatever you need here—binders, lab notebooks, pencils, pens, calculators. You can also get electronic parts like resistors and transistors and integrated circuits. Just pick out what you want and sign it out over there. If you don't see what you're looking for, they probably have it behind the counter."

I was incredulous. "You mean I can get anything—I just have to sign for it?"

"That's right," George replied. "Pretty neat, huh?"

I shopped around for a while before settling on some basic items: three-ring binders and paper, lab notebooks, pens and pencils. I figured I'd come back for electronic equipment and parts after I learned more about my assignment.

"George eyed my new blue Bell Labs binders and said, "You know, Dave, practically every kid in Holmdel is going to school with three-ring binders from the Bell Labs stockroom. And a lot of guys come down here to get supplies for their government projects."

"I don't understand," I said. "Government projects?"

"Yeah, you know—personal projects. Things they're building at home like a burglar alarm or a fancy stereo system. We call that kind of stuff government projects. If you're taking parts home for a personal project, and someone asks you about it, just say, 'It's for the government' and they'll leave you alone."

George chuckled. I could tell he was happy to relate some inside information I was unaware of. I chuckled too, uneasily. I intended to keep working on my music synthesizer project at home, but I had no intention of using Bell Labs parts, no matter how accepted the practice was. For a moment I imagined myself jailed for stealing parts from the stockroom. I imagined folks saying, "See, we hired this black guy and look what happened!"

After the trip to the stockroom, I decided to take John Colton's advice and wander the halls. I wanted to reintroduce myself to the group members I'd met on the way to lunch and find out more about them.

The first colleagues I stopped to see were Jose Garcia and Steve Sato. Jose and Steve looked a few years older than I. Jose was a lanky white guy, and wore a full beard. Steve was Asian, and was short and trim. Jose was the more outgoing one. When I asked where he was from, he said his family had emigrated from Cuba after being dispossessed by Castro. His tone and demeanor assured me he was no fan of the dictator. Steve Sato said he'd grown up in Hawaii and that his folks had emigrated there from Japan.

Next I stopped in to see Tony Genovese and Mike Tom. Tony and his family had emigrated from Italy, and he was by far the best-dressed guy in the group. His nicely pressed dress slacks, silk shirt, and designer loafers contrasted markedly with the khakis and cotton shirts worn by other group members. After we talked for a while, Tony said, "When you get settled, you have to come over for dinner with me and my wife Gianna." I was pleasantly surprised by the invitation, and told Tony I would definitely take him up on it.

Mike Tom was from Taiwan. I noticed that the nameplate on his desk said "M.K.K. Tom." I asked what the K.K. stood for, and he said they were the initials of his Chinese name, Kin Kui. I asked him why he didn't use his Chinese name, and he said "Mike" was easier for people to pronounce.

Next I looked in on Lloyd Ottesen. Lloyd was a white Brooklyn native, probably the oldest engineer in the group. At first he seemed wary of my presence, but after we'd talked for a few minutes he seemed to relax. Lloyd was working on RTS-5—the same project I was assigned to. He told me to come by if I ever needed information on the system or help with my designs.

The last stop on my get-acquainted tour was the lab, where I had earlier met the Jamaican, Earl Brown. Earl's desk sat in a corner, surrounded by impressive-looking electronic test instruments. I was eager to find out how Earl liked working in Colton's group.

When Earl saw me enter the lab he said, "So you're the new Mr. MTS, eh?"

"I guess so," I said. I was trying to sound nonchalant about the "Mr. MTS" title he bestowed on me.

"What project are you working on, Earl?"

"I'm a TA, mon," Earl said. "I do anything they tell me to do."

TA was short for technical associate, one step down from George Callahan's STA designation. Earl's job was to support the group by building and maintaining prototype equipment.

I lowered my voice and looked Earl in the eye. "So how do you like working here, man?"

Earl said, "This a nice place, mon. You need something, mon, you just order it. No problem! How you like this computer, mon?" He gestured toward the gleaming IMSAI 8080 microcomputer system next to his desk. "It's one of the first ones they make!" Earl was beaming.

I probed further. "So what is it with this group? Are all of the groups like Colton's, with such a mix of people?"

Earl chuckled. "No, mon. Dere ain't no other groups round here like Colton's. I dunno—maybe he try to make a name for himself. But he seem like a okay guy, mon. He treat me right, too."

I reflected on what Earl Brown said. Maybe Colton's group *was* unique, but that didn't have to be a bad thing. Maybe I got lucky and got a really progressive boss. Maybe Colton's group was an oasis of diversity in the Bell Labs desert. Whatever the case, I felt more comfortable after talking to Earl and my other colleagues.

Earl started packing his things. I looked up at the clock and realized it was time to go home. My first day at Bell Labs was already over. I went back to my office, packed up my own things, and prepared to return to the Molly Pitcher. I retrieved my car from the visitor's lot, but before driving back to Red Bank I drove slowly around the road that encircled the Holmdel building. It was my way of celebrating the fact that I had "arrived."

The next morning I awoke early. I didn't want to be late for my first full day of work. I made sure to pack my basketball clothes so I could join Colton at the after-work games. I arrived well before starting time, and pulled into one of the two vast employee parking lots. I was glad I had arrived early.

BELL LABS HOLMDEL, FROM ABOVE.

The spots nearest the building were already taken, so I had to park near the middle of the lot. Parking on the outer fringe would have made the walk to my office take almost as long as the drive to work.

George Callahan was already hard at work. We exchanged greetings, and I started poring over the project materials Colton had given me. I wondered if I would be able to meet his high expectations. I wondered if the group's pace was as relaxed as it seemed, or if I would soon find myself behind the eight ball.

I barely noticed when George left to do some work in the lab. I sat alone, alternately reading and daydreaming and fretting. A female voice broke my solitude:

"Hello."

The word was sung more than spoken. I looked up and saw two young black women in my doorway. One was light-skinned and had a large afro. The other was darker and wore neatly pressed hair. I hadn't seen either of them before.

The young lady with the afro spoke up first:

"So you're the new MTS everybody is talking about."

She spoke with a charmingly soft Southern accent. I didn't know how to respond. I looked at my guests and flashed a nervous smile.

She spoke up again.

"Dot wanted to meet you, but she was too shy to come up here by herself."

"Uh… I see."

Then the other young lady spoke up. Dot had a Southern accent too, but her tone was more formal, less flirtatious.

"Don't pay any attention to my friend. She wanted to meet you just as much as I did. My name is Dot Stokes, and my so-called friend here is Nettie Colquitt."

"Nice to meet you both," I said. "Come on in. Where do you guys work?"

Dot and Nettie came in and stood next to my desk.

"I work in PBX systems engineering," Dot said. "I'm in Building Two on the second floor."

"And I'm in the Operations Research Department," Nettie said. I'm in Building One on the fifth floor."

I was flattered that two attractive young professional women would seek me out.

"How did you find out about me?" I asked.

Nettie smiled and said, "Oooh—everybody knows about you, Mr.

Tarver. It's not very often a new black MTS starts here. The news travels fast. So... are you single?"

"Yes, I am." My answer was more emphatic than I'd intended.

"Ooh-wee," Nettie said. Every floozy in this building gon' be after you, including this one here!" Nettie gestured toward Dot, and Dot gave a look of mock exasperation.

For a moment, I felt as if we were in grade school, flirting on the playground.

"Like I said, don't pay any attention to Nettie. She's a little touched, if you know what I mean."

Dot and Nettie giggled. I tried to suppress it, but I broke out in an ear-to-ear grin. There was something endearing and innocent about these two Southern belles. Both were attractive and apparently technically minded, but they were also down-to-earth and silly. I felt as if I knew them already, and took an immediate liking to them.

"So, Mr. Tarver," Nettie said. "Whatcha doin' for lunch?"

"Nothing that I know of."

"Well, why don't you come down to lunch with me an' Dot? We can introduce you to some of the folks."

"That sounds fine."

Just then, George Callahan returned with a set of drawings and sat at his desk. I could tell he was suppressing a grin.

"George, I'd like you to meet Dot and Nettie."

George gave my guests a friendly glance and said, "Hello, ladies." Then he stuck his head into the drawings.

I agreed to meet Dot and Nettie in the atrium at lunchtime, and they left, talking and giggling as they sauntered down the hallway.

George looked up and smiled. "I see you're a popular guy around here already," he said.

I found myself feeling eager for lunchtime. I looked forward to talking with Dot and Nettie again and finding out more about life at Bell Labs. At 11:45, I met them in the atrium and they escorted me to lunch.

The cafeteria and a much smaller, more intimate area called the Grill Room were both located on the lower level. I assumed we would eat in the cafeteria, but Dot and Nettie steered me into the Grill Room.

"Most of the 'folks' get together in here," Nettie explained.

The Grill Room served made-to-order burgers and sandwiches of all kinds. As we waited in line, I saw one of the cooks carving slices from a long, reddish piece of meat.

"What's that?" I asked Nettie.

"Ummm, cow tongue," Nettie said. She stuck her tongue out and wagged it mischievously for emphasis. I retracted my tongue deep into my tightly closed mouth, as if to protect it from the carver.

We got our food and ventured over to a group of tables pushed together in one corner of the room, where ten or fifteen other black employees were seated. Dot introduced me.

"Hey, everybody. I want you to meet David Tarver. He just started as an MTS here yesterday."

The folks at the surrounding tables smiled and welcomed me. We sat down to eat, and before long the people nearest to our table began peppering me with questions:

"What organization do you work for?"

"What school did you go to?"

"Are you married?"

Nettie flashed a knowing smile at that last question, as if to say, *I told you those floozies were gon' be after you!*

After conversing with my new black colleagues for a while, I realized they held a variety of job titles. Bob had a PhD in mathematics, and his job involved something called queuing theory. John was an electrical engineer like me, and was designing electronic switching systems. Peggy was an administrative assistant. Willie was a janitor. Skip was a graphic design technician. It didn't take long for me to see that the primary forces binding the group were cultural rather than professional. The common bonds seemed to be race, youth, and relatively short tenure at Bell Labs.

At one point, Nettie offered an explanation: "Bell Labs just started hiring black folks a few years ago, after the Asbury Park riots."

I thought about what Nettie had said. Changes in law, the civil rights movement, and the social unrest of the 1960s created the impetus for large corporations to hire black executives in significant numbers. Bell Labs was at the forefront of that movement, and my black coworkers and I were products of it. We were the first wave of blacks at Bell Labs, and similar waves were washing over corporations all across America.

I wanted to know how my work situation compared with that of my black peers. I described the mix of people in Colton's group, and asked them what their groups were like. Most said they were the only person of color in their group. A few indicated that their group included one or two Asians. It appeared that Earl Brown was right—Colton's group was unusually diverse.

Even though there were maybe twenty blacks in the Grill Room that day, and perhaps a hundred in the building, we were a mere drop in the bucket. Bell Labs Holmdel employed more than 5,000 people. My black colleagues and I were sprinkled throughout the place like fleas on a very large dog. I understood immediately that getting together in the Grill Room was a way for the black employees to avoid isolation, compare notes, and get moral support and encouragement. Even though we were spread throughout the organization, it was great to have access to so many other talented black professionals. I had never experienced anything like that.

After lunch, Dot and Nettie and I walked back toward our offices. I was glad they had connected me with the other black employees—they made Bell Labs feel much more comfortable and familiar. My new friends and I parted ways at the elevators. I returned to my office reinvigorated and ready to tackle my job.

I spent the rest of the afternoon talking to Lloyd Ottesen about RTS-5. He told me about the work that remained, and showed me where my project fit in. Things were starting to come together, and my confidence, which had waned that morning, began to return.

Shortly after 5:00 p.m., John Colton stuck his head in my office. "So, Tarver, are you going to play basketball today or what?"

"Sure," I said. "I'll meet you over at the school."

I gathered my things and left the building. It was a gorgeous summer afternoon. As I descended the steps, three separate young women stopped me and introduced themselves. I was an unmarried black MTS at Bell Labs. I felt like a celebrity.

I arrived at the Holmdel Village School just as my coworkers were choosing teams. Colton invited me to play with his team, and the games got under way. As the evening progressed, I got to know some of the other players. They were mathematicians, engineers, psychologists, technicians, administrative assistants, drivers, porters, and janitors. The group included people of every education level, from high-school grads to PhDs. Most were white, but there were a significant number of blacks and a few Asians and Latinos.

Between games, in a moment of quiet reflection, my thoughts again drifted to the Bell Labs recruiting brochure and Leesing Pang: *Attractive, intelligent, fit, fun-loving, hard-working, cosmopolitan, and international—the best of the best.* It was only my second day on the job, and already I was working and playing with some of the world's best and brightest engineers and scientists.

I couldn't imagine a better place to work.

I couldn't imagine a better place to *be*.

22. Garden State

THOUGH I LOVED MY JOB AT BELL LABS, I encountered many reasons for hating New Jersey. The reasons went far beyond the turnpike smell and the tollbooths—and even beyond my new home, Monmouth County, being identified as a cancer hotspot. My discontent flowed from much more direct, personal, everyday experiences.

The Molly Pitcher Inn in Red Bank served as my temporary home while I looked for an apartment. The inn, named for the Revolutionary War heroine, was dank and uncomfortable, and the town of Red Bank had seen better days. The Rex Diner, open 24/7, was just down the street, but the food left a lot to be desired. Next to the Rex was a store that advertised Peachy Plumbing Paraphernalia. It, too, was long in the tooth, and its front window displayed plumbing fixtures that appeared to date from the early 20th century.

This dank little hotel and this dying little town were my introduction to New Jersey life, and it was not a happy introduction. Fortunately, my new job and my search for a permanent home left little time to bemoan the Molly and the Rex.

Bell Labs had a well-staffed, professional personnel department ready to help new employees find suitable housing. I met with Mr. Jones, a housing specialist, and he clued me in on the situation. Though Bell Labs was located in Holmdel, the town offered absolutely no housing suitable for an engineer like me. By "like me," I mean fresh out of college, with a fresh-out salary. There were no apartments in town, and the minimum lot size for houses was one acre. Needless to say, buying a home in Holmdel was out of the question. Mr. Jones quickly made me aware of that, and he deftly turned my attention to more realistic locales.

"Most guys like you end up looking at garden apartments. There are some nice ones available in Matawan, not too far from here. There are also some nice ones in Eatontown and Ocean. You might also want to consider Long Branch and especially Neptune."

Neptune? I thought. What the hell kind of place is called Neptune? I couldn't imagine living in a place with such a goofy name.

"Well, I guess I'll just have to take a look at what's available," I said, to appease Mr. Jones.

"Good. We'll have one of our housing consultants take you around to see some apartments. We'll definitely get you settled in a nice place."

After the meeting with Jones, I decided to look around the area by myself. On Saturday, I jumped into my trusty old Camaro and drove to some of the spots he had mentioned. Holmdel featured sprawling farms and pastures interspersed with subdivisions of large lots and huge homes. I didn't see a single apartment building. I couldn't imagine living there.

I drove ten minutes north to Matawan, a busy area with crazy traffic. I did see some apartment buildings there, but none looked nearly as nice as my place in Ann Arbor. I reasoned that as a working engineer I should have an apartment at least as nice as the one I'd had in college.

Next I drove south through Eatontown and Ocean, and I saw several large garden apartment complexes. I hadn't heard the term "garden apartment" before arriving in New Jersey, but I quickly learned it applied to low-rise apartment buildings with considerable landscaped area outside. Applied to the old and decrepit places I saw, the term was almost a euphemism. Nothing reminded me of a garden and nothing captured my imagination. I kept driving south to Neptune, and I realized why Jones had pointed me there: lots of black people lived in Neptune. Perhaps he thought I would feel more comfortable among "my kind," but I didn't see any apartments, or even any neighborhoods, that excited me. I crossed Neptune off my mental list and headed back toward Red Bank.

Before retiring to my room at the Molly Pitcher, I drove around Red Bank to get a better feel for the place. A railroad track bisected the town into east and west sections. To my amazement, the west side was almost entirely black, and the east side almost entirely white. The west side was the domain of the working poor and a dwindling middle class, while the east-side residents seemed considerably more affluent. Then it hit me: During my entire drive that afternoon, I hadn't noticed any thriving black neighborhoods. The black areas I saw seemed to be struggling. The white areas I saw were middle or upper class. My family's neighborhood in Flint was almost entirely black *and* middle class. Many neighborhoods in Flint and Detroit were majority black and economically well off. The lack of a sizable black middle class in my new home was startling, and it didn't bode well for my long-term prospects in the area. Disappointed and tired, I retreated to the Molly Pitcher.

At work the next morning, I met the housing consultant, a pleasant, youngish, attractive white woman named Kayla. I could tell right away she wasn't a Bell Labs employee because she wore a casual, loose-fitting dress and sandals, and her legs were bare. She looked ready for an afternoon at the beach rather than a day at Bell Labs.

Kayla and I spent a few minutes discussing areas where I might be interested in living. I described my Village Green apartment in Ann Arbor and said I might like a similar place in New Jersey. I expressed a desire to be close to Holmdel, because I didn't want a long commute. Kayla listened intently and made a few notes on a small pad, and then we set off in her car to look at apartments.

Kayla was warm and disarming, and I liked her immediately. In the car, she seemed even more relaxed and friendly, and we conversed easily.

"You married?" she asked.

"No. I just graduated from college, and I'm out here by myself. I just need a one-bedroom place, or maybe one bedroom with another room for an office."

"Well, David, there are lots of apartments around here, but most of them are garden apartments. It sounds like you're not crazy about those."

"That's right," I said. "Most of the garden apartments I've seen around here are old redbrick places. They actually look pretty depressing. I had a much nicer place when I was in college."

"I understand. We'll definitely find you something, and hopefully we'll have some fun in the process. Okay?"

Kayla flashed a big smile as we left the Bell Labs property in her Saab. Was she hitting on me, or was she just a friendly, fun-loving person? Could it be both?

"Okay," I said. I decided to test Kayla, to see how much she knew about the area and whether she was sensitive to my situation.

"Kayla," I said. "You seem to be pretty familiar with this area. Something is bothering me about the housing situation around here. Maybe you can help me understand."

"Sure, if I can," Kayla said. She looked more serious now, concerned and caring.

"Well, it seems that there are black areas and white areas. Driving around, I didn't see any integrated neighborhoods, and I didn't see any mostly black neighborhoods that were doing well. Mr. Jones seemed to be pointing me to Neptune and to the west side of Red Bank, but I think he was trying to steer me to the black neighborhoods. So what's going on?"

"Wow, David, you're very perceptive." Kayla still wore a concerned look. "The Bell Labs housing staff does usually steer black recruits to areas where, they think, you'll be more comfortable and more able to afford the rent. It sounds like you want a nice place in an integrated neighborhood, and I can't blame you for that. If that's what you're looking for, that's what you'll get. I'm going to stay with you until you find just what you want."

Kayla flashed that warm, radiant smile again, and I was smitten, convinced that she liked me and would do her best to find me a nice place. I believed she was on my side.

Kayla and I looked at places for a week, and when we didn't find anything suitable, Bell Labs extended my stay at the Molly Pitcher, and we kept looking for another week. Finally, I settled on a garden apartment in Long Branch. It wasn't what I had envisioned, but the Edgewater East complex was relatively new and nice. Also, it was only a block from the ocean. I saw whites, blacks, and a few Latinos in the complex. My apartment had a large bedroom and a dining room that could serve as a study. Kayla was true to her word—she had helped me find a decent place. It wasn't perfect, but it was pretty good.

"This is a great place," Kayla beamed. "And it's *sooo* close to the beach. I just *love* the beach. You're going to have a great time here. After you get moved in, I'm going to come back and we'll celebrate!"

I didn't know what to make of that, but I was happy to have found a place. I signed the lease, and the next week the movers arrived from Ann Arbor. I never heard from Kayla again.

Once I got settled into the new apartment, I started to get a feel for what life was like in the area. Edgewater East was nice enough, but the surrounding town of Long Branch was experiencing hard times. Many white residents had fled, leaving a devastated downtown and a city core populated mainly by blacks and Latinos. The physical infrastructure was crumbling. A sewage treatment plant two blocks from my apartment emitted an awful smell. In fact, when folks asked for directions to my apartment, I jokingly (and correctly) told them to "hang a left at the smell." It never failed.

It was hard for me to feel connected to the community. In Flint we had a local TV station to tell us what was going on in town—local politics, high school sports, social events, and such. Not so in my new home. Monmouth County contained an amorphous collection of little towns and villages, and places like Long Branch and Red Bank were just two among many. Almost all broadcast media came out of New York City. New York was only twenty miles away as the crow flies, but it might as well have been Tokyo. The news from New York had nothing to do with Monmouth County. With so many little towns, it was impossible for the local newspapers to adequately cover events. It was hard to get a sense of the overall area or know what was going on in the community. Perhaps the natives had ways of finding out what was happening, and perhaps they felt a sense of community. I felt like a stranger in a foreign land.

My Edgewater East neighbors were a big help. The white couple next door, Stan and Fran Pieczara, were several years older, and they immediately took me under their wing. I spent lots of time in their apartment. Stan was an unemployed stockbroker, and Fran was an operating-room nurse. We spent hours sharing meals, watching TV, and shooting the breeze. Stan regarded himself as an authority on every subject, but delivered his information with humor, so he didn't come across as obnoxious. Fran was a nice lady who was extraordinarily patient. I could tell she had heard every Stan story ten times, but she would listen intently as he told it for the eleventh. I never saw Fran display anger about Stan's lack of employment. Through Stan and Fran, I started to learn about the local scene—what to do, where to eat, whom to see to get things done. They were a big help, but New Jersey was still New Jersey.

A few months after arriving, I summed up my feelings in a letter to my mother.

> "As you know, I came out here for the experiences and opportunities that were available. While I don't know how long it will take me to accomplish my goals here, I do know that there will never be any place like home, and I look forward very much to the day when we're all together again. New Jersey is a very crowded, congested, polluted place. The per capita cancer rate for this county is the highest in the USA. The cost of living is 30 to 40 percent higher. In short, this is a terrible area. However, the job is very nice. I'm learning a lot and contributing a bit, and I feel that I'm starting to do some of the things I came here to do, so everything is not so bad."

I wanted to get the experience I needed from Bell Labs and leave New Jersey as soon as possible. The only things keeping me there were the Bell Labs job and the desire to start my own business.

The letter must have worried my mother, because the next thing I knew, she had arranged for my family to drive out from Michigan for Thanksgiving. Fred and his wife, their two kids (one of whom was an infant), Mimama, and my mother all piled into Fred's car for the long drive from Flint. After twelve hours, they arrived in the area, but my brother needed directions to my place. Around 9:00 p.m. he and the family pulled into the adjacent town, West Long Branch, and found a pay phone in a McDonald's parking lot. Fred parked and stepped out to call me. He noticed a McDonald's worker emptying a waste bin. When the worker saw Fred, she dropped her trash bag and ran inside.

Several West Long Branch police cars soon converged on the parking lot and surrounded my brother's car. The officers from one car got out and approached Fred. They wanted to know his name, why he was in West Long Branch, and where he was going. Fred explained that he was on a trip to visit his brother in Long Branch, and that the rest of the family was in the car. The officers looked inside and saw everyone sitting in silent fear and disbelief. Then the officers told Fred the McDonald's was closing, and he should be moving along. Fred explained that he had just stopped to use the pay phone, and that he would be leaving immediately. The officers didn't offer any directions or other assistance. Fred surmised that the police had received a call from McDonald's about a group of blacks sitting in the parking lot at closing time, and their only concern was to get rid of them. As Fred and the family proceeded toward my apartment in Long Branch, two West Long Branch police cruisers followed. When the family reached the Long Branch border, the cruisers turned away, their mission accomplished.

When Fred related the story to me, I was angry and disappointed, but not surprised. I had suspected that some in my new home were hostile to blacks, and the hostility had hit home. Despite their unpleasant introduction to the Garden State, my folks and I had a good Thanksgiving. Their visit helped ease the pain of being in New Jersey.

I settled into a pattern: I worked a full day at Bell Labs, then worked on my synthesizer until late in the evening, then went over to Stan and Fran's to watch the late news and the *Tonight Show with Johnny Carson*.

I hated my new state. I loved my new job. I was committed to my goals. I wanted to get the hell out of New Jersey.

23. A Nigger in the Woodpile

MY FIRST BIG BELL LABS PROJECT MEETING took place in the fall of 1976. I will always remember it, for reasons that have nothing to do with electrical engineering. John Colton asked me to attend a Switched Access Remote Test System (SARTS) planning meeting with him. I didn't know why he wanted me there, but I sensed it was an opening of some kind. I had been in New Jersey for only a few months, and didn't know much about the SARTS project or how my own work fit into it. I thought the meeting might be a chance to find out.

I walked up to the conference room with John. Some people from the systems engineering group and a few development engineers were there. The systems group leader was an older engineer, William "Gigs" Giguerre. Gigs was the chief system architect. Aside from me, Colton was probably the youngest person at the meeting. The rest seemed like crusty old-timers. The crustiest, and the scariest, was F. Gordon Merrill. I didn't know what the "F" stood for—everyone just called him Gordon. He was a tall guy with dark, deep-set eyes. When I had seen him in the hallways, he hadn't smiled or greeted me the way most other department members had. I didn't know Gordon, though, and was too excited about being at the meeting to worry about him.

Colton stood and started the meeting. He said the purpose was to outline the new architectural features for the SARTS system—in particular the new RTS-5 capabilities. Colton began his presentation, and I realized why he had asked me to attend. My assignment was to *design* the new RTS-5 capabilities.

I looked around while Colton spoke. John was clearly the whiz kid—everyone else was reacting to him. Each old-timer seemed to have a concern or a point of view born of his vast experience or past failures. Frank Dargue, for example, was the expert on electromechanical relays and resistance lamps, primitive but necessary components in a system that featured newer innovations such as microcomputers and memory chips. Gigs was a lot more sophisticated than the others, and raised some system architecture concerns, but in the end he deferred to John. After a while, it wasn't clear to me why most of the others were there. Most simply observed the interaction between John and Frank and Gigs, and piped in with safe and obvious comments.

Colton led everyone through the agenda, and finally the meeting approached a conclusion. I hadn't said a word because I knew little about SARTS and absolutely nothing about relays and resistance lamps. I was all set to go back to my desk and prepare to go home when I heard it. I couldn't believe what I had just heard, but I was absolutely sure I *had* heard it. F. Gordon Merrill had just said, "We have to make sure we clean up these issues—we don't want…"

Then he said it… "a nigger in the woodpile."

All of a sudden I couldn't think straight. I couldn't even see straight. Some of the engineers were looking at me as if expecting a reaction. Others showed no clue that they realized what Gordon said. John Colton turned beet red. I didn't know what to do or say. I didn't react. I just sat there, incredulous. Meeting adjourned.

When I got back to my office I was still in a daze. I didn't know what was worse: that Gordon had that phrase in his vocabulary, or that he felt comfortable uttering it with a black person in the room. I started to think about my options. I had harbored suspicions about Gordon, and those suspicions had been confirmed. *I knew it! What should I do? I can't let him get away with this.* I called my friend Dot and told her what happened. She was surprised, especially since I had told her several times how progressive my group was. In fact, I had bragged that my organization was far more progressive than hers.

Dot said I needed to do something—that I couldn't just let the incident pass. I already knew that, of course. I was definitely of the opinion that people who remained quiet about injustice usually got screwed, and people who stood up for themselves might or might not get screwed. I decided that if I were going to get screwed, I'd get screwed standing up. At least that would make it more difficult.

I went to see Colton. John seemed to relate to everyone—the eager new engineers and the crusty old-timers. He told me he felt terrible about Gordon's comment, and asked what I wanted him to do. I said I would have to think about it. John said, "Well, when you decide, just let me know. That bastard had no right to make a statement like that. I don't think he meant anything by it—he's just a stupid, old-fashioned bigot." It made me feel good to know John was on my side. At least I wouldn't be putting my future at Bell Labs in jeopardy by going after Gordon.

I went home and thought it over. If I did nothing, I would feel like a chump. On the other hand, I wanted to do something that would be effective, not merely vindictive. I wanted to make sure I wouldn't have to suffer an environment that would condone a statement like Gordon's, and that no one else would, either. I went to bed, but couldn't sleep. I kept thinking, tossing and turning, and thinking some more. Finally, in the wee hours, it hit me: I would insist on an apology from Gordon—a *written* apology. That way, if I did get screwed later, I would pull it out and illustrate the attitudes I'd had to deal with in my department. Yeah! Gordon's written apology would be my "get out of jail free" card, my insurance policy against racist asshole mischief. The decision made, I fell asleep.

When I got to work the next morning, I immediately went to see John. "I've decided what I want to do about Gordon," I said.

"What's that?" John asked.

"I simply want him to give me a written apology for the statement he made in the meeting."

John sat back in his chair and thought about it for a few seconds. He seemed to realize the implications of my request. We both knew that if Gordon gave me a written apology, he would be documenting his transgression and would be at my mercy if I chose to pursue the matter further. John thought some more, and then said, "Okay—let me see what I can do."

I went back to my desk, a happy camper. At lunch, I told Dot and Nettie what I had decided. I thought the plan was brilliant, and they thought it was pretty good, too. Both, however, doubted Gordon would ever give me a written apology. Nettie was very wise about Bell Labs politics, and she expressed her opinion in simple, graphic terms:

"That old boy ain't no fool! If he gives you that apology he might as well let you put some pliers 'round his balls. It ain't never gonna happen."

I didn't think it would be easy to elicit a written apology from Gordon, but I placed a lot of faith in John Colton. He was my boss, and a powerful manager in our organization. If anyone could force an apology from Gordon, John could.

That afternoon, I received a call from my department head, Ed Spack. Ed wanted to meet with Colton and me at the end of the day. I hadn't met with Ed Spack since the welcome lunch on my first day at work, so his invitation underscored the gravity of the situation. At five o'clock, I walked up the hall to Ed's office. John was already there. His face was red, his glasses were off, and he appeared to have been rubbing his eyes a lot. Ed seemed his usual stern self, but his face had a sympathetic cast I hadn't noticed before. When Ed saw me, he said, "Come on in, Dave, sit down." I walked over and sat next to John in front of Ed's rather large desk. "I understand Gordon Merrill made some pretty stupid statements in a meeting you guys had a couple of days ago."

"Well," I said. "He made one statement I thought was pretty offensive, but I don't know if he meant anything by it. Anyway, I've worked out a way to handle it that's fair to everyone." I went on to explain that I simply wanted a written apology from Gordon, and that as far as I was concerned, that would clear the matter up completely.

Ed Spack had obviously heard my request already. He paused a few seconds, and then he said, "I think it's really rotten what Gordon did. That kind of thing has absolutely no place in this department. I want you to know that we are happy to have you. We were lucky to hire you, and you have a bright future here."

Ed went on to say many other complimentary things, and John chimed in with a few of his own. Their theme seemed to be that I shouldn't worry—things would be great for me in the department and at Bell Labs.

When they finished, I said, "That's great. I really appreciate what you're saying. I just think it's important to get a written apology from Gordon."

Both Ed and John seemed disappointed. I could see they thought their assurances about my "bright future in the department" would be enough to put the matter to rest. Before ending the meeting, they agreed to "talk to Gordon," but that was the extent of their commitment.

I returned to my office to gather my things and go home. I was beginning to think Nettie was right—the Gordon situation wasn't going to be resolved easily. As I walked down the hall toward the parking lot I heard a lot of noise coming from Ed's office. His door was closed, but I could hear three voices—John's, Ed's, and Gordon's—engaged in heated discussion. I couldn't make out what they were saying, but I surmised that John and Ed were trying to convince Gordon to apologize. I left the building and drove home.

The next day, I got to work late, around 9:30. It was the third day of the "Gordon situation." A pink message slip from the director's secretary sat on my desk: "Bob Yeager would like to meet with you—please call me to set up an appointment." Things were getting really serious. Yeager was director of our entire organization—the Special Services Maintenance Systems Laboratory—and he never had time to meet with a lowly MTS like me.

I was getting worried. I needed to let someone outside the organization know what was going on. After all, Gordon, John, Ed, and Bob had probably been friends and colleagues for years, and I was brand-new. For all I knew, they could have been teaming up to screw me.

I decided to visit the Affirmative Action Center (AAC), which was there to make sure all Bell Labs departments adhered to the company's affirmative action and equal opportunity policies, educate the employees, and help resolve disputes. The AAC managers were company employees, so they were unlikely to make waves, but they were viewed as advocates for minority employees. They were also the "eyes and ears" of management with respect to minority concerns. Going to see the AAC folks seemed like a logical step.

I walked over to the Affirmative Action Center and met with a counselor. I told him about the incident with Gordon Merrill, and the counselor seemed quite sympathetic. He didn't seem surprised to hear my complaint, and said he would look into the problem. He asked for my department number and the names of the managers involved. He said he would contact my department and try to work things out.

I told the counselor my bottom line: "What I really want is a written apology from the guy who did this. If I don't get it, I think he and his friends will just be out to get me in the future, and I won't have any protection."

The counselor said he couldn't guarantee I'd get a written apology. He seemed to realize how powerful a document that would be. I decided to force the issue.

"I know folks at the NAACP," I said. "I'm prepared to go to them and let them know about this incident."

I was bluffing, but I knew the counselor would be sure to communicate my message to management. If he didn't, and the president of Bell Labs suddenly got a letter from the local NAACP chapter asking about my treatment, the Affirmative Action Center would not look good. I was making a serious threat, signaling that I was ready to pull out all the stops.

The counselor reiterated that he would contact my department and try to work things out. I left feeling confident he would give it his best shot.

On day four I had my appointment with Bob Yeager. Bob was a massive guy—not in an athletic way, but in a too-many-company-luncheons way. Personality-wise, Bob was unfailingly friendly in purely social situations, but he had a reputation for being direct and sometimes brutal concerning work issues.

Bob was cooler than Ed—almost matter-of-fact. Our meeting began with the usual small talk. Bob wanted to know how I liked living in New Jersey (I didn't), and what I thought of my project (it was fine until the incident with Gordon), and then he got down to business.

"David," he said, "you have a bright future in this organization. Everybody seems to like you, and you've gotten off to a great start.

This thing with Gordon isn't going to cause you any problems at all. I've known Gordon for a long time. He doesn't always say things in the best way, but he's a good guy at heart. He feels terrible that he upset you with his comment. It was a really stupid thing to say, and he realizes that."

I felt honored to be sitting there having a conversation with the director of my organization. And he was saying I had a bright future! What more could one ask? Still, I had a sinking feeling. Bob Yeager didn't say anything about a written apology from Gordon. In fact, all the nice things he said seemed to boil down to: *Look, I'm telling you everything is going to be okay. Forget about this thing with Gordon. If you don't want to believe what John and Ed and I are telling you about your future here, then tough luck.*

I knew the AAC folks had probably contacted Bob Yeager, and I was in no mood to relieve the pressure. Besides, I wanted to see just where the department stood. Nice words were one thing, but action and documentation were another. I spoke up, nervously.

"Mr. Yeager, I just want to get a written apology from Gordon. I'm glad the department thinks I can do well here. I intend to work hard and do a good job. But I was offended by what Gordon said. I don't know what he intended, and I don't plan to make a big stink about it—as long as I get a written apology."

At that moment, Bob Yeager made the switch. He changed from his social, affable self to his official, stern self. He said, "That's up to Gordon. I just wanted to let you know how the organization feels—that we support you one hundred percent. This incident is not going to have any negative effect on your career here."

That's funny, I thought. *There shouldn't even be a question about the effect on* my *career. The question should be what effect it will have on* Gordon's *career!*

Bob had covered the base he intended to cover. He had given me the official line: that his organization didn't discriminate; that I was well regarded; that my future was bright. Case closed.

Later that day, John Colton stopped by my office. He asked how the meeting with Bob had gone, and I said it went okay. Then he surprised me by saying, "Gordon would like to meet with you."

I said, "That's great. Is he going to give me a written apology?"

John said, "I don't know, but I know he wants to apologize. He wants you to stop by his office. Just talk with him and see where things go from there."

Late that afternoon, I walked over to Gordon's office. He seemed completely different from the sinister person I remembered from our hallway encounters. He even managed a smile.

"Dave," he said. "I understand I made a comment the other day that bothered you. I feel terrible about that. It was just an expression. I didn't mean anything by it, and I didn't intend to offend you."

I said, "Well, I never heard that expression before, and I *was* offended by it."

Gordon went on to say he was not a "prejudiced" person. He told me about a fellow he had met in the army who had become one of his best friends, a fellow who happened to be "colored, er, a Negro."

I said, "That's all fine, Gordon. Why don't you just give me a written apology and we'll consider it done."

Gordon's face turned red, and he seemed to get angry. "Why do you want a *written* apology?"

"Because I want to make sure this is documented so it doesn't happen again."

Gordon said, "Well, I'm telling you I'm sorry. I'm not prejudiced. That's all I wanted to say."

I said, "Well, I'd still like a written apology."

It was a tense moment. I got up to leave, and Gordon didn't say anything else.

At the end of the day, John Colton stopped by to debrief me. I said, "Well, he apologized, but I still want a *written* apology." I was candid. I told John how Gordon had related the story of his "colored" friend. I told him how Gordon was noncommittal about giving me a written apology. I also told John I had been to see an affirmative action counselor, and that I was thinking of involving the local NAACP. John seemed to understand. I could see that he was troubled, and I felt bad for him. He had been doing his best to run a top-notch group, one of the few really integrated groups in the whole company. Still, I felt I had to stand my ground. The "Gordon situation" seemed like a turning point in my young career.

On the fifth day I arrived at the office intent on getting back to my technical work. The whole situation had become far too distracting. I was tired of all the meetings about the incident. I wanted to refocus on my project and get back to work. When I sat down, I noticed a piece of folded typewriter paper sitting in the middle of my desk.

Slowly, I unfolded the paper and looked inside. To my great surprise, it was a note from Gordon:

> David
>
> In a recent meeting, I used an
>
> unfortunate euphemism that
>
> troubled you. For that I am sorry.
>
> F. Gordon Merrill

It didn't matter to me that the note wasn't written on Bell Labs stationery. It didn't matter that it didn't specify the meeting date, or what the "unfortunate euphemism" was. It didn't matter that it looked like Gordon had typed the note on his home typewriter so that no secretary would see it. It didn't even matter that Gordon's "unfortunate euphemism" was actually not a euphemism but a racial slur. The typewritten message still felt like victory. I folded Gordon's note and placed it in my briefcase. I was satisfied. I was an MTS at Bell Labs, the world's leading electronics research and development organization. My project was going well. My future in the organization was bright. Best of all, I had stood up for myself and served notice that I would allow no one to disrespect me.

I didn't get invited to any big project meetings for quite some time.

24. Breakthrough

BY THE SUMMER OF 1977 I HAD A GOOD HANDLE on my Bell Labs work assignment, which was to design sophisticated electronic instruments for RTS-5, the measurement subsystem for the Switched Access Remote Test System. SARTS would enable technicians to detect and diagnose AT&T network problems from a central location. When completed, SARTS would eliminate thousands of hours of service-technician travel and labor and save AT&T millions of dollars.

Most of my assigned instrument designs were straightforward. The main challenge was to employ microcomputer technology that would allow the instruments to be remotely controlled. In the late 1970s, microcomputers were new to many Bell Labs engineers, but they weren't new to me. I had worked with the very first Intel microcomputer at AC Spark Plug in 1972, and was comfortable with the technology.

I had access to earlier Western Electric and Hewlett-Packard instruments that, although they didn't employ microcomputers, provided useful benchmarks for my designs. One of my assigned designs presented an unexpected challenge, though, and I couldn't find any prior examples to follow. That surprisingly difficult assignment was to design an accurate, precise, wide-ranging tone generator.

A tone generator was one of the most basic tools in the communications technician's kit. To determine the frequency response of a communication link, the technician would inject a series of tones at one end of the network, and measure the level of each tone at the opposite end. The procedure is analogous to a hearing test. The bandwidth of many communication networks, particularly digital networks, far exceeded the audible frequencies, so a wide-ranging tone generator was required.

When John Colton outlined the assignment, neither of us thought it would present any special problems. In fact, Colton proposed several solutions. As I delved into the details, however, I grew concerned. None of Colton's suggested solutions could generate tones with the required range, accuracy, and precision. After a few months of thinking and tinkering, I worried that I might not find a suitable approach. During a one-on-one meeting in which John and I reviewed the status of my designs, I broke the news.

"I'm having trouble coming up with a design for the tone generator. I looked at VCOs, phase-locked loops, and off-the-shelf chips, but nothing seems to satisfy the specs. All my other designs are coming along fine, but I'm still looking for an approach on the tone generator."

John was patient and understanding, and above all, optimistic.

"I'm sure one of the approaches I suggested will work," he said. "You might have to sharpen your pencil and tweak one of the designs, but just keep working at it."

Colton was confident I would produce a workable design if I merely followed his guidance. I left the meeting disheartened that he didn't accept my assessment. I spent several weeks further reviewing and tinkering with his ideas, and arrived at the same conclusion: they just wouldn't work. To solve the tone generator problem, I needed to come up with something neither John nor I had thought of before.

I started spending my days in the Bell Labs library. I pulled every book I could find on the subject of tone generation or frequency synthesis, and pored over each one looking for a solution. After a few days, I found a description of a technique called direct frequency synthesis (DFS). On first inspection, the technique seemed to be the solution I was looking for. It was so elegant and simple that it hardly seemed workable, but the more I analyzed it, the more suitable it seemed. DFS relied on digital computing techniques to generate tones accurately and precisely. The digital electronic "chips" required to make the technique practical had only recently become available, and that was why my research hadn't turned up any commercial implementations.

I spent a few days producing a rough DFS tone generator design, and couldn't wait to spring it on Colton. I knew he would be pleased. I had overcome the last hurdle in my assignment by devising an approach that was novel and effective, and that met all requirements. I had arrived at a solution neither John nor I envisioned when he gave me the assignment.

I presented the solution to Colton, and at first he was unconvinced. His first question was, "How can this design produce the range of frequencies you need to generate?"

I eagerly took John through my calculations and showed that the frequency range was more than adequate.

"How can you generate the step sizes you need?"

Once again, I showed him how the design exceeded the requirements.

"How can you..."

John asked several more questions, and I grew more confident with each answer. I wasn't cocky—just proud and sure of my design.

When John ran out of questions, he just sat there, his face reddening, his eyes fixed on the tablet of paper that bore my design. Finally he looked up at me and said, "Why didn't you think of this before?"

That's when I knew I had a solid design. I had a lot of respect for my boss. He was one of the brightest engineers I'd ever met. If he couldn't fault my design, it was solid. I had come up with a solution he hadn't thought of, and that meant I was a pretty good engineer, too.

I felt I had earned John Colton's respect, and that I had finally earned my title: Member of Technical Staff, Bell Telephone Laboratories.

25. A Dream Dies

———

I DESIGNED AN AWESOME MUSIC SYNTHESIZER, a work of art full of microcomputers and other advanced electronics. It made familiar sounds and far-out space-age sounds. Its elegant black control console oozed technological sophistication. Polished cherry wood framed its gleaming piano-like keys. A small panel left of the keyboard revealed a set of controls that suggested underlying complexity.

Musicians learned about my synthesizer by word of mouth, and top players in the industry called to inquire about it. I was flattered, but I was interested in having one and only one person play my fantastic creation. That person was none other than the genius, my musical hero, Stevie Wonder.

Then one day it happened. Stevie Wonder visited Bell Labs scientists in Murray Hill, New Jersey, to check out their experimental music synthesizer. One of the scientists informed Stevie about my invention, and that very afternoon my phone rang. I was amazed to hear my idol's voice on the line.

"David, I hear you've designed a really cool synthesizer. I would love to see it. Might I come by?"

"Mr. Wonder, I would be honored if you would play my synthesizer. I've followed your career since we were both just kids, and I can sing every one of your songs."

That evening, Stevie showed up at my apartment. My synthesizer and piano and other instruments were arranged around the living room. I took Stevie by the arm and helped him to the piano bench in front of the synthesizer.

Stevie ran his hands over the keyboard and cooed, "This is cool, man. You have a really nice place here, and this synthesizer is just the bomb! The keys feel so nice. May I play it?"

"Of course, Stevie! Go ahead!" I was beaming with joy and pride as Stevie started playing Superwoman... *on my synthesizer! I couldn't believe my ears and eyes. Stevie finished playing* Superwoman *and immediately lit into* Tuesday Heartbreak *and then* Livin' for the City. *He played and I sang, then we sang together. On some songs, Stevie sang lead and I harmonized, and on others we switched it around. We played song after song until the wee hours. When we were both exhausted, Stevie got up to leave. He said, "David, this was great, man. You are a wonderful soul. I want you to come to Los Angeles and work with me. I want you to sing with me and build synthesizers for me. We'll put your synthesizer on the map, and we'll have fun, man!"*

That was the best evening of my entire life.

The next morning, I awoke in a daze, wiped crusty sleep from my eyes, and stumbled into the former dining room I had converted to serve as my electronics laboratory. My eyes focused slowly on my synthesizer, crude but complete, sitting on a makeshift plywood lab bench. Gradually, I realized that no famous musicians were clamoring to play my synthesizer, and that there had been no magical evening with Stevie Wonder. The jam session with my hero was just a dream.

My synthesizer was no dream, though. It was very real and very much in need of marketing. On that spring morning in 1978, after six years of dreaming and learning and designing and building, my invention was complete. I had finally created the object of my dreams, but only a few family members and close friends had seen it.

It was time to present my invention to the world, but I was apprehensive about showing it to professional musicians. I wasn't sure what would happen when my dreams met reality, but I knew if I was ever going to turn my synthesizer project into a business, I had to act.

I had no contacts in the music business, and didn't know where to start. My uncle, Bill Hayden, knew some musicians and was willing to help. Uncle Bill worked in a Detroit Ford factory by day, but by night he was a jazz aficionado, raconteur, and man-about-town who hung out at Baker's Keyboard Lounge, a nationally known jazz spot. In his younger days, he was so handsome and debonair that his friends called him "The Dark Clark," as in the melanin-enhanced version of Clark

Gable. Uncle Bill was an amateur photographer whose photos of touring jazz artists were posted on a wall just inside the Baker's entrance. Bill's outgoing nature and photography skills allowed him to develop relationships with many jazz stars. One was pianist Les McCann, perhaps best known for his musical collaboration with Lou Rawls and for his *Compared to What* recording at the 1968 Montreux Jazz Festival. He was still a popular figure on the jazz circuit ten years later, and he and Uncle Bill were pals. I figured if I could get Les McCann to play my synthesizer, I would be on my way.

I called Uncle Bill, who told me that Les was scheduled to perform at Baker's the very next weekend. I thought it was divine providence, or at least serendipity. Whatever it was, I was determined to take advantage of the opportunity.

Les McCann did indeed perform at Baker's, and Uncle Bill asked him if he would take a look at my synthesizer. The next day, Uncle Bill called.

"He said he would be glad to meet you," Uncle Bill said. I could feel his jovial grin beaming through the phone.

"Les said he'll be in New York City next week and will meet you there." Uncle Bill gave me the name of Les's New York hotel, and the meeting was set. My synthesizer was going to be seen, and maybe played, by a great jazz artist! I was giddy with excitement and shaking with nervous energy. This was my big break.

During the week leading up to Les McCann's New York appearance, I went over my synthesizer with a fine-tooth comb. I cleaned it up to make it as presentable as possible, tuned it so the pitch of every note would be exact, and programmed the best sounds I could create into its memory chips. When the weekend rolled around, the synthesizer and I were ready.

Les was performing Friday night, so I waited until Saturday morning to call. I asked for his room, and the operator put me through. My palm was sweating so much I could barely hold the receiver. The phone rang and rang, but no one picked up. I was disappointed, but relieved to have more time to get control of my nerves. I called again later, but again no answer. As evening approached, I called once more. This time the hotel operator told me Les McCann had checked out.

I called Uncle Bill, who said Les probably slept late after his performance and then got up and hit the road. Uncle Bill asked if I had left a message. I felt stupid—I hadn't. I had missed my big chance, and it was my own fault.

"Don't worry," Uncle Bill said. "Les told me he was going to Washington, D.C., after his New York gig, and he told me the name of his hotel. Maybe you can catch him there."

New York was only an hour away from my home in New Jersey, but Washington was a four-hour drive. I didn't care—I was willing to go there to catch up with Les. I took the hotel name from Uncle Bill, and late on Sunday morning, I called Les again, determined, this time, to leave a message. The hotel operator switched me to Les's room. I was prepared to hang up after a few rings and call back to leave a message, but to my surprise, a sheepish voice came on the line.

"Hello?"

"Hello, is this Mr. McCann?"

"Yeah, who is this?"

"This is David Tarver. You know my uncle Bill Hayden in Detroit. He told me I could call you about my music synthesizer."

"Bill... oh, yeah, Bill."

"Yes, Bill Hayden."

"Yeah. Now what is it you want?"

"I have this music synthesizer that I designed. It's digital. My Uncle Bill... uh, Bill Hayden, said you might be interested in seeing it. I tried to reach you in New York but there was no answer on your phone. I could bring it down to Washington if you want... if you would be willing to see it."

"That sounds cool, man, but uh... well, I'm leaving here after tonight. I mean I don't have time to be lookin' at nuthin' while I'm here. I'll be back soon, though, and maybe I can take a look at it next time I'm in town. Maybe next time I'm in New York."

"When will that be?"

"I don't know, man—I just follow the schedule. Check out the schedule and let's get together next time, okay? Bye."

Les McCann hung up. I had no idea when he would come back around, and besides, he sounded flaky. I don't know what I was expecting, but talking to Les certainly wasn't like talking to a Bell Labs colleague. He seemed unfocused and distant, and above all, unconcerned. Our telephone encounter was far from serendipitous—it was heartbreaking. Most disappointing, I learned nothing about my synthesizer. I got no feedback, good or bad. Pursuing McCann further would do no good. My synthesizer and I weren't important to him.

I became even more determined to show my creation to a musician, *any* musician. I decided to lower my sights and take the synthesizer to a local player. There were sure to be keyboardists hanging out at a music store, and maybe the sales staff could give me ideas about marketing my invention. One Saturday morning, I took my "baby" to the largest musical instrument store in Red Bank.

When I told an older male employee, apparently the manager, I had designed a synthesizer and that it was out in the car, he said: "Well, uh, I guess we could take a look at it, but we're not looking for any more products to sell right now." He was bristling, already in rejection mode.

"That's okay," I said in my most reassuring voice. "I'm not trying to sell anything. I just want some musicians to look at this thing and tell me what they think. I've been working on it for a while, and I need some help to know where to take it from here."

The manager seemed to relax a bit. He said, "Okay, bring it in the back door there, and I'll get a couple of guys to look at it."

I was ecstatic. Someone, a musician, was finally going to see my creation. Finally I would get some feedback.

I rushed out to the car and pulled around behind the store. I unloaded the synthesizer—first the control unit, then the cherrywood-clad keyboard—and took it inside. The manager had corralled a couple guys—musicians, I assumed—and they stood eyeballing my invention.

"Looks pretty impressive," musician #1 intoned. Where did you get it?"

"I designed and built it myself. I've been working on it for about five years, ever since I was a junior in college."

"Well, hook it up. Let's see how it sounds," said musician #2.

That was fine with me. I set the control unit and keyboard on a big counter, plugged in the keyboard, and plugged the control unit into a wall socket. I flipped the power switch, and the whole thing came to life.

"Can you play it?" musician #1 asked.

"No, I only play trombone. I was hoping one of you guys would play it."

Musician #1 sidled up to the keyboard and hit a few notes.

"Mono, eh?"

He meant my synthesizer was monophonic rather than polyphonic. That meant it didn't play chords—it played just one note at a time. That was how the early Moog and Arp synthesizers worked, but newer designs from outfits like Yamaha and Kurzweil could play chords.

"Yeah, I designed it to be monophonic. The new thing about this synthesizer is that it's computer-controlled, and the waveforms are stored in ROMs. The waveforms are digital, so you can make them whatever you want."

"Cool," croaked musician #2.

"How much does it cost?" asked musician #1.

"Well, this is just a prototype, so it's hard to say. I guess it'll cost maybe two or three thousand when it goes into production."

"Wow, that's a lot," said musician #1. I don't know any musicians who'd be willing to pay that much for a mono synthesizer, computer-controlled or not. There's a lot of stuff out there cheaper than that, and there's more coming out every day. It may not be as sophisticated as this, but musicians don't care—they just want something cheap and rugged. Besides, this thing is too big. It might work in a studio, but no musician is going to drag this around."

"This is just a prototype," I said. "The final unit will be smaller. I'm just trying to get some feedback so I'll know where to take it from here." I could hear myself sounding defensive, and worse, hurt. These guys were attacking my baby, but I understood what they were saying. Scanning the room at the various keyboards for sale, mine was bloated and fragile by comparison. For the first time, I started to feel I had missed the mark.

Musician #2 jumped on the bandwagon. "This is basically an expensive low-end synthesizer—it costs too much. Musicians want low-end synthesizers, but they don't want to pay a lot. If you want to sell to high-end musicians, you'll need to have more features. The high-end guys can get synthesizers from Yamaha or Kurzweil—they don't need something like this."

My bubble burst, I thanked the guys for taking the time to look at my synthesizer. Then I loaded it back into the car and drove home.

It was a bitter realization. I should have been talking to musicians much earlier, before I'd invested so much time and money designing and building my synthesizer. I had patterned the design after the Moog synthesizer I saw at GMI in 1972, but didn't finish mine until 1978. Times had changed, and the music industry had moved on. Far from being the music of Stevie Wonder's mind, my synthesizer couldn't even impress two geeky musicians at a local store. I had developed a cool but obsolete product.

I was working alone with minimal funds and no music industry connections. I had good, fresh technical ideas, but better-funded developers had good ideas, too. Many had the music industry experience and ties I lacked. I had been fighting a losing battle.

That evening, I realized I had reached the bitter end of my synthesizer journey. For years, I dreamed of starting my own music synthesizer business. That evening, the dream died.

26. An Idea is Born

LATE ON A FRIDAY AFTERNOON, I lay on a lab floor in Holmdel taking measurements and jotting results in a notebook. It was early 1979, and I had been working at Bell Labs nearly three years. A few days earlier, I had received some disappointing news from Flint. I was burying myself in my work to avoid thinking about that, so I didn't notice Bob Heick peering over my shoulder.

"Did you check the frequency accuracy? What number did you get?" Bob stood poised with pencil and lab notebook, ready to record my response.

"Yes, Bob," I sighed, "I checked the frequency accuracy, and it's right on the money. It has to be, because the system is digital—that's how I designed it."

"What about phase noise—did you check the phase noise?" Bob was starting to get on my nerves, but I did my best to remain calm.

"Yes, Bob, the phase noise is within limits. Everything is checking out just fine. Believe me, I want this to work just as much as you do. After all, it *is* my design!"

I could hear myself sounding perturbed. I turned to look at Bob and noticed his wounded expression. "Okay, Dave. I'll let you get back to it," he said, before retreating to another part of the lab.

I felt bad. Bob Heick was a decent guy. I knew his constant checking on me wasn't a "white thing," but more of a "German thing." Bob was much older and much more experienced. He and I were responsible for different parts of RTS-5, but he seemed to feel responsible for the entire system, hence his constant checking on my work. I didn't know and didn't care if he checked on the other engineers' work as much as mine. My frustration that afternoon was not about Bob Heick—he was merely a catalyst for the angst building inside me. I was frustrated with myself. My ideas for starting a business were going nowhere. My Bell Labs career was not advancing according to my schedule. I felt trapped, and for that reason I lashed out at Bob.

After my synthesizer idea went up in smoke, I cooked up all sorts of new ones. I designed a small factory-automation computer, dreamed up a coinless Laundromat, toyed with designing an automatic hearing-test machine, and considered creating a combination computer/telephone that could access all sorts of useful information. In the end, I abandoned each idea.

Before my synthesizer project crashed, I'd pursued another business idea. In the fall of 1977, I had been at Bell Labs for just one year. My U-M bioengineering studies were still fresh in my mind, and I thought they might lead to an exciting and lucrative business. Hospitals were beginning to deploy advanced electronic instruments—EKG machines, fetal monitors, and such—but they often lacked the capability to properly use and maintain such devices. My idea was to form a company that would help hospitals evaluate, select, use, and maintain new high-tech equipment. I called the idea the Healthcare Information Service.

I decided to launch the venture in Flint. Hurley Hospital was the natural place to investigate my idea's possibilities. Hurley was owned by the city and was the largest hospital in the area. Not only had my mother served as the first black administrator there—assistant director of nursing during the 1960s—but my sister Bernice graduated from Hurley School of Nursing and worked at the hospital, and her husband, Evans, worked in the medical lab. I developed an extensive proposal, clearly detailing the services my company would provide and the operational and financial benefits. I felt like I was completing a circle. I was born at Hurley, educated at a great university just down the road, and after gaining experience at the world-renowned Bell Labs, was prepared for a return to my hometown to apply all I had learned.

I finished my proposal in just a few weeks, and then went straight to the top to present it. Hurley's second in command was Phil Dutcher, who, like me, was a GMI alumnus and former AC Spark Plug employee. I knew Phil was interested in improving operations and reducing costs, so I was sure he would be receptive. I contacted Phil by phone, introduced myself, and told him what I was trying to do. He expressed interest and arranged for me to meet with the Hurley CEO, Richard Schripsema.

A few weeks later, I traveled from New Jersey to Flint, nervous but confident. I had a good idea, and knew I could carry it through. Things were falling into place. I felt my association with Hurley was meant to be.

The meeting with Schripsema lasted about thirty minutes. He listened intently and seemed genuinely interested. He said he recognized the need for the service I proposed. He said he wasn't sure whether it made more sense to hire an outside company or to hire someone in a staff role. He asked if I would prepare a job description for a staff biomedical engineer who would provide the services I proposed, and then sent me to see the director of the physical plant department for a detailed review of my ideas.

I thought it odd that I should be asked to meet with the person responsible for maintaining the hospital's heating and cooling systems and such, but later that same day I stopped by Richard McCormick's office. I was surprised to learn that he was the father of one of my high school basketball teammates. That seemed like a good omen. I presented my ideas to McCormick. I told him Richard Schripsema liked my proposal and had asked me to write a job description for a staff person who would manage the service.

McCormick didn't seem particularly interested. My words about calibration and electrical safety and software standards seemed to fly over his head. That was okay, though—he didn't have to understand. If Hurley approved my proposal, I would handle all the technical details so he wouldn't have to worry about them. That was the whole idea. I wasn't crazy about making my Healthcare Information Service an in-house operation, but if that's what it took to get the idea off the ground, I was prepared to do it. I could always turn it into a standalone business later, after gaining experience at Hurley's expense.

McCormick said he would consider my proposal and talk to Schripsema. When I got back to New Jersey, I immediately prepared and mailed the job description Schripsema had asked for. I expected I would soon get a call asking me to come to Flint and establish the Healthcare Information Service as a Hurley department.

Weeks and months went by and I heard nothing. I placed calls to McCormick and to Schripsema, but their secretaries referred me to personnel. After a few months, disgusted and disappointed, I gave up. I moved on to the last lap with my music synthesizer project, experimented with other business ideas, and pretty much forgot about Healthcare Information Service. Then, nearly two years after my visit to Hurley, my brother-in-law Evans told me Hurley had established a new position: staff biomedical engineer. I contacted Hurley's personnel department and was told the position had been filled. I asked for and received a copy of the hiring notice. The duties described in the notice bore an eerie resemblance to the job description I'd written. Concluding that my services were not wanted at Hurley, I was extremely disappointed. What had at first seemed preordained turned into another dead end.

So there I was at Bell Labs late on a Friday afternoon, nearly two years after making the proposal to Hurley, tediously testing and retesting my RTS-5 designs. The bad news from Flint hung over me like a thick fog. Time was running out on my entrepreneurial dreams, and I didn't even have a viable idea. I tried to ease my despair by focusing on my job and trying to create a perfect product that my Bell Labs colleagues couldn't criticize. I was at the lowest point in my career that Friday afternoon, when Bob Heick entered the lab and began peering over my shoulder, double-checking my work.

After the encounter with Bob, I resumed the tedious process of testing my designs. I generated each kind of noise or interference that might be found on a telephone connection, and verified that my instruments could accurately measure each one. Then I compared the value my equipment measured to the value measured by my Hewlett-Packard "reference" measuring instrument. I used something called a Bradley Box as a noise and interference source. The official name was "Bradley 2A/2B Impairment Simulator." Despite its high-tech-sounding name it was a pretty rudimentary device. The enclosure looked like a large lunch pail, not unlike my father's old Morse code practice machine. Ancient toggle switches and knobs controlled the distortion levels. As I sat looking at it, I chuckled. I knew that, as primitive as the Bradley Box was, Bell Labs paid $6,000 for each one, and we owned a lot of them. On top of that, I needed a $10,000 HP measuring instrument to tell me what the Bradley Box was generating. *What a racket*, I thought.

Then it hit me: This Bradley Box was nothing but an expensive synthesizer! My failed music synthesizer was more sophisticated than the Bradley Box, yet I couldn't sell it. The Bradley Box sold because it filled a well-defined niche in the telecommunications industry, and because companies like Bell Labs could afford to buy it. In fact, with all the money floating around Bell Labs, the Bradley Box seemed inexpensive.

Just like that, I had a new business idea. I would make a better Bradley Box, an impairment simulator that was easy to use, precise, accurate, and reliable. I would use all the computing and digital signal-processing techniques I had developed for my synthesizer, only this time I would make a product that would sell. Not only did I completely understand the technology, I understood how the impairment simulator would be used, because I was already using it. I was familiar with the customer base, too, because they were engineers like me. The impairment simulator would be marketed to well-funded corporations, not to financially struggling musicians. My plan seemed to make perfect sense. A ray of hope burst upon what had been a miserable spring afternoon.

Several months later, I had an unexpected opportunity to test my new idea. My Bell Labs colleague Jim Ingle, an expert in telecommunication measurement techniques, said his friend Frank Bradley was coming over from Fort Lee, New Jersey, for a lunch meeting. Frank owned Bradley Telecom and had invented the Bradley Box. Jim wasn't aware of my entrepreneurial interest in the device—he just knew I was using it and thought I might have questions for Frank. Jim invited me to join him and Frank for lunch, and I cautiously accepted.

I was cautious because I thought Frank Bradley might debunk my new business idea. Perhaps he was already planning to improve his Bradley Box, or maybe my suggestion of an improved version would prompt *him* to produce one, but I knew it was better to go ahead and meet Frank. If my new business idea was dead in the water, it was better to find out sooner rather than later.

Jim, Frank, and I met for lunch in the Bell Labs cafeteria. Frank was a crusty fifty-something fellow who wore thick glasses and carried a plastic pocket protector in his shirt. His appearance and his gruff, no-nonsense character made me think of Vince Lombardi. Frank didn't seem the least bit interested in me. His objective was to bend Jim's ear about his new-product ideas. I listened attentively, waiting for a chance to ask Frank about the 2A/2B. When the conversation waned, I jumped in and explained to him that I had been using his product heavily and that it was crucial to my work. Frank seemed flattered. I went on to explain that, as an analog product, the 2A/2B required an expensive external instrument to confirm its settings. Silently hoping his answer would be no, I asked Frank if he had ever considered developing a newer version, one that incorporated digital precision and accuracy.

"I would only do a redesign if one of our customers paid for it," Frank said. "The 2A/2B does a good job, and it's an industry standard now."

Jim Ingle concurred with Bradley's view. He said the 2A/2B did things "just right," and that engineers were already familiar with its capabilities and limitations.

I felt a massive rush of relief. Frank wasn't inclined to develop a more modern version of his product, and I knew no one was going to pay him to do it. I had just gotten a green light.

The Bradley 2A/2B was similar in concept to my beloved but failed music synthesizer. It sold for thousands of dollars to Bell Labs and other corporations. My knowledge of the Bradley Box, my access to key players in the telecommunications industry, and my synthesizer experience would enable me to produce a much better product.

After lunch, bursting with excitement, I nearly ran back to my office. I had found the idea that would launch my company, and couldn't wait to start working on it.

27. Love and Technology

I BECAME A REGULAR PARTICIPANT in the after-work basketball games at Holmdel Village School almost as soon as I arrived at Bell Labs. Just a few weeks later, in August 1976, I was surprised to see a young Asian girl on the court. She wasn't merely participating in the mostly male games but was one of the most aggressive players. She had no qualms about setting a hard pick or driving to the basket. At the end of the game, she untied an elastic band and revealed a flowing mane of shiny black hair. She was cute! Within a few days this uniquely attractive Asian point guard walked up to me in the cafeteria and asked, "Are you going to be playing basketball again tonight?" I was surprised she would strike up a conversation so directly. She seemed so outgoing, so friendly, so *free*.

"Yes, I plan to be there. How about you?"

She grinned and said, "Yep, and I'm going to take you to the hoop again!"

I was so caught up by her outgoing, playful nature that I didn't even ask her name. That night between games I had another chance to talk with her. Diane Sheng was her name, she was raised in Connecticut, and her parents were from China. She was studying at Stanford University for a PhD in applied mathematics, and was visiting Bell Labs for a few months to work with her thesis advisor. Like that image of Leesing Pang in the Bell Labs recruiting brochure—intelligent, fit, the best of the best—Diane Sheng seemed to embody the allure of

Bell Labs. I admired her immediately.

A few weeks later, Diane left Bell Labs and returned to Stanford. I didn't expect to see her again, though I hoped I would.

As the summer of 1977 approached, I remembered Diane and wondered if she would return to Bell Labs. The summer came and went and I didn't see her, so I figured she wasn't coming back. I was disappointed, because this cute, intelligent Chinese girl had intrigued me, but I concluded that another meeting just wasn't in the cards. As the summer progressed, thoughts of Diane receded. Then, in September, I received a big surprise. I walked into the office of my friend Nettie Colquitt, and there was Diane! I was happy to see her. The way she smiled and played with her hair said she was happy to see me, too. Diane said her thesis advisor had returned to New Jersey, but this time he had decided to visit during the fall rather than summer. Diane would be in New Jersey until December.

The wheels in my head spun rapidly. I had no idea whether Diane had a boyfriend, and certainly didn't know how she would feel about dating a black guy, but every time I saw her she would run up to me like a kid and ask about playing basketball. I was sure no one could be *that* excited about playing pick-up basketball after work. Still, we had never talked about anything serious or in-depth—we were just basketball buddies. I resolved to ask Diane out on a date, and I enlisted Nettie as a surrogate suitor.

One day at lunch, I approached Nettie and said, "I'm interested in going out with your new officemate. Do you think she would date a black man?"

Nettie giggled. "So, you're interested in some oriental stuff, eh?" Nettie had a way of breaking things down to their basest elements.

"No, I'm serious. I kind of like her, and I'd like to get to know her. What do you think she would say if I asked her out?"

"I don't know," Nettie replied. "But I know she has some black friends at Stanford. She seems pretty cool, so you should just ask her and see what happens."

"Well, I don't want to just ask her," I muttered back. "Would you just tell her I like her, and see how she reacts? Just tell her I'd like to get to know her better."

"Okay, I will," Nettie said. "But you're on your own after that."

Back in my office, I tried to work, but all I could think about was what Nettie was saying to Diane and what Diane might say in response. I imagined Diane might "freeze up" and become distant once she knew I had more than a basketball-buddy interest in her.

Late in the day, I called Nettie and invited her for a coffee break. As we strolled toward the cafeteria, Nettie said, "Diane said she's interested in getting to know you."

That was all I needed to hear. I followed Nettie back to her office, beckoned Diane into the hallway, and walked her to the balustrade overlooking the atrium. Gazing down at the garden, I popped the question:

"How would you like to go out and get some dinner with me?"

Diane's reply was immediate and enthusiastic: "I'd love to!"

That Friday evening, I picked Diane up and drove her to What's Your Beef, my neighbor Stan's favorite eatery in Rumson, New Jersey. Diane and I had a good time getting to know each other over dinner. We did talk about basketball, of course, but also school, Bell Labs, and career aspirations. I told her about my music synthesizer project and my plan to start a business. She told me about her doctoral studies and her desire to teach at a university. I was amazed someone so bright and beautiful could be so down-to-earth and nice. My admiration for her grew deeper.

After dinner, we got into my Scirocco and headed back to her apartment. We had traveled just a few blocks when the car suddenly stalled and stopped. It was out of gas. In my excitement to go out with Diane I had forgotten to fill the tank. I was totally embarrassed, because I knew Diane would think I'd planned to run out of gas as a prelude to groping and kissing. I must have apologized at least ten times, insisting that I wasn't about to try anything. Diane laughed it off, and I walked back to the restaurant to call Stan for help. After thirty awkward minutes, he arrived with a gas can, and we resumed the trip to Diane's apartment. We sat on her sofa and talked for quite a while, and I felt the romantic tension rising between us. Finally, I grabbed Diane awkwardly and kissed her on the lips. To my great relief, she kissed me back.

That autumn, Diane and I spent a lot of time together. We still played basketball with the guys, but we also went to movies, enjoyed long weekend breakfasts, and stopped after work for drinks with my Bell Labs friends. During that blissful season, we concentrated on our work but also focused on our blossoming relationship. When Diane returned to Stanford at the end of 1977, we were committed to continuing our relationship from opposite ends of the continent.

I flew out to visit Diane a few times, and each time we grew closer. She acquainted me with the San Francisco sights, and I treated her to relaxed breakfasts and sumptuous dinners at Palo Alto restaurants. By the time summer 1978 rolled around, Diane and I were deeply in love, and we were thinking seriously about adjusting our careers so we could be in the same place. Diane was due to finish her PhD program in December, and had begun interviewing for a job. Bell Labs was prepared to offer her a position, but she preferred to work at a university. When Diane landed what she thought was the perfect job—a UCLA faculty position—I was happy for her, but also worried whether our relationship could survive an extended East Coast/West Coast separation. I realized that I might have to relocate to keep the relationship alive.

I arranged an interview with the Rand Corporation, a think tank based in Santa Monica. Rand did classified research for the U.S. government, and had become famous in 1971 when one of its employees, Daniel Ellsberg, released the Pentagon Papers analysis of the Vietnam War.

The first phase of the interview took place at an office in Washington, D.C. The two interviewers, one of whom was a Rand director, were serious and secretive. I wondered what I was getting myself into. At the conclusion of the interview, the director invited me to Santa Monica for a second interview. Intrigued, I made the trip.

The Santa Monica interview went extremely well. I was impressed with Rand, though it was very different from Bell Labs. Rand didn't develop products, at least not in the traditional sense. Rand's "products" were ideas and analyses, and the place was filled with people who sat in well-appointed offices and traded in both. I met only one black staffer, but he was a memorable character. The idea he was working on was a magnetic levitation train that would travel through a tunnel between New York and Los Angeles. It would require no external power because the tunnel would be dug in such a way that the lowest point would be under the center of the country, and the highest points would be at each end. The train would leave New York and gain speed as it traveled down the tunnel's arc toward the middle of the country. Then, after passing the low point, it would gradually lose speed as it coasted to a stop at Los Angeles.

I couldn't believe someone would get paid, and paid well, to sit around and think up such ideas. The intellectual environment at Rand seemed freer, and definitely wackier, than Bell Labs. The Santa Monica beach was only a block from Rand, and at lunchtime many employees walked over and played volleyball. Santa Monica weather was perfect year-round, and the thought of playing volleyball in the California sun contrasted nicely with the nasty, slushy New Jersey winters.

Rand wanted me. In fact, they wanted me so much they were willing to sponsor me to study for a PhD in public policy at UCLA. The salary offer was more than I was making at Bell Labs. I could have an ideal job, get my PhD and the alluring "Dr. Tarver" title, and be with Diane. All I had to do was leave Bell Labs and put aside, at least for a while, my ideas about starting a business. It was an easy decision. Back in New Jersey, I thought about the Rand offer for one day, and then accepted it.

Diane was as excited as I was. She would be a professor at UCLA, and I would be a researcher at prestigious Rand Corp. We would be together in beautiful, sunny Southern California. I fully understood that the move might spell the end of my long-held business dreams, but I was ready. I had grown weary of my synthesizer project, and the prospect of being in California with Diane altered my priorities. I envisioned a career in public policy, and perhaps even a run at politics. I went to see John Colton and told him I loved my job at Bell Labs, but that I had accepted a position with Rand. I told John about the situation with Diane and her offer from UCLA. He knew Diane from the basketball courts, and knew I was dating her, but he didn't know how serious we were. John displayed no anger or irritation or even surprise. He said he understood and wished me well. He said I was one heck of an engineer, and that if I stayed, my future would be bright. Then he issued the standard line: "If for some reason things don't work out at Rand, you'll always be welcome here at Bell Labs."

I was relieved. Quitting wasn't as hard as I had anticipated, and John's reaction was much more understanding and supportive than I'd expected. Now I could begin in earnest to plan the move to California, my work at Rand, and a life with Diane.

In July 1978 I traveled to Palo Alto again to visit Diane. My plans for our relationship were unfolding, and I wanted to let her know what I was thinking. We spent an afternoon at a nearby beach. It was a clear, sunny day, but the air was cool, and the occasional wind gust made it seem even cooler. The beach was nearly deserted. We walked along the rocky shore holding hands, and when we found a sandy spot we put down our blanket. As we cuddled, I could barely contain my emotions.

"Diane," I said. "I can't believe we're going to be together every day!"

"I can't believe we'll be together and I won't have to work on my thesis!" Diane smiled broadly, and I realized the tremendous weight her studies must have been. I appreciated that she could give our relationship so much attention as she toiled on her PhD.

We were both ecstatic. Diane was about to achieve her longtime goal, and I was moving to a great position at a prestigious think tank. A warm feeling came over me. *We're good for each other*, I thought.

The time had arrived. I thought back to when I first met Diane on the basketball court—what an extraordinary, wonderful girl! I recalled the nervousness I'd felt more than a year later when I invited her on that first date. That nervousness had returned. Just as before our first date, I thought I knew what Diane felt, but I couldn't be sure. This time, Nettie wasn't there to speak for me. I had to express my feelings for myself.

"Diane, I want you to marry me."

There. It was out. My heart was pounding and tears welled up in my eyes. Then I blinked, and the tears came streaming down my cheeks. Diane looked at me and grinned.

"Why are you crying?" she asked.

"Because I love you so much, and I want us to spend our lives together."

I was hopelessly sappy and trite, but couldn't help it.

Diane was still grinning. "Okay, I'll marry you, but you have to promise to stop crying."

I chuckled, wiped my face with my palms, and said, "Okay."

I took Diane in my arms and we fell onto the blanket. We stayed there and hugged and kissed and whispered to each other for the rest of the afternoon. Our two lives were becoming one.

Soon after that blissful afternoon, things started falling apart. First, Diane called with news that UCLA was rescinding her job offer. She didn't know why, but she knew it was true. Diane was crushed. I was pretty upset, too, and not only because I felt badly for her. I was already set on working at Rand, and had given notice at Bell Labs. Now I would have to mentally shift gears, forget Rand, and try to get my job back.

I couldn't help but wonder why UCLA had taken the unusual step of rescinding Diane's job offer. Could someone have found out she had a black boyfriend? Bell Labs and academia were closely connected, so it wasn't inconceivable that our relationship could fuel gossip in UCLA's System Science Department. Besides, when I interviewed at Rand and met with the folks at UCLA about their PhD program, I mentioned that my close friend was going to be a professor in the System Science Department. I told Diane she should make noise about what was happening—even raise the issue of discrimination—but she was disinclined to do so. She just started looking for jobs elsewhere.

Then things got worse. In August, Diane traveled to China with Suzanne, her only sibling, and their parents. I hadn't met Diane's family, so I had no idea how they felt about her dating, much less marrying, a black man. Diane's parents may have known she had a boyfriend in New Jersey, but until the trip to China, they didn't know how serious the relationship had become.

The trip was an emotional experience for the Sheng family. It was the first opportunity for Diane and her sister to meet their relatives in China. Diane's parents had grown up in China and had emigrated to the U.S. after World War II. Diane and Suzanne were born in the U.S.

and had never been to their parents' homeland.

Diane revealed to her parents that she was seriously involved with me and that we were planning to marry. Instead of responding with joy and good wishes, her parents spent the remainder of the trip berating Diane about our relationship and trying to convince her to change course. Her mother told her it was "strictly forbidden" for her to have a relationship with a black man. They seemed to feel that Diane had become too liberal, too free in her thinking, and that she needed to be "deprogrammed." Whatever Diane's parents said to her in China was persuasive. When she returned to the U.S., she broke off our engagement.

By September I found myself alone in New Jersey once again. There would be no glamorous cross-country trek to Rand. There would be no wedding. There was only my job at Bell Labs and my badly tarnished dream of starting a business. I embraced both with renewed vigor. I told myself that everything happened for a reason, and that I was not *supposed* to go to California. I became even more convinced that my rightful place was in New Jersey, preparing to start my business.

That fall, I participated as an on-campus recruiter for Bell Labs at the University of Michigan. At the time, the Bell System was still a regulated monopoly that included Bell Labs, Western Electric, AT&T Long Lines, and most local telephone companies. The Ann Arbor visit brought together executives from several of the Bell companies for a week of reviewing and discussing résumés, interviewing students, and conferring about employment decisions. There were lots of opportunities to socialize with the other team members, and I met and became romantically involved with one of them. She was attractive, charming, intelligent, and black. Our budding relationship helped me get over the pain of losing Diane. I was moving on.

Diane had not moved on. She was deeply conflicted between her feelings for me and respect for her parents. She was enduring a lot, finishing her thesis and examinations, licking her wounds over the UCLA job fiasco, and deciding where to start her career. It was too much to deal with at once. In breaking off the engagement, Diane wasn't saying she didn't want to be with me. She was saying she wasn't sure, and needed more time to process what happened in China. I was too hurt by what I perceived as her rejection to see shades of gray. In my mind, Diane was either in or out, and I decided she was out.

Diane called me a few times that fall, but I found it painful to talk to her. One of her Stanford classmates, Amy, also called. Amy was a white Tennessean, a fireball of positive energy. On this occasion, though, she seemed more upset than I was.

"You know Diane really loves you."

"Is that so?"

"Yes, she really does. This thing is tearing her up. You should see her moping around out here. Just give her some time, she'll come around."

"Yeah, maybe when her parents tell her it's okay to be with a black man."

"I don't know why they're so stupid. They're just idiots—they haven't even met you. Trust me—everything is going to work out. You guys are too good for each other to break up."

I found Amy's call ironic and sweet—Diane's white college buddy trying to push us together as her parents were trying to pull us apart. Amy's appeal was deeply moving, but it was too late.

In November, Diane called again. She told me she had accepted a job offer from Bell Labs and would be moving to New Jersey in December. I congratulated her and said it was great that she had finished her PhD, but I was noncommittal about our relationship. I didn't say, "I can't wait to see you," or "I hope we can start our relationship again." I didn't want to set myself up to get hurt again. I really had moved on.

When Diane arrived she called to let me know she was staying at the Ramada Inn in West Long Branch, and asked me to come by so we could talk. I made the five-minute drive across town. As soon as I saw her face, all my apprehension about our relationship melted. It was as if we had never split up. That night, we decided we would stay together no matter what.

Diane rented a second-floor apartment in Eatontown, ten minutes from my place in Long Branch. Her downstairs neighbor was a young engineer in my group at Bell Labs, a recent transplant from South

Carolina named Steve Moore. I was soon spending practically all my spare time at Diane's place. We worked hard at our jobs during the day and brought work home at night, but we still found lots of time to enjoy each other's company. We decided that when my lease was up in August we would rent a place together. More important, we decided we would get married in November.

Diane and I rented a townhouse at a place called Hilltop Terrace, on the east side of Red Bank. The townhouse had three levels: bedrooms on the top floor, living room/dining room/kitchen on the ground floor, and a full finished basement. I was most excited about the basement, where I immediately set up my electronics lab. Now that I knew I would be staying in New Jersey, I planned to get back to work in earnest launching my electronics business. I realized that moving to California would have dealt a severe blow to my entrepreneurial dreams, and didn't regret staying in New Jersey.

As our wedding day approached, Diane and I focused intently on our plans. I told her that starting a business was one of my main goals in life, and that I didn't want anything to get in the way of reaching it. I made it clear to her that the last thing I wanted to be was a Bell Labs "lifer." The vision held by many Bell Labs engineers consisted of enjoying the job, getting married, buying a nice house, having a couple of kids, and living a life of suburban bliss. That wasn't my vision, and I wanted to be sure Diane understood that before we got married. Living a comfortable suburban life was not my priority—starting a business was. Diane seemed to be okay with that, probably because she was so focused on her own job and her mathematics research. She valued hard work and seemed satisfied with life's simple pleasures. Ultimately, I felt assured our goals were compatible.

On November 17, 1979, Diane Dwan Sheng and I were married at the United Methodist Church in Oakhurst, New Jersey. Fifty-five of our closest family members, friends, and colleagues were there. My brother Fred was best man. Diane's sister Suzanne was maid of honor. Diane's parents did not attend.

28. Movin' on Up

WHEN DIANE'S UCLA JOB OFFER was mysteriously rescinded, and I ended up staying at Bell Labs, I had no choice but to refocus on my previous goals. While my ultimate aim was to start my own business, my immediate objective was to be promoted to Bell Labs management.

An ambitious MTS might wait six to ten years for promotion to group supervisor. I wasn't prepared to wait that long. When I joined Bell Labs as a brash twenty-three-year-old, I thought I would get promoted after one year and leave to start my company a year later. Even two years felt like a long time, and six to ten seemed an eternity.

The year 1979 was my third at Bell Labs, but I wasn't particularly worried that I had exceeded my "expiration date." The work was still satisfying, and my project was going well. John Colton and others had assured me my future was bright, so I expected to be promoted any day.

The opening came suddenly, unexpectedly. One afternoon Ed Spack called an impromptu department meeting and announced that John Colton had been promoted and would be leaving the department immediately. The news filled me with excitement and anticipation. My armpits were sweating, and my arms and legs were tingling. I waited nervously for Ed to announce my promotion to group supervisor. Was I ready for the responsibility? What would the other guys in the group think? What would I say when Ed announced my promotion? I decided to keep it short and sweet. I would say I was happy to have the job and ready to get started. I was so busy scheming and daydreaming that I didn't focus on what Ed Spack was saying. I should have.

"I'm pleased to announce that, effective immediately, Jim Ingle will be taking over John Colton's group. We're lucky to have Jim, because he's an expert in data parameter testing techniques."

I couldn't breathe. I was devastated, but tried not to react. It was as if someone had punched me in the stomach, but I had to absorb the blow and pretend nothing happened. I had noticed Jim Ingle when I entered the meeting, but I certainly didn't think he was there to take over the group—*my* group. John Colton had all but promised me the job. I hoped the announcement was some kind of prank. I hoped Ed would say, "Just kidding!" and announce my promotion, and everyone would laugh and congratulate me. That didn't happen.

Jim Ingle had been at Bell Labs a long time. I had worked with his group to establish the measurement methods I employed in RTS-5. Jim *was* an expert on telecommunication measurement techniques, but he had no business running my group. He was a theoretician, not a product development guy. I was certain Jim's appointment was a mistake, and I couldn't believe Ed had chosen him over me.

Everyone filed out of the lab, and I trudged back to my office. A few engineers gathered in an office down the hall and spoke in hushed tones. I wondered whether they felt sorry for me, or were glad I didn't get the job. I sat at my desk and tried to understand. Why would Ed Spack bring in a pencil-pushing nerd to run our cutting-edge development group? I convinced myself Ed had brought in Jim because he didn't want *me* to have the job. Ed knew I was the best candidate, perhaps the only candidate, in our department, so he appointed someone from the outside. Even though Jim Ingle had no product development track record, his expertise in telecommunication measurements allowed Ed to say Jim was more "qualified" for the job.

Did Ed not want a black supervisor in his department? And what about John Colton? He knew I desperately wanted to be a supervisor. Now he was going off to a cushy department-head job and leaving me to fend for myself. Did John recommend me for the supervisor position? If so, did Ed Spack disregard the recommendation? The best way to get to the bottom of things was to talk to John. I spent the rest of the day and night fuming and scheming, and the next day I went to see him.

I walked into Colton's office and plopped onto his guest chair. "What's up?" he asked. He appeared nonchalant, but he had to know I was pissed.

"Well, first of all, congratulations on your promotion," I began.

"Thank you," John said in his best casual, upbeat voice. "I'm going to be running the DACS department. I'm excited because I came up with the DACS concept myself a few years ago."

DACS was the acronym for Digital Access and Cross-Connect System, but John's new assignment was the least of my concerns. I just wanted to know who was responsible for Jim Ingle getting the supervisor job I was supposed to have.

"Well, that's great—it sounds like a great position. I was just wondering, though. I mean I kind of expected that if you got promoted I would get your job. I don't understand why Jim Ingle is being brought in. I thought I might get the job after all the work I've done in the group. You told me you expected me to be promoted in two years, and now it's been nearly three."

"Well Dave, Ed Spack felt he needed to have someone more experienced running the group." John's reply was direct, unsmiling. "Jim has a lot of experience with communication measurements. You know that—you dealt with his group while you were doing your designs. Just work with him for a while and your promotion will come, you'll see."

"Jim doesn't have experience developing equipment—he's a pencil and paper guy. You know I can do a better job running this group!"

"It's Ed's call, Dave. Your name came up for the position, and Ed assured me you'd get promoted. He just feels Ingle is the guy he wants to run the group right now."

So that was it. My name "came up." I was disappointed, angry, and hurt. I didn't know what else to say to Colton, so I just left his office. Since John said it was Ed Spack's call, I went to see Ed.

In my early days at Bell Labs, I would have been nervous about confronting Ed Spack. Back then I was awed by his big office, his power, and his supposed wisdom. With time and experience, I came to realize Ed was just a guy. He didn't have special powers. He merely had a higher position, more experience, access to more and better information, and a legion of people working for him.

I hadn't spoken to Ed at length since the "nigger in the woodpile" incident. Back then I was an innocent victim and he was a powerful and caring executive. Now I saw him simply as an older engineer, wiser than me in some ways, less so in others. I realized our relationship was forever changed. I was no longer an unsure, green engineer. I was a confident, angry young man stepping up to insist on my promotion to management.

Ed rose to shake my hand and offered me the chair opposite his big wooden desk. I sat erect and prepared for confrontation. I noticed he wasn't the sympathetic, genial man I remembered. He wore a stern look, and his manner was tense, as if bracing for a challenge. I sensed the hard core that probably got Ed his department-head position. He expected a challenge, and I was going to give him one.

Ed started the discussion. "So, Dave, you wanted to see me. What did you want to talk about?"

"Well, Ed, I just wanted to talk to you about my position in the department. I mean, you said yesterday Jim Ingle would be taking over John's group. When I started here, John told me I was supervisor material, and I thought that with him being promoted I would take over the group."

"I see," said Ed.

"So I wanted to know why you're bringing Jim Ingle in to run the group. I mean, I don't understand."

Ed said, "Well, Dave, Jim is an expert on communication measurement techniques. He's got a lot of experience. I think you can learn a lot from him."

"But Ed, John told me I would be a supervisor in two years. Now it's been three, and there's an opportunity right here in your department, and I don't get the job."

"Dave," Ed said sternly, "I don't know about any promises John might have made. I know you're well regarded here, and if you keep doing a good job you'll probably get promoted at some point. There are no guarantees, though."

That answer wasn't good enough for me. "How much longer do you think I'll have to wait?"

"Like I said, David, I can't make any guarantees." Ed seemed to grow impatient. Calling me "David" instead of "Dave" was the signal that he was ready to end our meeting.

"Well, I'm going to finish my project and make sure the field trial goes well. I hope I'll be in a position to get promoted by then."

"Okay, Dave, that's good. You do that and we'll see what we can do."

Ed didn't promise me a promotion or shed any new light on why Jim Ingle got the supervisor job. He only repeated the vague promise "your future is bright," which was becoming tired.

A few days later, Jim Ingle began managing our group, and I sat down with him right at the start to get a few things straight. I knew Jim would want to know the status of my project, so I opened with that.

"Everything is going well with my hardware and software," I said. "We should be ready for the San Diego field trial right on schedule. I'm doing some final tests and documenting the designs. I'm documenting everything thoroughly, so if I get promoted or leave the department, someone else can pick up where I left off."

That was a not-so-subtle hint that I expected to be promoted or leave the company, but if Jim picked up on it, he didn't let on.

"Uh-huh," Jim replied. He didn't seem particularly pleased with my report. "I'm going to be checking all the measurements to make sure they conform to PUB 41009. Your measurement methods are consistent with that, aren't they?"

"Of course they are." I was indignant, because Jim knew full well the methods I used were proper—I had consulted his previous group to *determine* the methods. This was Jim's way of asserting control. It was as if we were playing a game of horse, and he started the game with an easy lay-up. He didn't expect me to miss; he just wanted me to follow his lead.

We discussed the project status a few minutes more, and then I came straight to the point:

"Jim, I think I'm ready to be promoted. John Colton told me I was ready, and in fact I thought I might get this group when he was promoted."

"Uh-huh, I see," Jim said. He didn't seem concerned that I coveted his job, he just continued methodically: "Let me see the schematics you've done up to now. Give me a copy of your documentation, too. I want to make sure all the measurements are being done properly before we go to field trial."

I decided the best strategy for dealing with Jim Ingle would be to do my job to perfection. I would produce an outstanding product that he couldn't possibly criticize. Then I would clearly and completely document my work in the best technical memorandum (TM) Bell Labs had ever seen. If that got me promoted, fine. If not, the excellence of my work would shout to my bosses, "Shame on you!"

It wasn't easy to persevere. Jim questioned every aspect of my work—the designs, the measurement techniques, and the results. I worked extra-hard to be ready for his questions. The last thing I wanted was for him to catch me in an error. I spent hours and hours documenting my work in a TM with the intriguing title "SARTS/RTS-5A Data Parameter Tests." It didn't exactly read like a novel, but it did meticulously describe the system I had devised. My tome also pretty much guaranteed that no one else would be able to take credit for my work.

In the spring of 1980 I was due to receive my first Jim Ingle-authored annual performance review. I had received glowing reviews from John Colton, but wasn't sure what to expect from Jim. I was pleasantly surprised: Jim rated my performance excellent, and then he told me what I wanted to hear—that I had been placed on the "promotion list." I believed the recognition was overdue, but it still felt good.

William Tarver — from MTS (59232), to Group Supervisor, Diagnostic Control Devices Group (43422), HO

Bell Labs News

PROMOTED – AT LAST!

Several weeks later, I accepted a group supervisor position in the Data Communication Laboratory. I was eager to prove I could manage high-powered Bell Labs engineers, because I would need that skill to run my own business.

I had achieved one of my major career goals, slightly behind my schedule. None of my plans to leave and launch my own business had changed, but meanwhile I was determined to be a good manager and contribute to the success of my new Bell Labs organization. My synthesizer business plan was dead, but the epiphany born of the need for a better Bradley Box remained very much alive. I was confident that a little time and management experience would prepare me to make my move.

29. NERD WORLD WAR

MY NEW BELL LABS ORGANIZATION, the Data Communication Laboratory, was responsible for developing modem networks and other business telecommunication products. The Western Electric division of AT&T manufactured the products. Our target customers were banks, governments, and retail chains—any commercial entity that needed to move data from one place to another.

I joined DCL as Group Supervisor—Member of Technical Staff, and was keenly aware that I was DCL's only black technical manager. I was undaunted, though, because I was eager to prove I could manage top-notch engineers at a time of enormous technological change. In 1980 the telecommunications world was rapidly evolving from analog to digital, and AT&T was dabbling in new, unregulated markets. DCL management had declared the analog modem dead, leaving that still profitable field to other companies while focusing instead on emerging digital communication technology. It was an exciting time to be in the telecommunications business.

My group was assigned to develop the Graphics Workstation (GWS), a device for transmitting images (pictures and graphics) and voice simultaneously over the telephone network. GWS would facilitate a brand-new concept called teleconferencing, and would mark our organization's first foray into the freewheeling office-equipment market. Like most new Bell Labs projects, GWS technology was on the absolute cutting edge. The system used the newest high-speed custom logic chips, advanced optical scanner technology, a custom-designed software operating system, and newly developed high-speed digital communication links. GWS was an ambitious and impressive project, even by lofty Bell Labs standards.

The leader of my new department was Dennis Morgan, a tall, fortyish man with a neatly trimmed mustache and thick-rimmed glasses. Upon meeting Morgan, I realized DCL was going to be quite different from my old department. My first boss, John Colton, had been casual and improvisational. Morgan was formal and structured. The reception he gave me wasn't hostile, but he seemed indifferent to my presence. Colton had stressed that he wanted me in his group and that he valued my expertise and experience. Morgan conveyed no such message. I wondered if he had chosen me for his department or if someone else, perhaps *his* boss, was responsible.

Five supervisory groups, a total of about forty-five people, were assigned to GWS. That seemed like overkill to me. In my old department, one group of a dozen people developed a fairly complex hardware and software system. I wasn't excited about being just one of five group supervisors on the GWS project, but I was excited to be developing a product for businesspeople instead of telephone companies, and to be working with brand-new technologies.

The social climate in my new department was disappointing. The managers displayed little camaraderie. We never did anything together away from work—in fact we rarely ever ate lunch together *at* work. We knew little about each other's lives outside the glass walls of Bell Labs. We never got together informally to discuss GWS or to brainstorm. In short, we didn't seem like much of a team.

I occasionally longed for my old department. Our relationship had been far less than perfect, but my former colleagues and I talked about things other than work. We sometimes ate lunch together. We played softball and basketball together. We were a team, and we produced results. Some of the alienation I felt in my new job came from having just arrived in DCL, but much of it seemed endemic to the organization.

The other four GWS supervisors were white males. Al was responsible for overall system design, and was de facto project leader. Tom managed software development. Fran was responsible for project management. Larry led system test. My group was assigned development of three hardware modules. One, already designed by Al's group, was nearly complete. Another was considered childishly simple. The third was a diagnostic module that would not be shipped with the final product. My group's assignments didn't seem glamorous, but I had learned in my old department to turn "lemon" projects into lemonade. Working with my group, I redefined our assigned subsystems to make them more functional, more elegant, and more cost-effective. Then I turned my attention to making the overall system simpler and less costly.

During staff meetings, some of the other supervisors challenged my group's technical ideas and designs by saying things like "Is that really necessary? Why don't you just go along with the program?" They seemed to resent my "ownership" of the project. I resented their resentment—the project was just as much mine as theirs. Moreover, I knew my engineers would be disgruntled and unmotivated if they didn't feel their work was interesting and challenging.

I didn't mind the other supervisors raising specific technical issues about my group's work—that's what team members were supposed to do—but when I turned the tables and challenged their technical ideas, they were indignant. Soon we were barely speaking outside of staff meetings. Rather than deal with me, the other supervisors would go directly to one of my people or would send one of their subordinates. Dennis Morgan tended to ignore the conflict brewing within his management team. Meanwhile, he gave me little positive feedback.

Amid this cauldron of contention, our organization's annual performance reviews took place. All DCL supervisors and department heads participated in ranking the performance of non-management technical staff. The review meetings spanned several days, and the resulting rankings determined each employee's salary and promotion eligibility. We spent hours upon hours arguing the relative merits of people working on different projects in different departments.

The review was a zero-sum game. Salary increases came from a fixed pot of money, so a larger raise for one engineer meant a smaller raise for another. Supervisors were encouraged to play up differences between engineers, and to reward outstanding performers with more money. In theory the approach was logical; in practice it became a huge fight, with each supervisor praising or defending his or her own people. The process was highly political, as supervisors and department heads made side deals and secret pacts to support or bash particular engineers. For most participants it was a grueling, draining, wholly unsatisfying process.

For me, one astounding aspect of the review was that several top performers—all of them black or Latino or Asian—were considered "not ready" for promotion. Victor Lawrence was a prominent example. Born in Ghana, Victor had received his PhD from London University. Every supervisor acknowledged that Victor was a top performer who had made important fundamental contributions to DCL, yet he was deemed "not ready." One supervisor commented, "Victor is a great guy—he's just not management material." Another said, "I don't know if he's so interested in managing other people—he doesn't seem to have that fire in his belly." No one gave specifics as to why Victor wasn't "management material," and no one suggested asking him if he was interested in management.

The vague comments about Victor reflected a familiar Catch-22 concerning people of color: If cooperative and non-confrontational, they were deemed "not management material." If bold and assertive, they were considered "angry" or "not a team player."

Ultimately, Victor was not placed on the promotion list. His rejection triggered an epiphany for me. I decided that if I were to go down, I'd go down swinging. I refused to sit back and let the other supervisors dictate GWS project direction. I felt I had to assert myself or risk being categorized "not management material." I knew that by speaking out I might rub some people the wrong way, but realized I would not be successful otherwise.

I redoubled my efforts to address the bloated Graphics Workstation design. I believed simplifying and solidifying it was the most important contribution I could make. Some of my colleagues acted as if we were still in the "good-old days" of Bell Labs, performing a research project instead of designing a product that would live or die in the competitive marketplace.

Brimming with confidence, I decided to meet one-on-one with Al to discuss my concerns about GWS complexity and cost. I documented my ideas and compiled a list of questions before meeting him in his office. I was nervous but determined to get to the bottom of the issues.

Al was the prototypical engineering nerd. Tall and slightly oafish, he resembled a cross between Clark Kent and the Pillsbury Dough Boy. When he talked, his eyes blinked rapidly behind thick glasses. He carried the requisite collection of pens and pencils in a pocket protector ensconced in a neatly pressed white shirt. Though nerdy, Al seemed like a nice guy. He was polite and articulate, always ready to expound on any aspect of the project.

Al and I sat on opposite sides of a small table, and I began by asking why we were using such expensive, high-powered components in GWS when simpler, less costly components were readily available. Al acknowledged that the expensive parts were overkill, but he insisted that they might be needed in the future. I was dumbfounded. Al's response confirmed that the GWS design was unnecessarily complex and costly. I presented my ideas for simplifying the system. Al listened, but it was clear he didn't want to change the design. Moreover, he seemed to resent the implicit challenge to his project-guru status.

After making no headway with Al, I presented my ideas and concerns to Dennis Morgan and the supervisor team at one of our regular staff meetings. Everyone more or less lined up behind Al. No one wanted to consider simplifying the Graphics Workstation.

More convinced than ever that GWS was headed down the wrong path, I continued to assert that we were creating an expensive, over-designed product. Our staff meetings became so heated that, on a few occasions, I thought actual blows might be thrown. The situation was so bad that I decided to take my case to Data Communication Laboratory director Don Hirsch. I detailed my frustration with the GWS project direction and my conflicts with the other supervisors. Also, I expressed concerns about the recently completed performance-review process, especially the treatment of Victor Lawrence and other nonwhite engineers. Don listened carefully and said he would get back to me.

A few days later he responded, through Dennis Morgan. Dennis called the supervisors together and told us Don had decided our management team needed an "intervention." He said the director had chosen Dr. Clifton L. Smith to work with us.

Dr. Smith was a psychologist and organization development specialist. I had often seen him in the building but didn't know what he did. A tall, slender man with dark brown skin and a full crop of wavy, jet-black hair, Dr. Smith was always immaculately dressed and supremely poised. He looked like Billy Dee Williams and sounded like Barry White. Though most Bell Labs staffers thought it en vogue to drive a sporty European car, Dr. Smith arrived on the Bell Labs campus each morning in a huge Lincoln Continental. He was like no other Bell Labs employee. Dr. Smith was a unique character.

Dr. Smith's assignment was to meet with our unruly supervisor team and implement something called "team building." Though he was the leader of our "team," Dennis Morgan didn't take part in the intervention—only the five group supervisors participated.

We met with Dr. Smith at a hotel about twenty minutes from our Holmdel workplace. We were directed to stay at the hotel day and night until we worked out our problems. Removed from day-to-day distractions, we could concentrate fully on the issue at hand—our team's severe dysfunction.

We met in a regular hotel room converted for our purposes into a far-from-glamorous conference room. The four other supervisors and I arrived before Dr. Smith. While we waited, the others engaged in small talk. I sat alone, trying not to look like the kid no one wanted to play with. After a while, Dr. Smith entered the room with regal bearing. His lanky frame was clad in an expensive suit, his hair perfectly coiffed, his cologne too strong.

Dr. Smith plopped his large, fancy leather satchel onto the table and began extracting his materials. He casually introduced himself and told us to call him Cliff. He engaged us in small talk about the hotel setting, pro basketball, and other low-intensity topics, and then pronounced, "Oooh-kay, time to get down to business."

Cliff started by talking about his role at Bell Labs. He said he was a member of the Organization Development Department, tasked to provide management consulting services to any Bell Labs department that needed or wanted them. I had no idea that such an organization even existed, but as Cliff described it, I couldn't help but be more impressed with Bell Labs.

Next, Cliff gave us a brief summary of his academic background: PhD in psychology from the University of Chicago, assistant professor at Indiana University, postdoctoral research fellow, and so on. Cliff seemed supremely confident, extremely competent, and utterly prepared—and he was *black!* I especially appreciated that last fact, because I thought Cliff's professionalism would send an implicit message to the other supervisors. I wondered if Don Hirsch was sending a message by picking Cliff to lead our team-building effort.

Cliff asked us to introduce ourselves. As each supervisor spoke, Cliff scratched notes on a yellow legal pad. After the introductions, he asked each of us why we needed the team-building session.

Fran spoke up first: "I really don't know, except it seems like Dave has some problems with the way we're doing the project, and he seems to have some problems getting along with the rest of the team."

Then Larry chimed in: "Yeah—I think things are going along pretty well on the project, but Dave really seems to have some problems fitting in."

I tried to keep my emotions in check. I knew Fran and Larry were trying to get under my skin, and I didn't want them to know they were succeeding.

Al had the same view as Fran and Larry, but he cloaked his answer in project-specific jargon: "Things are going well on the project. We've had some issues with Dave's group getting the image storage module working, but overall things are okay. Dave has some problems with the system architecture and components, but the rest of us are comfortable with it."

Tom was the only one who seemed to take a balanced view of the situation, and I was heartened by his comments: "Look—for whatever reason it's clear we haven't been operating as an effective team. I assume we're here to find out why that is and to see what we can do about it."

I waited until last because I wanted to see what the others would say. I said I agreed with Tom, and added that I hoped the team-building session would help us work together more effectively so we could build a better Graphics Workstation.

Cliff concluded the introductions by giving us his understanding of the meeting's purpose. He said Don Hirsch was concerned about our inability to work as a team, and that he wanted us to solve our problems and move forward. Dropping the director's name certainly got everyone's attention, so when Cliff asked us to commit to being open and honest during our counseling session, we all readily complied.

Over the next day and a half, we did some things I had never done in the context of a business meeting. For example, Cliff started our first working session by asking each of us to draw a picture that represented our project team. Some of the other supervisors seemed quite uneasy with the task, and kept insisting that Cliff tell us what he was "trying to get at." They seemed to view the exercise as a trap, as if by participating in an abstract activity, they might unconsciously reveal something about themselves. Cliff managed to work through their apprehension and we completed the exercise. I noted with interest that all our pictures showed the other supervisors more or less together, and depicted me as isolated from them.

The remainder of the intervention continued in a similar pattern. Cliff would have us do an exercise to stimulate thought and discussion. Then, based on our responses, he would identify a problem our team was having. After we acknowledged the problem, Cliff had us negotiate agreements that we were to adhere to when we returned to the workplace. I found the exercises illuminating, and hoped they would reveal the real reasons for my colleagues' hostility. Cliff's intervention yielded an uneasy truce. None of us believed our problems were solved, but all of us were ready for the sessions to end.

Back at work, things seemed better on the surface, but the underlying problems remained. Al was still opposed to simplifying the system, though he was less dismissive of my concerns. Fran and the other supervisors began to bring their issues directly to me instead of going to my subordinates. I made efforts to be more responsive to their requests. We were like parents trying to make our marriage work for the sake of our bloated, expensive, complex "child," the Graphics Workstation.

After numerous technical challenges and many staff-years of effort, GWS limped into operation. We negotiated with our manufacturing division, Western Electric, to produce the system, and they estimated the product cost. The marketing folks at AT&T met with potential customers to gauge the GWS's sales potential.

The closer we got to actually manufacturing the GWS, the more it looked like a white elephant. The AT&T marketing staff concluded that, based on the Western Electric cost estimates, the product would have to be priced at $100,000 or more to produce a profit. Marketing staff said customers who were willing to pay anything close to that were few and far between. A crisis was brewing, and it had the potential of making someone look very bad. Why hadn't we determined the market potential for the product before we started developing it? Why was the cost so damned high? We had to get answers, and we had to decide what to do with our bloated creation. If we didn't solve our problems quickly, there was real danger that the GWS project would be canceled. Such rumors were already floating through the corridors of DCL.

Faced with an intractable dilemma, we did what every red-blooded Bell Labs manager of that era did—we kicked the problem upstairs. We kicked it all the way up to our vice president, Tommy Thompson.

Dennis Morgan arranged a meeting with Tommy, attended by all the GWS supervisors. Our director, Don Hirsch, was there too, as was John Sheehan, our executive director. The meeting was our last opportunity to save GWS from the ax. We all knew our dog-and-pony show would have to be really good.

Al got up and began to describe the Graphics Workstation. He talked about all the great technology in the system, and how we were laying the groundwork for a potentially lucrative AT&T teleconferencing business, one that would enable executives around the world to meet and discuss important business issues without the cost and inconvenience of air travel. Al was asking for permission to proceed despite the product's high cost, even though we would likely lose money on each system we sold. The essence of Al's argument was: *I know this thing costs a lot more than anyone thought it would, but it's so important for AT&T to establish a teleconferencing business that we should go ahead anyway.*

Tommy Thompson listened intently. He didn't ask many questions. When Al finished, Tommy thanked us for coming. Then he rendered his decision: "What I want you boys to do is put all the bells and whistles on the system. Don't worry about the cost. Make the best system you can, and we'll take it out to field trial. Then we'll see what kind of response we get."

Everyone was relieved. Tommy had spoken, and we still had a project. Moreover, he was telling us we didn't have to worry about cost. All we had to do was make the system as good as we could make it. Surely we knew how to do that. Al, above all, was elated. He was vindicated in his "pull out all the stops," high-cost approach to design, and he didn't have to spend time trying to simplify or cost-reduce the system. For the moment, it seemed as if the old Bell Labs mantra, "Just build it and they will come," was alive and well.

We didn't realize at the time that Tommy Thompson, in his wisdom, had already canceled the GWS project. He hadn't done it officially or overtly; he did it by converting the Graphics Workstation into a market research project. He realized the product was hopelessly expensive and ahead of its time, so he decided to cut his losses and get what he could from it. He reasoned that if a hopelessly expensive Graphics Workstation with "all the bells and whistles" could generate some excitement among the field trial customers, AT&T might have a future in the teleconferencing business. If, on the other hand, we produced the absolute best Graphics Workstation we could, regardless of cost, and it still didn't generate much customer interest, the whole idea of an AT&T teleconferencing business was dead on arrival. In effect, we had spent many man-years of effort and millions of dollars creating a product prototype for a few trial customers.

The trial customers didn't get excited about the Graphics Workstation. It was too cumbersome and wasn't interactive enough to beat the combination of a low-cost fax machine and a standard telephone line. Shortly after the trial, management officially pulled the plug on the project.

I was amazed that AT&T and Bell Labs would spend so much money to develop a product without first understanding what customers wanted and how much they would pay for it. Bell Labs people were among the world's brightest, yet our team had failed to recognize the most basic business principles. I resolved that I would never again work on a project without first understanding its market potential.

In the end, the fact that our project was too costly didn't turn out to be the deciding factor in its demise, but if its features had been marketable, the GWS would have stumbled because of high cost that resulted from over-engineering.

I wondered if the project would have failed had we supervisors worked more effectively as a team. All of us had been busy trying to prove ourselves, to show how smart we were, to get ahead. Dysfunction similar to ours was occurring throughout Bell Labs, and projects were being canceled left and right. Part of the problem was that AT&T had not learned to operate in a competitive world, but another part was Bell Labs engineers' tendency to treat every product development like a research project. We engineers thought we would be judged solely on how cool our technology was, rather than how well we solved problems for our customers. None of us was working toward the right objective—to understand what the customer wanted and provide it at a price that would yield a decent profit for AT&T.

After canceling the Graphics Workstation, management threw a party for the troops at Jumping Brook Country Club in Neptune. We had great food, a live band, and an open bar. We danced and talked and reminisced and joked about GWS until the wee hours. It was the best party I'd ever attended, but somehow it didn't seem right. We had failed, and we had blown 27 million dollars of AT&T's money.

That party, that celebration of failure, is my fondest memory of the Graphics Workstation project.

30. Last Stand

THE GRAPHICS WORKSTATION'S 1982 DEMISE left me disappointed, disheartened, and thoroughly disgusted with Bell Labs. I had never put so much effort into a project only to have it flushed down the toilet. On the other hand, the cancellation presented an opportunity. I had not wanted to quit in the middle of the project, even though I had been preparing for some time to start my own business. The abrupt termination of GWS had me thinking it was time to pursue my dreams full-time.

Before I could act on that thought, I received a call at home from the Data Communication Laboratory director. He had just been promoted to executive director in a brand-new organization—AT&T Consumer Products.

"Dave," he began. "This is Don Hirsch."

"Hello, Don."

"Dave, I've got a new project in Consumer Products you might be interested in. It involves a new type of touch-screen-based telephone, and it's exciting. We're calling it Touch-Tel. A fellow from Murray Hill—Rich Thompson—has been working with a new kind of touch screen, and he's agreed to transfer to Holmdel to work on the project. He would be in your group. You can select whatever other engineers you think you'll need. We're going to assign the marketing and physical design folks to your group, too. I want this to be a small, self-contained team—no AT&T bureaucratic bullshit—and I want you to run it. What do you think?"

Wow! Don's totally unexpected offer felt like vindication. He was asking me to manage a new project exactly the way I thought the Graphics Workstation project should have been managed—a small, close-knit team working without the usual hassles. I was flattered and wanted to accept, but I was cautious.

"Let me think it over for a day or so, Don," I said. I'll get back to you by day after tomorrow."

"Fine. Just let me know what you want to do, and remember, this will be your baby."

I hung up the phone and nearly exploded with excitement. Don Hirsch thought I was a good manager after all! And he was willing to assign Rich Thompson, the PhD champion of an important new technology, to work for me. He trusted me to handpick my own team and run my own project. It wasn't the same as running my own business, but it was close.

The next day, I hunted down Neville Chandler, one of my lead engineers on the Graphics Workstation project, and invited him to lunch. Neville, a Barbados native, was one of DCL's few black engineers. I really liked working with him, and had tremendous respect for his computer hardware skills. I had decided overnight that if Neville would join me on Don Hirsch's new project, I would take the job.

Neville asked a lot of questions, but I could see he was nearly as excited as I was.

"Who is this Rich Thompson guy?"

"He's a PhD from Murray Hill, and he's going to be responsible for the overall system design. He's not a hardware designer—you would be the lead hardware guy on the project."

"What are we going to do about software development?"

"I don't know yet, but Don told me I could pick my own team, so I'm sure I can get somebody good."

Neville was sold. I was sold.

I called Don Hirsch that afternoon and accepted the job. I couldn't wait to get started and prove to my Bell Labs colleagues that I could manage an entire project.

Don arranged for me to meet my new boss, Dick Frenkiel. Dick's claim to fame was impressive: he and his partner Joel Engel had pioneered the cell phone. As soon as I met Dick, I knew I had made the right decision. He was the total opposite of my old GWS department head, Dennis Morgan. His dress was casual, his hair a bit messy. He looked more like a cleaned-up former hippie than a Bell Labs executive.

"Come in Dave. Have a seat," Dick said.

His voice was soothing, disarming.

"This is a new department," he began, "and we have two main jobs. One is your project, Touch-Tel. The other is cordless phones. Touch-Tel is going to be your baby, and I'm going to focus on cordless. I'm here to support you, but this is your project. I don't know much about what you guys are going to be doing, and I don't want to know. Just tell me what you need and I'll try to get it for you. You can use me to shield you and your guys from all the AT&T bullshit."

Dick Frenkiel seemed too good to be true.

The next order of business was to meet Rich Thompson—Dr. Richard Thompson. I was apprehensive because I had never managed a PhD engineer. Would Rich respect me, or would our association be a replay of my difficult relationship with Al, the Graphics Workstation technical guru? Murray Hill, New Jersey, was Bell Labs headquarters, and was the place where most of the really cool scientific research was done. Most Murray Hill PhDs would sooner cut off an arm than move to Holmdel and work on developing an actual product.

Rich came to Holmdel for our initial meeting. The first thing I noticed was that he seemed to exude happiness. Rich looked to be in his mid-thirties to early forties, and his hair was prematurely white. He looked fit and trim, and he flashed an easy smile.

Rich and I got to know each other by sharing our backgrounds—where we went to college, what we had done at Bell Labs, how we came to be assigned to Touch-Tel. We had a good, relaxed discussion. Rich repeatedly said he was happy to work with me and excited to see touch-screen technology used in an actual product. Rich said Don Hirsch had spoken highly of me, and was confident we would build a great product.

We turned our attention to assembling a Touch-Tel product development team. I told Rich that Neville Chandler was a first-class team player and the best microcomputer system designer I had ever met. Rich was sold on Neville right away. Then Rich told me about his friend Dewayne Perry, a computer science PhD who was teaching at Carnegie-Mellon University. Rich assured me Dewayne was a great software designer, and was willing to take a leave from Carnegie-Mellon to join our team. Dewayne sounded terrific, and I wasn't about to look a gift horse in the mouth. We added him to the team.

We assembled our core development team in just two days. Rich was convinced Touch-Tel was going to be an easy job, and that he would head back to Murray Hill in a few months. I was confident we would get the job done, but not quite as confident as Rich. I had just been through product-development hell on the Graphics Workstation, so I was prepared for another rough experience—at least I *thought* I was prepared.

Our team got right to work. Sure enough, in just a few months we had a working Touch-Tel prototype. Neville's hardware design was simple and elegant, and Dewayne's software was state-of-the-art. Rich mostly concerned himself with the design and performance of the touch screen and with the overall user experience. We were really rolling.

Realizing that Bell Labs had blundered on projects like the Graphics Workstation, my marketing person suggested we get early feedback to determine whether customers would actually buy Touch-Tel. She arranged a focus group at a hotel in Queens, New York, not far from Kennedy Airport. I had never heard of a focus group, but I was certainly eager to find out what everyday people thought of our product.

The focus group subjects were seated around a conference table. Rich and I were behind a wall of one-way glass. We could see and hear the potential customers, but they couldn't see or hear us. We placed several manufacturers' latest telephones, answering machines, and automatic dialers in front of our subjects. Then a facilitator asked the subjects which devices they used and how they used them. We were surprised to find how absolutely frugal our focus-group subjects were. Most didn't own any kind of fancy telephone, and some even refused to pay one dollar per month for Touch-Tone® service. They were content with their archaic rotary dial phones. Our subjects clearly considered a computer-based touch-screen telephone extravagant. I worried that Touch-Tel might be in trouble—big trouble.

We returned to Bell Labs and racked our brains over the focus-group results. We decided we would have to sell Touch-Tel for $350 or less and market it to "high-end" consumers. Our target was near the price of the other high-end devices we'd shown the focus group, but none of those products used touch-screen technology. We thought Touch-Tel would prove viable because it would perform a host of useful household functions: phone number directory and automatic dialer, appointment calendar, memo recorder, and more. We were creating something the consumer had never seen, and we were confident it would create a real splash in the market.

Around that time, we learned that AT&T Information Systems (ATTIS) was also developing a touch-screen telephone/data terminal, a product that targeted business customers. It seemed like a good idea, but when we heard their target price, we were dumbfounded. Information Systems planned to sell their terminal for $3,500. At that price, the ATTIS engineers wouldn't face anything like the cost pressures we faced on Touch-Tel.

Our team was undaunted. We believed we could drive the Touch-Tel price to under $500, and had a reasonable shot at hitting our $350 target. The single most expensive system component was the cathode ray tube (CRT) display. Rich Thompson scoured Asia for a low-cost CRT, and Neville devised an elegant, low-cost microcomputer to control the system. Our focus group experience told us where our costs had to be, and we were determined to get there. The experience reminded me of the Graphics Workstation, only now there were two teams: one building a bloated product that would never sell, and the other building a product people would actually buy. The challenge was invigorating, and our team was up to it.

Our project was going well, but then something I hadn't considered occurred—corporate politics kicked into high gear. We got word that ATTIS management wanted to meet with us to discuss the differences between their product and Touch-Tel. They were embarrassed by Touch-Tel's low target price, and worried that business customers would buy our product instead of theirs. Their systems engineers insisted that Touch-Tel adhere to design standards intended to govern business terminals. Business customers required lots of costly, complex features. We felt that Touch-Tel, a consumer product, should not have to comply with the bureaucratic ATTIS requirements. Many businesses, especially small ones, didn't care about ATTIS standards, anyway. Given the choice, they would buy the much less expensive Touch-Tel.

We shrugged off the ATTIS requirements and remained focused on Touch-Tel. We developed a physical mockup. We got Neville's hardware and DeWayne's software to mesh smoothly. We successfully tested a CRT display with Rich's patented touch screen. Rich negotiated a ridiculously low CRT display price with an upstart Korean company called Lucky Goldstar (LG). In short, we were on an even bigger roll.

Then the hammer dropped.

One afternoon, Dick Frenkiel called me to his office. He was seated behind his desk wearing his usual matter-of-fact expression. I didn't see it coming.

"David, I hate to tell you this, but the powers-that-be have decided to cancel Touch-Tel."

"But why?" I was incredulous.

"Well, it seems there was a battle between the higher-ups in Information Systems and Consumer Products, and Information Systems won."

So our marvelous little Touch-Tel was canceled, and our tight, talented team disbanded. The super-low CRT price Rich obtained from Lucky Goldstar was never confirmed. Later, we learned why—the fellow he negotiated the deal with was killed when a Soviet Union fighter jet shot down Korean Air 007 on September 1, 1983. That tragic news made our project woes seem puny by comparison.

I was extremely disappointed by Touch Tel's demise, but wasn't despondent. I would have felt much worse had our team not produced such an outstanding product. The decision to cancel Touch-Tel had nothing to do with its quality or market potential. It had everything to do with corporate politics in a company undergoing wrenching change.

Touch-Tel was dead, and I had a decision to make.

31. Basement Brigade

Base·ment (bās′mənt) *n.* 1) The lowest habitable story of a building, usually below ground level. 2) The lowest or fundamental part of something.

"WHAT'S YOUR NAME AND WHERE DO YOU LIVE?"

The large, ruddy-faced policeman barked the question from the driver's seat of his cruiser. Diane and I were walking our new collie, Wilson Wong, within a block of the home we had just purchased in Little Silver, New Jersey. I had no idea why the policeman stopped us.

We had spent little more than a year in our Red Bank townhouse before buying the three-bedroom brick ranch in Little Silver. The price was $123,000, more than twice our combined annual income. Our mortgage interest rate was 16 percent, but we easily qualified for a loan because we both had great Bell Labs jobs. We loved our house and the grassy, nearly two-acre property. I was especially pleased that the wide driveway apron accommodated a sizable basketball court, but the feature that excited me most was the large finished basement. I envisioned it as the subterranean launch pad for my electronics company, and realized that I was again completing a circle. My father had started his neighborhood radio and TV repair business in the basement of our home in Flint, and in the process had ignited my love of electronics. Now I was about to start a business in the basement of my own home, an endeavor that might one day inspire my own, as yet unborn, children.

When Diane and I had gone house shopping, I completely missed one significant fact: Little Silver had no black residents—absolutely none. That was hard to fathom, because our new home was less than a mile from our Red Bank townhouse. Roughly 3,100 of Red Bank's nearly 12,000 residents—about 26 percent—were black. Little Silver was a more affluent town with only half the population, but I had assumed that, given its proximity to Red Bank, Little Silver had at least *some* black folks. I should not have assumed. Diane and I were integrating Little Silver, and it didn't take long for that demographic fact to rear its ugly head.

"State your name and where you live!"

The officer's tone was rude and insistent, and his face was burning bright red. He addressed the command to me, not to Diane. Just then I thought of a recent news story. A black California man who often walked through all-white neighborhoods had refused to identify himself after being stopped by police. The case went all the way to the U.S. Supreme Court, which ruled that the man did not have to identify himself if police didn't suspect him of a crime.

Recalling that case, I said, "I don't have to identify myself to you. Am I doing something wrong?"

The officer didn't back down one inch. "Look, don't make me bust your chops. Just tell me who you are and where you live. I don't want to have to take you downtown."

I turned to Diane. She didn't say anything—she was frozen to her spot on the sidewalk, looking both frightened and embarrassed. I felt intense anger rising inside. We had enjoyed a perfect afternoon, laughing and joking and frolicking with Wilson Wong at the high school fields. I wanted to get back to normal, to get this belligerent policeman out of our lives. I decided to comply with his demand.

"You see that big house up there at the top of the hill, the brick ranch with the acre of grass in front? My name is David Tarver, and that's where I live."

I said it just for spite, to rub the officer's nose in the fact that my house was probably much nicer than his.

The officer said nothing more. He watched until we reached the house, then sped off. I tried to forget the incident, but I couldn't shed my anger and dismay. I couldn't let the officer's behavior go unchecked. This was 1981, not 1955. I was a taxpaying homeowner, not a trouble-making interloper. I recalled how West Long Branch police had surrounded my folks when they stopped at a McDonald's to get directions. I didn't want anyone else to experience that humiliation, or the humiliation I had just experienced down the street from my own home.

The next week, I went to the Little Silver council meeting and insisted that the town enact an ordinance that would preclude police from stopping someone simply to determine his or her identity. On the day of the meeting, Diane worked late. I suspected she didn't come along because she hated to make waves. Following my council appearance, a reporter from the local newspaper, the *Asbury Park Press*, wrote a story about the incident. Several people called and expressed sympathy. My real estate agent, whose husband was a Bell Labs director, said what had happened was terrible and that the officer was a "dope." The mayor, also a Bell Labs employee, hosted me for a nice lunch. He apologized on behalf of the town, and said he would ensure that nothing like that would happen to me again.

The council never passed the ordinance I requested or otherwise adjusted the town's policies. Because of that, I eventually regretted identifying myself to the officer. I should have let him arrest me and "bust my chops," then sued the town for false arrest. I wouldn't have charged racism or otherwise ascribed a motive to his actions. I would have left it for Little Silver officials to explain in court why one of their policemen arrested a young Bell Labs executive for the crime of strolling along the sidewalk with his wife and dog. The lawsuit might have resulted in a hefty financial settlement. I could have owned that town!

Notwithstanding that episode, I was becoming satisfied with my career and life circumstances. Diane and I didn't live extravagantly, but we were far from struggling. We owned an impressive home in an affluent, quiet town. I had achieved one of my main goals: promotion to Bell Labs management. My business ownership dream was intact, and I had a great idea for my first product. My plans were on track. Then one afternoon at Bell Labs, something shattered my budding complacency.

I was walking past a modem-testing lab near my office and, briefly peering inside, I saw a Bradley Box and other familiar instruments. The lab belonged to a different department, but I stepped inside anyway and noticed, perched atop a lab bench, an instrument I hadn't seen before. Its metal housing was dark blue, and the front panel was a dark acrylic plate that bristled with toggle switches and red alphanumeric displays. Small, white letters labeled each switch and display, and larger letters announced the instrument's distressingly ominous name: "AEA S-3 Telephone Channel Impairment Simulator."

I froze in front of the device, and an odd sensation gripped my stomach. Was this what I thought it was? Had someone else implemented my idea? I noticed a technician on the other side of the lab, and I called him over and asked, as calmly as I could, "What is this?"

The tech looked a little startled—he must have wondered why a stranger from another department was interrogating him. He answered the question, though.

"This is the new simulator from AEA. "It's going to replace our Bradley Box and a bunch of other stuff."

The tech's answer confirmed what I suspected: Someone had already built a better Bradley Box. This "AEA" outfit had beaten me to the punch, and my business opportunity seemed doomed.

I muttered a disingenuous "thanks" and dragged myself from the lab. I spent the remainder of the day holed up in my office, despairing over yet another crushed dream. I told myself it might be time to forget about starting a business and concentrate instead on building my Bell Labs career.

For the next several days I buried myself in work and tried, unsuccessfully, to forget my dream. The memory of the AEA product kept gnawing at me until I found myself back in the lab where I'd first made the devastating discovery. I had to know more about it, had to see if it rendered my latest and most promising business idea obsolete.

I examined every switch, every display, and every label on the AEA. My up-close investigation yielded some surprising discoveries.

The first was that the AEA simulator's construction seemed shoddy and unprofessional, as if produced in someone's garage. I spoke with a second technician, who was actually using the AEA, and he expressed disappointment with the product's accuracy. He said he was using the AEA only because, unlike the Bradley Box, he could control it with his computer and automate his tests.

The technician's assessment offered a glimmer of hope, and my confidence slowly returned. There was a market for an accurate, computer-controlled telephone channel simulator, but the AEA product's flaws left the door wide-open for a better product. Unfortunately, my product remained no more than an idea. AEA had a big head start, and was probably already designing its next-generation model. I would be left in the dust if I didn't produce something soon. Time was running out.

I would never catch AEA if I kept working alone—my music synthesizer project had taught me that lesson the hard way. I needed to assemble a product development team. I couldn't recruit DCL engineers, because I was new to the organization and had no track record. Engineers there would think I was crazy if I asked them to join me in my basement business venture. Moreover, I wasn't getting along with my management colleagues, and if they learned I was surreptitiously planning to start my own company, they would surely use that information against me. I would have no credibility in the organization, and might even be fired.

Who would be willing to work with me on this long-shot project? I pondered that question for days before settling on two people: Steve Moore and Charles Simmons, engineers from my old SARTS organization. I had known Steve and Charles since they'd arrived at Bell Labs. In fact, I had helped John Colton interview and recruit them. We were members of the same exclusive club—black engineers with impressive résumés. Steve and Charles both had master's degrees in electrical engineering—Steve's was from Rutgers and Charles's from Stanford. Their undergraduate records were just as impressive. Steve was from a little town called York, South Carolina. He was a no-nonsense guy who had graduated top of his class from the University of South Carolina, a school not known for coddling black students. I figured if Steve could do well there, he could do just about anything. Charles was from Virginia Beach, Virginia, and had received bachelor's degrees in physics and electrical engineering from a Morehouse College/Georgia Tech dual-degree program. Steve and Charles had been my lieutenants on the SARTS project, and that allowed us to become well acquainted, personally and professionally. Both men had proven themselves extremely capable, hardworking engineers.

I hadn't worked with Steve and Charles for nearly two years, but often ran into them around the Holmdel building. We hardly ever socialized outside work, but we maintained a good relationship. We respected each other, and we shared the pride of being top performers in our organization. Steve and Charles had assumed leading roles on a new SARTS project. When I called and invited them to lunch they had no idea of the agenda.

I was a bundle of nervous energy as I met Steve and Charles in the atrium. I knew my "better Bradley Box" project would get a big boost if I could recruit them, but I wasn't sure how they would respond. I felt almost as if we were prison inmates and I was proposing an escape. I didn't know if they would go along with me or laugh and turn me in.

As we approached the cafeteria, I instinctively dispensed with details and made a visceral appeal. I blurted out, "I have an idea that will make us millionaires."

I hadn't planned to say it that way, but it worked—Steve and Charles paid rapt attention throughout our lunch. As we huddled at a table devoid of other employees, I revealed my intention to build an accurate, computer-controlled telephone channel impairment simulator. I explained that the required technology was similar to what we had used in SARTS and what I had used in my synthesizer. I told them I had spent the past several months designing the system, and that I needed their help to build the prototype. I said we would sell our product to engineers like us who needed to develop and test communications equipment. I even had a name for our company: Telecom Analysis Systems—TAS for short.

Steve asked good, detailed questions about the system design, and he seemed ready to dive in. Charles was apprehensive. He wanted to know if Bell Labs would object to us developing a product based on ideas we'd acquired there. I told Charles that I had been concerned about that too, but was pretty sure we were okay. I explained that AT&T was clearly not in the test equipment business, because that market was too small to justify their attention and resources. I said if we asked permission to develop the product, we might be refused simply because Bell Labs corporate bureaucrats were cautious, so our best strategy was to go ahead and develop our product and assume Bell Labs would not intervene. I said the same corporate bureaucrats who would be too cautious to explicitly permit our project would also be too cautious to stop it. I said our product would *help* Bell Labs by providing their engineers a superior piece of test equipment.

Later that week, Steve and Charles came by the house for further discussions. I offered more project details and presented a high-level design. Then I asked them to join me—not to leave Bell Labs, but merely to help me complete the prototype. If things didn't work out, they could bail out and concentrate on their "real" job.

Steve and Charles agreed to sign on, which was a huge, encouraging step. I finally had a technical team, and I was confident we would design and build a great product.

From 1981 to 1983, we worked feverishly to build the prototype. We worked at Bell Labs by day and in my basement by night, often until 1:00 or 2:00 a.m. Typically, we worked together all day Saturday, and on Sunday we usually worked at our respective homes. We couldn't afford commercial lab equipment, so we designed and built our own. We couldn't afford to buy computers, but John Yaman, a generous manager at a local Heath/Zenith electronics store, gave us three and told us to pay him when we could. John was more than a store manager—he was an angel.

Completing the prototype proved extremely difficult, especially because our long TAS work night followed our long Bell Labs workday. To ease the burden, we enlisted two Bell Labs technician friends to help assemble and test the prototype. They were excited to join us, but they quickly tired of the long hours and uncertain rewards. Several weeks later, Steve, Charles, and I were back to doing all the work ourselves. It was grueling, but each technical milestone we achieved encouraged us. We felt we were building something special, and we truly believed what Steve often said: "If it was easy, everybody would be doing it."

By summer 1983 our prototype was nearly finished. I was satisfied with our technical accomplishments, but realized that, to build a successful business, we needed to address other functions. We needed to write a business plan. We needed to develop marketing, finance, and sales strategies. Engineers and technicians couldn't help us accomplish those tasks.

I didn't know many people with business management expertise, and knew still fewer who might be inclined to assist our project. I remembered a fellow U-M student, Kedrick Adkins, who had studied engineering and had later earned his MBA degree. I also remembered a General Motors Institute classmate, Mary Powell, who had since gotten a Harvard MBA. I figured either or both could inject some business expertise into our team, or at least give us sound advice.

I did a bit of research and discovered that Kedrick was working at a Big Eight accounting firm in Detroit, and Mary was working for a different Big Eight firm in New York City. I called Kedrick and told him about my business plan. We had a nice conversation, but I could tell he was too busy to be enticed into our project. Then I called Mary, who seemed genuinely interested in further discussing our fledgling enterprise.

Mary and I had dated briefly during our days at GMI, but hadn't spoken since. I was excited to see her, and even more excited about persuading her to join our team. We met at the Lobster Pot, a little Jersey Shore restaurant, and after we reminisced, I laid out my nascent business plans. Mary asked good questions, many of which I couldn't answer. I tended to view our business opportunity from a technical standpoint, but Mary's questions focused on marketing and finance. I quickly concluded that we needed someone with her expertise and point of view.

Mary said she would write the business plan for our venture. As with Steve and Charles, Mary wasn't committing to leave her day job, but she did agree to devote her spare time to our project. Her commitment was another important step. The next time I saw Steve and Charles, I described my meeting with Mary, and they readily accepted her participation. In fact, they seemed relieved, as though Mary's involvement validated their own. They seemed to be thinking: *Maybe we're not so crazy after all!*

Unfortunately, Mary didn't produce the business plan. I never knew why. Perhaps her day job became too demanding, or maybe she simply lost interest. I spent several weeks waiting and then wrote the plan myself. Steve and Charles pitched in with valuable research on actual and potential competitors. I had never written a business plan before, but I knew the business better than anyone, and no one was more motivated to make it succeed.

By fall 1983 our venture was coming together. We finished the prototype and completed the business plan. Steve, Charles, and I were a solid and capable team, but we were still full-time Bell Labs employees. I needed to devote my full attention to the business, but as a Bell Labs manager, I had important responsibilities that I couldn't just dump. I was approaching a moment of truth. I had a decision to make. It was do-or-die time for TAS.

Then Touch-Tel was canceled. I took that as a sign that TAS was my destiny and that it was time to leave Bell Labs. Even so, the prospect of quitting my only job since college was daunting.

An unexpected setback helped push me over the edge. In October, I contracted a bad throat infection and spent a few days in the hospital. My condition wasn't life threatening, but the hospital stay allowed me to stop and ponder the future. I imagined a much older version of myself lying on my deathbed, looking back at my life. At that moment, I sensed the deep regret I would feel if I didn't pursue my dream. I resolved once more that I would devote myself to building Telecom Analysis Systems. I resolved to leave Bell Labs and never look back.

32. Leaving Bell Labs

IN THE END, IT WAS SURPRISINGLY EASY TO LEAVE. The Touch-Tel project was dead, and my disgust with Bell Labs had me feeling headaches and nausea and dizziness by the end of each day. My body was telling me it was time to go.

In August 1976 I had started my career at AT&T Bell Labs, one of the most respected research and development organizations in the world. By October 1983 I was working for AT&T Consumer Products Labs (still referred to internally as Bell Labs), a confused outfit that developed overpriced telephones. I accomplished all that without even changing buildings. The landscape simply changed under my feet. My employer had become a far different entity from the one I'd joined seven years earlier, straight out of Ann Arbor.

Touch-Tel was canceled for stupid political reasons, so I didn't feel guilty about leaving. Bell Labs had lost its most attractive qualities. In the old days, you knew you wouldn't get rich working for Ma Bell, but you got to work on exciting projects, and nobody cared about deadlines or costs. It was like being paid to pursue your favorite hobby and play with the latest gadgets.

The "new" Bell Labs was all about costs and deadlines. Most employees were confused about the company's direction and strategy. Top management made completely illogical decisions about which projects would go forward and which would be canned. In fact, it seemed that most projects were being canned. Canceling a project took away an engineer's greatest motivation. For most of us, going to work wasn't about money or fame, it was about the sheer joy of creating a new, unique gadget and seeing other people use it. Taking that joy away was like severing the engineer's right arm.

By the time Touch-Tel was canceled, I had come to realize that AT&T's management structure was inverted. The people who needed to be on top in the new, competitive world were near the bottom. The people at the top had come of age in a regulated monopoly and were not prepared for competition. They also knew little about microprocessors or communication protocols or digital signal processing. The landscape had shifted beneath their feet, but they were too far above ground to notice.

Although we were entering a new, competitive age, the incentive systems for engineers were held over from the old days. Everyone was paid basically the same salary. We managers would argue for days in performance review sessions over the right to pay a top engineer 2 percent more than a bottom feeder.

When you couldn't get paid for extraordinary results, and you couldn't be sure the project you were breaking your neck for would see the light of day, you just felt disheartened. It was no wonder that many people who stayed with Bell Labs settled into a conformist, confrontation-avoiding stupor.

When I made the decision to go, it was easy. I just walked into Dick Frenkiel's office and laid it on him. Dick seemed mildly surprised, but, as was his way, he calmly reviewed the situation with me. When he saw my reasoning was sound, and that I wasn't leaving just because my project was canceled, he just looked at me and said, "Okay."

Dick later told Don Hirsch I was leaving, and the executive director made the obligatory phone call. After I explained my plans, Don said I would have a bright future if I stayed with AT&T, and that he already had another assignment lined up for me. I had no idea what the assignment was, and Don didn't elaborate, but the next time I spoke to Dick Frenkiel, he intimated that I was being considered for a department-head job. Dick and Don didn't understand that the last thing I wanted was an even higher position as a corporate functionary within AT&T. My soul and mind had already left Bell Labs, and my body would soon follow.

My coworkers held a farewell luncheon for me. It was a nice affair and an opportunity to reminisce with my Touch-Tel, Graphics Workstation, and SARTS colleagues. Friends from other departments were there, too. It was a nice sendoff.

At the luncheon, and later in the corridors of Bell Labs, many people wondered what had caused me to leave. Had I been fired? Had I gone crazy? It was as if no one in his right mind would leave a cushy Bell Labs job to start his own business. Truth was I never felt I'd received what I was worth at Bell Labs, but it wasn't only about the money, it was about recognition and appreciation. I felt a lot of folks at Bell Labs believed the company was doing me a favor just to let me work there, as if a black guy should be happy just to *have* a job with Bell Labs; as if affirmative action was the only reason I'd gotten the job in the first place.

A major goal in leaving was to determine my true value. I wanted to use *all* my skills, not just my engineering skills. I wanted to apply my ability to dream up new products. I wanted to fully employ my people skills, my writing ability, and even my singing talent. I felt corporate life had stifled the abilities and interests I'd acquired during my youth. I longed to feel like a whole person again.

Despite my resolute posture, on moving-out day I had to pause. It was a big moment, and I wanted to savor, if not reconsider, my decision. Years earlier, I had quit General Motors Institute to attend MIT, only to return. I had nearly left Bell Labs for Rand Corporation, but had ended up staying. At least a few minutes of reflection were warranted.

I reminded myself of all the attractive options I had turned down in the past—MIT, the University of Illinois, IBM, NASA—to pursue my dream of starting a business. Then I thought about the endless possibilities inherent in what I was about to pursue. Once again I recognized the regret I would feel later if I foreclosed on my dream. Then I picked up the box that contained my personal belongings and walked out of Bell Labs. For the first time in months, I didn't have a headache.

Part 4: Venture

TAS FOUNDERS
(L-R) STEVE MOORE, CHARLES SIMMONS, DAVID TARVER

33. Money

MY FIRST MAJOR TASK UPON LEAVING BELL LABS was to raise enough money to get Telecom Analysis Systems off the ground. Diane and I had financed the prototype development with our savings and credit cards, but fully establishing the business would require much more cash. We needed to design and test the production version of the product and build the first production run. We needed to move the company into a suitable commercial space. We needed to hire someone to answer phones and manage the office while Steve, Charles, and I focused on product development. We were traversing an exciting but dangerous period. Lots of money was going out, but none would be coming in until we started shipping products to customers and getting paid. We needed to finance that transition or TAS would be sunk.

We didn't need to borrow a huge sum. We had finished developing our prototype while still employed at Bell Labs, so that significant expense was already covered. Diane and I didn't have many expenses beyond our home mortgage. We lived frugally, and Diane's Bell Labs earnings covered our living costs. My VW Scirocco was six years old. Diane drove an eight-year-old Chevy Nova. We rarely bought new clothes and didn't take vacations. We seldom went to the movies. The business was everything, and all my energy and resources were directed toward its success.

I assumed it would be easy to get startup money from a bank or the government. I had heard of programs designed to help blacks and women launch businesses and win government contracts. I learned about outfits called SBICs (Small Business Investment Corporations) and MESBICS (Minority Enterprise Small Business Investment Corporations), supposedly designed to fund small businesses and minority-owned companies. I was confident I would have no problem securing the money we needed to get TAS rolling.

I was a young electrical engineer of considerable accomplishment, more than able to handle—right from the start—the science and technology of my business vision. My knowledge of the financial side, however, was much more limited, and my first foray into business finance betrayed a lack of financial expertise.

I strapped the TAS 1010 Channel Simulator prototype into the passenger seat of my VW and proceeded to a local bank's drive-in window. I was reluctant to go inside, partly because I didn't know if this bank made business loans, but also because I was afraid of being rejected. When a bank clerk came to the window, I said, as matter-of-factly as I could: "I need some money to start a business."

The clerk looked at me as if I had just rolled up to the window with a carload of watermelons.

"What?"

"I need some money to start a business, and I have my product right here," I said, motioning to the passenger seat.

The clerk looked puzzled. "Wait a minute. Why don't you park your car and come in?"

I looked at her earnestly and said, "Should I bring my prototype with me?"

"That won't be necessary," she said.

Inside, a young man about my age ushered me to his desk. Since he was wearing a suit, I assumed he was a manager. He looked at me earnestly and said, "How can I help you today?"

I told him I was there to get a loan so I could start my business. I said my prototype was right outside in the car, and that I would be glad to get it and show it to him.

The bank employee said, "I'm afraid we only do home loans here."

Chastened and embarrassed, I drove home with the TAS 1010

prototype sitting beside me. Getting money from a bank was going to be harder than I thought.

I needed a financially astute person to show me the ropes, someone who would discuss my business with me and tell me how to make it successful. I was looking for someone to lend me money based solely on the fact that I was enthusiastic and willing to work hard. I wanted someone to believe in me and my business idea. I had heard that countless people and agencies were itching to help black folks start businesses. I wanted desperately to meet some of those people.

I decided to go straight to the source, so I drove to Small Business Administration headquarters in Washington, DC. I didn't know anyone at the SBA and had no contact names. I went in cold. I met a staffer who told me about SBA programs that might fit my situation. There appeared to be two: the Guaranteed Loan program and the Small Business Innovative Research grant program. The staffer gave me some pamphlets and referred me to the Newark SBA office for further information.

That was progress, but as a young black entrepreneur and former Bell Labs engineer, I had expected to be greeted enthusiastically by a crack team of experts just waiting to analyze my business plan, advise me on the next steps to take, and write a check for a startup grant. I was wrong. I saw no evidence of any concerted effort to help someone like me start a business. I saw no one who could *understand* my business plan, much less get excited about it. All I got was a pile of pamphlets and forms, a few encouraging words, and that referral to the Newark office.

I met my accountant, Bert French, and told him about my disappointing experience at SBA. When I mentioned the guaranteed loan program, his face lit up. He said he had a friend at a North Jersey bank who could surely get me an SBA-guaranteed loan. Things definitely seemed to be looking up.

The SBA application required several pages of financial disclosures and detailed financial projections, which Bert gladly prepared, for a fee. Armed with the TAS business plan and the SBA loan application, we drove to the North Jersey bank to see Bert's friend. I thought, after weeks of effort, that I was about to hit pay dirt.

Bert greeted his banker friend with a hearty handshake and a slap on the back. Then he gestured toward me and said, "This is the fellow who needs the SBA loan—David Tarver. He's got quite a good idea for a business. I think you'll like hearing about it."

Bert's friend looked like the generic small-town banker: white, middle-aged, middle-market suit, white shirt, and bland tie. As soon as Bert introduced me, I noticed a telltale contortion on the banker's face. It was a look somewhere between anger and embarrassment, accompanied by a reddish hue. I'd seen that look while apartment hunting in New Jersey, when the landlord wasn't expecting a black tenant. I saw it on the first day of a certain class at the University of Michigan when the professor didn't expect to see a black student. I saw it in some of my early Bell Labs meetings when the other participants didn't expect to see a black engineer. I knew that look on the banker's face meant trouble.

After a palpably awkward silence, the banker walked over and sat behind his desk, and Bert and I sat in front of it. Bert sensed something was wrong, and he spoke cautiously:

"Uh, uh, we have a, uh, business plan here. We also have, uh, the uh, SBA application completed."

I was amazed. Bert seemed nervous, even though the banker was supposedly his friend. Meanwhile, the banker's expression had morphed into a scowl.

"You know, Bert, we're not doing much SBA work anymore. I really don't think I can do anything with this, but I appreciate you coming up here."

The banker wasn't looking at me—he was addressing his comments only to Bert. I wanted to blurt out: You sorry-ass motherf*****! Why don't you just come out and say why you don't want to give me a loan?

I wanted to say that, but I didn't.

I glanced at Bert. He looked embarrassed and downtrodden. I felt sorry for him, because he really hadn't seen it coming.

Bert and I shared a quiet ride back to his office. He punctuated the silence by saying, "I just don't understand. He knew why we were coming, and he told me it would be no problem to do an SBA-guaranteed loan. This is really rotten. I'm sorry, Dave."

Bert learned a lesson that day, and it weighed heavily upon him. I learned a lesson, too: I learned that *someone* would probably give me an SBA loan. I just needed to find the right banker.

Bert and I next tried MidLantic Bank, a regional lender whose ads portrayed them as "The Hungry Bankers." I was anxious to see if they were hungry enough to finance an aspiring black entrepreneur. I visited the folks at MidLantic with my business plan and SBA application documents. This time, the reception was friendlier. A bank officer named Frank Anfuso met with me and reviewed my plans. A few weeks later, MidLantic agreed to lend me $150,000 if they could obtain the SBA guarantee. The interest rate would be "prime plus two"—two percentage points above the rate most banks charged their best customers. Obviously TAS, a tiny startup company, was not considered one of MidLantic's best customers. They probably felt we should have been grateful just to *get* a loan.

In addition to the SBA guarantee, MidLantic required—and got— personal guarantees from Diane and me, and we had to put up our home as collateral. The loan was a great deal—for MidLantic. They had virtually no risk, since the SBA guaranteed repayment, yet they were getting an above-prime interest rate. Prime rate at that time was 11 percent, which made our interest rate 13 percent. I belatedly realized that banks don't lend you money the way you might lend five bucks to a friend, knowing that you might or might not get your five bucks back. Banks *rent* you money the way Hertz rents you a car—they fully expect to get their money back, and the interest rate is the fee they charge for *using* their money.

MidLantic wasn't going out on a limb, but that didn't bother me. We got the same loan terms the bank would have provided any similarly situated client. We certainly didn't get any extra assistance or privileges as a black-owned business. That didn't bother me, either. I was willing to assume the financial and business risks because I, not MidLantic, would reap the rewards if the business succeeded. If TAS failed, it wouldn't be because we paid a slightly higher interest rate. If the company succeeded, we would grow rapidly and pay the loan off early, so the interest rate wouldn't matter.

The SBA-guaranteed loan from "The Hungry Bankers" provided the money we needed to build our product and start selling it. The rest was up to us.

34. THE GAUNTLET

THE TAS 1010 TELEPHONE CHANNEL SIMULATOR PROTOTYPE was an ugly beast. Our beige "better Bradley Box" was nineteen inches wide, more than a foot deep, and nine inches high. A series of status lamps protruded through holes positioned across the black front panel.

Far more important than its looks, the prototype still needed thorough testing. The TAS 1010 performed many complex functions that could only be analyzed with expensive, specialized test instruments we didn't have. Without that testing, we couldn't be sure we had a viable product. I strongly believed our product was better than the Bradley Box, better than AEA, better than anything else on the market, but we had no way to prove that. I had to find a way to get it tested by someone with the required equipment, expertise, and credibility.

I pondered the situation for a few days, and then a solution hit me like a slap on the head. I decided to enlist a former Bell Labs colleague—not just any colleague, but the one who could be expected to be the most severe nitpicker and naysayer. If he gave our product the seal of approval, we would have a real shot at success. I knew we would have to deal with indifferent, antagonistic, and even hostile engineers to establish the TAS 1010 in the marketplace, so I decided that our path to success would have to run straight through my old boss, Jim Ingle.

On the surface, Jim and I couldn't have been more different. Jim was old and white; I was young and black. Jim could be maddeningly careful; I was daring. Jim was an old-timer nearing retirement; I was a kid with my whole career in front of me. Jim was a Bell Labs lifer; I was a budding entrepreneur who had left the company to start my own business. Jim seemed satisfied with his position and accomplishments; I was struggling to find my place in the business world.

Beneath the surface, though, Jim Ingle and I had some important things in common: We both loved telecommunications and valued technical excellence, and where technical matters were concerned, we both wanted to get to the truth.

Jim was honest as a Boy Scout. In fact, he'd *been* a Boy Scout, and as an adult he served as a troop leader. Nothing was more important to him than getting technical details right. That's why he was a perfect test customer for the TAS 1010. He would give us exactly what we needed—a tough, brutally honest assessment.

I called Jim, described our business venture, and asked him to look at our product. He responded tentatively, but as I described our box and its unique features he began to show interest. He asked questions about the techniques we used and why I thought ours was a better unit than Bradley and AEA. He grew more and more interested, and finally agreed to take a look at the TAS 1010. I sensed that he was licking his chops for a chance to point out everything we had done wrong, but I didn't care. If we had done everything right, great; if we'd made mistakes, we would fix them and come back again. Either way, we couldn't lose.

I made the ten-minute drive to Jim's brand-new lab in Lincroft, New Jersey. The place had a modern, almost clinical feel. Jim was ready. He had arrayed a set of measuring instruments in one corner, and had cleared an adjacent table for the TAS 1010. He looked like a mad doctor who couldn't wait to start dissecting his latest patient. I was to leave the unit with him for a few days, but after seeing his elaborate setup I wasn't about to depart without watching him perform a few preliminary tests.

Jim needed me to show him how to operate our box. I fired up the prototype while Jim got his instruments ready. He asked me to turn on the noise so he could measure it. Success! He then asked me to turn on another function, and then another. When we were satisfied that the box was basically working, and Jim knew how to operate it, I left him to do his detailed analysis. I was relieved that the TAS 1010 was ready for its acid test, and didn't feel like pushing things any further. I would let Jim dig in and find whatever problems he could. Our product was in Bell Labs being analyzed to death by the world's leading authority. Whatever the result, we would learn important things.

Leaving our precious prototype to be evaluated at Bell Labs was a milestone. On the way home, I stopped at a local watering hole called the Pour House and had a glass of wine. I wanted to savor the moment.

A few days later, Jim called. "I found some problems with your box," he announced. He sounded almost gleeful beneath his clinical German facade, and my heart sank. "I think you should come out so I can show you what's wrong."

I was filled with anxiety. What had Jim found? Would the problems be too difficult for us to fix? Would my beloved simulator be sunk? Would we be the laughing stock of Bell Labs? I was a trembling, nervous mess as I drove to Jim's office. My armpits were soaked even before I arrived, but I was determined to persevere.

Jim greeted me in the lobby and escorted me to his lab. The bounce in his step was disconcerting. The TAS 1010 and Jim's array of test instruments still sat where I had last seen them.

"As I said, I found some problems with your box," he began. "A couple of them look pretty serious."

"Is that right?" I asked. I was trying to remain cool, despite the jugs of sweat I was generating. "Let's take a look."

Jim started by showing me that the levels of nonlinear distortion indicated by the TAS 1010 were wrong compared to what his instruments measured. I saw right away that it was an easily fixable calibration issue.

Then Jim delivered his coup de grace. "Now this is the major problem," he said. "The frequency spectrum being generated by your frequency shift function is all wrong. If you look at the AEA simulator's output, you can see that the Bessel sidebands should be here." Jim pointed to the spectrum analyzer display, and he seemed proud of what he had found. "I think this is a fundamental problem with the way you're generating frequency shift. It looks all wrong."

I couldn't say our Bessel sidebands were the same as AEA's—they clearly weren't. I *could* say that we generated frequency shift by the book, using the exact mathematical formula dictated by theory. I really didn't see how we could be wrong. I could have argued with Jim, but I didn't. I listened, and took notes, and took a printout of his spectrum analyzer traces. I wrote down the numbers indicated by his test instruments, and noted all of his comments.

As we reviewed the remaining TAS 1010 functions, I was relieved that, according to Jim, we seemed to have done everything else correctly. Still, he was fixed on the notion that our frequency shift function had a fundamental, perhaps fatal flaw. Just when Jim seemed most negative, and our meeting was ending, he surprised me with an offer.

"Why don't you borrow my test instruments and check out the problems in your box? You can use them for a couple of weeks if you like."

I couldn't believe my ears. My biggest technical antagonist at Bell Labs was offering to lend me $50,000 worth of test instruments so I could fix the "problems" with my product. He even let me borrow the Bradley simulator so I could compare it with the TAS 1010. At first I couldn't understand why he was making the offer. Then I realized it was Jim's inner Boy Scout emerging. It was the honesty and commitment of the techie fraternity, the desire to pursue technical excellence at any cost. I immediately accepted his offer. We did the necessary paperwork for security, and I left Jim's building carrying not only my TAS 1010, but also a carload of Jim's most expensive and sophisticated test gear.

Steve, Charles, and I immediately put Jim's instruments to work in our basement lab. In less than a week we resolved the issues Jim had raised. Best of all, we were able to confirm that we were doing the frequency shift function exactly right, and that AEA was doing it wrong. I couldn't wait to see the look on Jim's face when I showed him the proof. I called to set up another appointment.

Jim answered with his customary, "Ingle here."

"Jim, this is David Tarver. I just wanted to let you know we finished doing the measurements on our box, and I think we resolved the issues you pointed out. When can I come out to bring your equipment back and let you look at our box again?"

"Did you fix that nonlinear distortion problem?"

"Yes. It was a calibration issue. Now our box agrees with your test instruments."

"How about the frequency shift? That looked like a pretty serious problem. Your Bessel sidebands looked to be in the wrong place compared to AEA."

I didn't want to drop the bomb on the phone—that AEA was doing frequency shift incorrectly. I just said, "Yes, we figured that out, too."

On the appointed day, I arrived at Jim's lab with the TAS 1010 and all of Jim's equipment. He set everything up again, and a short time later we were making measurements. Jim examined each function and confirmed that our measurements looked correct. When he got to frequency shift, he frowned. He looked at me severely and said, "Your frequency shift function still looks wrong. The Bessel sidebands are still in the wrong place."

I said, "I know our box is doing something different from the AEA, but I can prove that our box is right." I took out my lab notebook and showed Jim the equations I'd used to develop our frequency shift function. Then I showed him the Bessel sidebands from his beloved Bradley Box, and they agreed with the TAS 1010, not the AEA. I told Jim I surmised that AEA was taking a shortcut that resulted in their Bessel sidebands being in a different place.

All Jim could say was, "Hmm." He knew I was correct—the proof was right there in my lab notebook. That realization was like a fly in Jim's meticulously prepared soup. The AEA, a product he had been using for months without question, appeared to have a design flaw—its results did not agree with the theory.

"I'm going to call AEA and let them know about this," Jim announced. "It does look like your box is doing things right, though. On the other hand, if your box doesn't agree with AEA, that could be a problem, because so many people out there are using it."

I was taken aback by Jim's assessment. We could be right, but still in effect be wrong because our box didn't agree with the *incorrect*

AEA. That seemed unfair. I couldn't help but feel that if *our* product had proven theoretically wrong it would be dead.

Jim did report his findings to AEA, and they confirmed that their product was doing frequency shift wrong. AEA had indeed taken a shortcut to make implementation easier. In the end, it didn't matter much, because AEA's approximation was good enough for most "real-life" situations. Still, it was comforting to know that our product was doing things exactly the right way. It gave us bragging rights.

Our tango with Jim Ingle couldn't have turned out better. Jim gave his blessing to our product. We could proceed with confidence to manufacture and sell the TAS 1010 Telephone Channel Simulator. We had the best product, and Jim Ingle, my former nemesis, was our ally.

35. 1984

NO MATTER HOW WELL IT PERFORMED, no one was going to buy a product that looked like our TAS 1010 prototype—it was just too ugly. We needed to do something about that.

An industrial design firm crafted the look of every Bell Labs consumer product, and I was always amazed to see them turn a drab prototype into something sexy and cool. I wanted to apply that same expertise to the TAS 1010 and make it the best-looking product of its kind. Hewlett-Packard was the benchmark for professional instrumentation design. We wanted the TAS 1010 to perform to that standard—and look as good, too.

AT&T's New York City industrial design firm, Henry Dreyfus Associates, was far beyond our financial means, so I asked some of our suppliers to suggest other firms. One of them pointed me to a small New Jersey company. On a drab morning, a few days before Thanksgiving 1983, I put the TAS 1010 in the back of my VW and headed up to Morristown.

James Wickstead Design occupied a small space in a nondescript industrial park building. Myra Tang, a designer, met me in the lobby and escorted me to a small conference room where James Wickstead joined us. Tall and slender, with prematurely gray hair, Jim's appearance, soft voice, and serious demeanor befitted a college professor. He didn't seem alarmed that his newest potential client was black. That was a relief, because I didn't want to write Jim off before we got started.

1984

After Jim showed me examples of his work, I lifted the prototype from the floor and placed it on the conference table. "Our task," I said, "is to make this product look cool. When people see it, I want them to think of Hewlett-Packard. I want the box to look like it's full of action, even though there's nothing on the front panel but lights and labels."

As I spoke, Jim and Myra stared pensively at the TAS 1010 enclosure. Both seemed to understand what I wanted, and they seemed interested in the assignment. Jim asked lots of questions: What dimensions did we want for the final product? What was going to be inside? What was our target cost? Who would be using the product, and where would it be used? Jim spent nearly an hour poring over the prototype, and then said, "I think we can create a really good design that meets your objectives." Myra concurred.

A few weeks later, Jim and Myra visited my Little Silver home and presented their TAS 1010 proposal. Steve, Charles, and I sat at the dining-room table enthralled by a design that transformed our ugly, rectangular prototype into an exciting instrument. The enclosure design incorporated slick front-panel graphics and a coordinated black-and-beige color scheme. We had achieved our objective of a great-looking, HP-like product, and we were ready for production.

"BEFORE": TAS 1010 PROTOTYPE

The TAS 1010 enclosure was a custom-painted Zero Halliburton box, a high-quality piece. Our product had no built-in controls or displays, aside from status lights. It was designed to be controlled by an external terminal or computer—an unusual feature in those days. I was determined not to reinvent the wheel by incorporating a complicated, expensive user interface. The advent of the microcomputer

"AFTER": TAS 1010 PRODUCTION UNIT

and the personal computer made it possible for us to take a bold new approach, and we did.

"Ready for production" meant it was time for Steve to leave his prestigious, lucrative job at Bell Labs and join me in the basement. Though we had agreed on the timing much earlier, I was nervous about Steve joining the venture. TAS had always been *my* dream, and starting an electronics company in the basement was *my* crazy idea. Now my dream was going to have a tangible, significant impact on someone other than Diane and me. Steve had gotten married the previous summer, so his decision affected his wife, Kimberly, too. Her job wouldn't support them both, but Steve never wavered on his decision. He often said, "You've got to be in or out, and I'm in. I'm doing this because I want to get rich!"

Steve's high expectations were intimidating, but I had promised that we would be millionaires. It was time to put up or shut up.

Steve left Bell Labs in February 1984, making TAS not just "Dave Tarver's crazy idea" but a real company. Steve and I took Diane and Kimberly to dinner at the West Long Branch Ramada Inn to celebrate, but nerves overtook my appetite. I looked at Kim's sweet, innocent face and realized she was ready to make whatever sacrifice Steve wanted. I hoped I wasn't ruining their lives.

I couldn't ask Steve and Charles to risk their careers without defining their reward. Hard work and dedication entitled them to an ownership stake, so I needed to incorporate the company and distribute stock. If I failed to offer them a fair share, the whole venture could fall apart. Entrepreneurship books were full of advice, but in the end I had to create an ownership formula. The TAS 1010 Channel Simulator was my idea. I had designed the prototype, raised the startup capital, and left Bell Labs to get the business started. I felt my contributions entitled me to a controlling share.

On the other hand, Steve's and Charles's shares needed to be large enough to motivate their continued hard work. I figured we could quickly build a business worth $10 million, and I wanted their equity to make them millionaires, at least on paper, at that level. I decided to offer Steve 20 percent of the business, and to offer Charles 14 percent. Steve's share was larger because he had developed more of the prototype and had joined the company full-time. I planned to raise Charles's share to 16 percent when he left Bell Labs to join our team.

I didn't want to present the ownership proposal in my basement. It was a crucial issue that required discussion outside the daily routine, so we held our summit conference at a little pizza joint in Middletown. We were all nervous, and it showed. Steve and Charles wanted to talk about product design issues, pro basketball, Charles's project at Bell Labs—anything but the subject at hand. After several minutes of small talk, I forced the issue.

"I've been looking at dividing up TAS stock among the three of us. I don't think it's fair to ask you guys to continue to work without knowing what you're working for. I promised you we would all be millionaires, and I think this stock breakdown will get us there."

Steve and Charles listened intently as I explained the ownership breakdown and the underlying rationale. I was sweating profusely, and had to blink frequently to keep my eyes from watering. I thought the proposal was fair. I hoped they would agree.

I felt relieved to have finally put the proposal on the table. We could discuss it, negotiate it, and, if needed, modify it. Then we could move forward as shareholders in our new corporation. Steve was the first to respond.

"That sounds good, man. That's more than I expected. You've done so much on this project, and you put up all the money."

I was relieved by his response, but I felt the need to reiterate my rationale.

"I know I've done a lot up to now, but you guys are going to be working hard for years to come to make this thing work. I think this breakdown is fair."

Charles was less vocal, but seemed comfortable with the arrangement. His main concern was his impending decision about leaving Bell Labs. He was on the verge of promotion to group supervisor, and was planning to get married that summer. Steve and I were both worried that Charles was having second thoughts.

Despite any doubts about Charles's full-time involvement, we left our pizza summit on a high. TAS had three owners, a solid product, and a viable business plan. On February 17, 1984, we formally incorporated the business: Telecom Analysis Systems, *Inc.*

We spent the next few months building the first TAS 1010 production units. I ordered the parts from a spare bedroom I'd converted into an office. Delivery vans came fast and furious with electronic components and other hardware. Charles and Steve toiled in the basement to assemble and test the products. In March we completed our first production "run," which yielded an inventory of seven units: two demonstration models and five available for sale.

It was time, finally, to make some money. I needed to put on my sales hat and visit potential customers, something I hadn't done since my childhood days selling handmade potholders and Wallace Brown greeting cards to my Sixth Street neighbors. Selling the TAS 1010 would be very different. First, our product's price was $15,950, so no one would be buying it on a whim. Second, our potential customers were engineers—usually white, often conservative, always nit-picky. They would buy our product only if they needed it and it was better than the alternatives.

The high price had a positive side. The TAS 1010 only cost about $3,000 to make, so our gross profit totaled roughly $13,000 per unit. We wouldn't have to sell many to achieve success. Our product cost as much as a nice imported car, but three guys in a basement could make and sell it. The moneymaking potential seemed enormous.

My first major sales task was to decide whom to call on. Our Bell Labs colleagues were a natural and legitimate target, a swarm of potential customers just a few minutes away in Holmdel. Many Bell Labs engineers did work that required a product like the TAS 1010. Sales prospecting therefore simply required getting on the phone and calling them. My old boss, Jim Ingle, was now an ally (or at least a decent reference), so I called him first and he gave me some leads. So did other former colleagues. I called each prospect and offered a demonstration to anyone who sounded the least bit receptive. My method was simple: set up an appointment, demonstrate the product, get the sale.

My first demonstration resulted in our first sale, but only after weeks of painstaking effort. Jim Ingle had told me that Rick Tauson was equipping a new test lab and had a pressing need for the

TAS 1010. When I spoke to Rick, I sensed he was sympathetic to our plight as struggling young entrepreneurs, but he turned out to be anything but an easy sale. Rick didn't want to get involved in the details of evaluating our product, so he delegated the assignment to an Indian émigré named Rao. Rao was a nice enough fellow, but knew absolutely nothing about our product or its intended application. Every sales visit became a training session: I had to bring Rao up to speed on what he was doing, why he needed our product to do it, and how to perform his tests. Rao wasn't confident enough to approve the TAS 1010, and he didn't have authority to buy it. I spent weeks alternately befriending and training Rao, and calling Rick to beg for the order.

The first order finally came in late May. By industry norms, that wasn't a long time for winning a $16,000 sale, but it seemed like an eternity. Once the purchase order came through, the turnaround to delivery was about twenty-four hours. We had five units just waiting for customer orders, so all we had to do was box one up and deliver it.

I put a TAS 1010 in the back of my VW and drove out to Bell Labs Holmdel. I pulled into the receiving dock, a part of the building I had never seen, parked next to a huge tractor-trailer, opened the rear hatch, and put the TAS 1010 on the loading dock. I gave the packing slip to the dock supervisor and had him sign a copy for me, then called upstairs and spoke to Rick Tauson to let him know I'd made the delivery and was available to help Rao set it up. On the ride home, I was grinning from ear to ear. I, chief executive of my own high-tech company, had just delivered our first customer order. I was also chief engineer, chief salesman, purchasing manager, manufacturing manager, deliveryman, and customer support technician. The feeling of accomplishment was beyond anything I had ever experienced.

Of course, Rick's wasn't the only order I was working on that spring. Orders began trickling in from other Bell Labs departments. We even got one from a Bell Labs facility in Indianapolis, based on a recommendation from someone in Holmdel. The Indy folks ordered a TAS 1010 sight unseen. Steve, Charles, and I were incredibly proud of those first orders. We had created a sophisticated product with our own minds and hands. It would have taken a whole department of Bell Labs engineers to make an item like that, but we had done it with just three engineers—three *black* engineers. We were able to do it because we were totally focused, fully committed, and extremely motivated. I felt that if we could just maintain our drive, nothing could stop us.

Not many days after parking my VW at the Holmdel loading dock, I flew to Flint and bought a brand-new Pontiac 6000 station wagon from the Superior Pontiac dealership near my mother's house. I drove it back to New Jersey ready to make the next deliveries in a vehicle from the city and the company that had made my success possible.

The Bell Labs Indianapolis engineers were astounded to learn that the guy in the Pontiac was the TAS president, and that he had driven all the way from New Jersey to personally deliver their new test instrument. I hoped that first impression would benefit us in the future, but I drove the unit to Indiana mainly because I didn't want a shipping company to lose or damage it. Also, I wanted to set up and demonstrate the product so there would be no delay in payment.

Meanwhile, activity at TAS was really heating up, and we needed an office assistant to handle the mail, answer the phone, and pay the bills. Steve, Charles, and I needed every minute we had to develop, manufacture, and sell our product.

I knew nothing about hiring an administrative assistant, and worried that interviewing candidates at my home wouldn't seem professional. I turned to friend and former colleague Peggy Austin for help. Peggy managed the Bell Labs secretarial pool, and was well versed in vetting candidates for office jobs. I would have hired *her* for the job, but she was way out of our price range. Peggy was a tall, beautiful, articulate black woman, and I was sure her presence would set a professional tone for the interviews. She and I formulated a job description and an interview procedure, put an ad in the *Asbury Park Press,* and fielded calls from applicants.

I expected most candidates would be white women, and was curious to see how they would react. I was sure some would think the whole thing was a scam—I probably would have. As it turned out, *all* the candidates were white women, but none seemed surprised or taken aback to discover they were interviewing with an African American-owned electronics company. The country was emerging from the depths of the "Reagan recession," and I concluded that applicants were much more concerned about having a job than working for a black boss.

One candidate stood above the rest.

Maureen Zamorski was a recent high school graduate. She was an attractive young lady: medium height, pale white skin, rosy cheeks, and long black hair. Maureen seemed totally at ease during the interviews. She was cheerful and friendly, and displayed no reservations about my race or the fact that she was being interviewed in my home. She seemed capable, and was enthusiastic about the opportunity. Peggy and I agreed that Maureen was the best candidate. I offered her the job, and she accepted.

In June, TAS moved from our basement to an unlikely new "office building"—the Red Bank Mini-Mall, a small, low-end retail complex in the heart of downtown. Our 800-square-foot office was downstairs, sandwiched between a travel agency and a tanning salon. The opposite end of the building housed the Barbizon modeling school, a magnet for starry-eyed high-school girls. It was an odd place to find a high-tech electronics manufacturing business, but the Mini-Mall had a few key things going for it: the rent was cheap, lunch and office supplies were within walking distance, and best of all, it was not my home.

By the summer of 1984, our little business was rolling. Maureen and a manufacturing technician had joined our staff. Charles was the one missing piece of our business puzzle. He had not yet left Bell Labs, and seemed to have growing doubts about doing so.

At Charles's July wedding, Steve was best man and I was the vocal soloist. Marlene was a radiant bride, and I could see she was looking forward to a blissful married life. Charles, on the other hand, was "sweating bullets." He had to choose between a promising career at Bell Labs and a roll of the dice in the Red Bank Mini-Mall. It was quite a dilemma.

By late summer, Steve and I needed Charles to make a decision. He had continued to work with us evenings and weekends, but our exploding workload required his full-time participation. If Charles wouldn't come aboard, we would have to hire another engineer, so we needed to know where he stood. One evening, Steve and I met

Charles in our tiny conference room seeking the answer to one simple question: *Are you in or out?*

I started the meeting by exulting that we had sold five TAS 1010s and had more orders on the way. Our initial success erased any doubts about our ability to produce a salable product, but Charles remained apprehensive.

"We sold a few units to the people we know around here," he said. "But what about the rest of the country? What if these are the only ones we sell?"

Charles had stated that concern several times, and he had a good point. We hadn't sold much outside our immediate area, and there was no guarantee we would have a viable business once we ran out of friends and former colleagues. On the other hand, we were selling to the most important customer in the industry, Bell Labs, and there was no reason to think we couldn't sell to others. We had no guaranteed business, but we were making progress. I tried to convey that to Charles.

"Look, Charles, in just a few months we've gone from an ugly prototype to a finished product. The people we're selling it to are buying it because they think it's the best. I believe people in other companies around the country will think the same. At any rate, we've gone too far to turn back now, so we'll just have to see. I'm willing to work day and night to make this business succeed, and I hope you and Steve are too."

I knew Steve was committed—he'd never have left Bell Labs otherwise. Steve was putting in as many hours as I was—more on some days. Still, it was reassuring to hear him affirm, "Yeah, I'm in. I'm either gonna get rich or die tryin'."

Charles wore a pained expression as he listened. I knew the decision to leave Bell Labs was more difficult for him than it had been for Steve, but Charles had been with us from the beginning, and I didn't see him turning back.

"This is a big move," Charles said. "I just want to make sure we're not fooling ourselves. I want this to work, and I'm ready to sacrifice everything except my family to make it happen." I didn't quite know how to respond, because Steve and I weren't planning to sacrifice our families, either. We ended our discussion that night still wondering what Charles would do. The next week, he resigned from Bell Labs, and the transition from *my* dream to *our* reality was complete.

As 1984 drew to a close, TAS faced new challenges. We had shipped several units, and we had no trouble collecting payment, but we still found ourselves in a cash crunch. We needed to pay Steve and Charles a modest salary, and we needed to pay our employees. (I was earning next to nothing—Diane and I were still surviving on her salary and our savings). TAS needed to pay its rent. We needed to buy materials for the next batch of TAS 1010s. We were making money, but not fast enough to pay our rising bills. We were hardly millionaire entrepreneurs. We were living hand to mouth, on the verge of having to reduce salaries and drop one employee. Things looked bleak, and we began to wonder if failure was such a bad option. If the business failed, we could return to Bell Labs, and our TAS exploits would provide fodder for years of great Grill Room stories.

Just before Christmas, I received a call from Dr. Victor Lawrence, the Bell Labs engineer from Ghana whose promotion I had advocated years earlier. Victor was now a group supervisor, and he was calling with a message I would never forget.

"Dave, this is Victor Lawrence."

"Hey, Victor! How are you?" I tried my best to sound confident and upbeat.

"I am well, but I think I need some of your equipment. Please tell me what I need to order."

That was encouraging. Victor's department hadn't ordered anything from us, and getting our foot in the door with his group could lead to further sales. The near-term income couldn't hurt, either. I figured I would ask Victor to buy a TAS 1010 and hope to secure more business later.

"Well, Victor, you could order a TAS 1010 Channel Simulator. I think that will fit your needs."

To my surprise, Victor acquiesced right away.

"Yes, yes, Dave, I will order one. Now is that all I need—just one? My department has a lot of money, and we must spend it by the end of the year, so please tell me everything I need. I think I may be able to use more than one channel simulator."

By the end of the conversation, Victor had ordered *five* TAS 1010s. It was by far the largest order we had received, and would definitely ease our cash problems. Having our products in Victor's test lab was sure to generate many future orders, because test engineers inside and outside Bell Labs looked to his group for guidance.

I thanked Victor profusely and put the phone down. Then I cried. Victor had come to our rescue. I didn't know if he really needed the products or was repaying a good deed or was trying to help some struggling black entrepreneurs. I had stood up for Victor years earlier because I felt it was the right thing to do—not just because he was black, but because he was an outstanding engineer who wasn't being treated fairly. Now Victor was standing up for us, and I like to think it was because he saw three black guys with an outstanding product—guys who were trying to find their way in a highly technical, highly competitive industry.

We ended our first year in business with sales of $100,000, a slim $2,000 profit, and, thanks to Victor, an order backlog. It wasn't much, but it was enough. We were on our way.

36. World's Oldest Profession

I WAS IRRITATED WHEN CHARLES first asked his skeptical question, "How are we going to sell anything outside Monmouth County?" I knew, however, that he had a point. I was the only TAS salesman, and there was no way I could sell enough products to sustain our company, let alone make us millionaires. Charles and Steve weren't interested in selling, and in any case were consumed with product development. I needed help to boost sales, but we couldn't afford to hire more employees.

In the spring of 1984 I received a call from Pete McLaughlin of Scientific Devices Inc., a sales representative firm in Boston. Pete said he had heard about TAS from Jim Ingle and others at Bell Labs. Pete wanted to meet and talk about selling our products. It sounded like a potential solution to our problem, and an answer to Charles's question, so I gladly agreed to meet.

A few weeks later, a Scientific Devices delegation showed up at our Red Bank Mini-Mall headquarters. With Pete were Rick Alexander, John Sobolewski, and Bill Graham. Rick was president of the company. Pete, John and Bill were salesmen. All were middle-aged white guys.

After greeting the SD group and introducing them to Steve and Charles, I ushered the visitors into our tiny conference room. It was just the SD guys and me; Steve and Charles returned to their work. We sat down at the makeshift TAS conference table and proceeded to break the ice. After joking about Jim Ingle and other mutual Bell Labs acquaintances, the SD guys started joking about each other.

Pete said, "If you sign us up, whatever you do, don't get Rick mad at you. He's got connections, ya know?"

John Sobolewski chimed in, "Yeah, some guy named Guido will come see you!"

Bill Graham saw my puzzled expression, so he said to John and Pete, "Come on, you're going to scare the poor guy off before we ever sign him up."

At that point, Pete decided to venture further down the ethnic path.

"We call John 'The Polish Prince,' yah know."

John deadpanned and said, "Yeah, and we call Pete 'The Professor.' He likes to think he's the expert on everybody's product. I'm Polish, so I don't give a damn about how the product works. I just sell it."

John paused for a moment to gauge my reaction, then said, "Now Billy Graham here—we call him 'The Preacher.' He'll smooth-talk those customers into buying anything, just like he's selling Jesus or something!"

The meeting was taking a weird turn. The SD guys were nothing like my Bell Labs colleagues. They seemed like an old white fraternity out to initiate their one new black member. Rick Alexander, the frat president, sat there looking cool and saying nothing. If he was trying to intimidate me, it was working. The talk about some guy named Guido breaking my legs if I crossed Rick was making me think maybe I didn't want sales reps after all.

Finally Rick said, "Okay, guys, let's get down to business."

Just like that, his guys got quiet and serious, and Rick took the floor. He gave a brief history of Scientific Devices and portrayed it as the only firm with enough experience to do the job for TAS. He stressed that his reps were dedicated professionals paid strictly on commission—"Oh, say fifteen or twenty percent"—on anything sold in their territory. Rick insisted SD reps were beneficial to small companies because they provided "feet in the street" without the cost of an "in-house" sales force. Because they were already selling complementary products from other companies, they had relationships with customers who would be interested in our products. They were pros well versed in the entire selling process: qualifying customers, closing sales, getting orders, and getting paid.

Rick's presentation sounded good. I liked the idea of more feet in the street. I also liked the idea of putting some people who looked like our customers (i.e., middle-aged and white) in front of our customers. When Rick finished his brief presentation, he looked around the table at his guys. Pete McLaughlin spoke up as if on cue.

"Look, Dave, Rick is giving you the official line about sales reps, and all that high-minded bullshit sounds really good, but just remember this: The bottom line is that reps are just like whores. When you come to town with your product, we love you, and treat you like you're the only one we work for. Next day, when the next guy comes to town—BAM!—we work for him, and we forget all about you."

When Pete said "BAM!" he slammed his fist into his open hand for emphasis. Then he continued: "We don't have to like you, and we don't have to like the customer. We just want to get in, get out, and get paid. You see?"

Pete laughed and the other guys laughed. I laughed too. Sales whores, or as the brothers would say, "sales *hoes*." It was an interesting metaphor, and not altogether inappropriate.

Despite the crass talk, I kind of liked the SD guys. They seemed honest and straightforward—they just wanted to make money. That was okay with me, because if they made money selling TAS products, we'd make money too. We agreed that Scientific Devices would be the TAS sales rep in the northeast U.S. I didn't have a contract to give them, so Rick Alexander said he would draw one up and send it to me.

A few weeks after signing Scientific Devices, I had an experience that drove home the meaning of "sales professional." Bill Graham invited me up to Boston to make a sales call at Concord Data Systems, a small modem manufacturer. Bill had arranged the call because he knew the guys at Concord were looking for a telephone network simulator, and he thought the TAS 1010 would fill the bill. I drove up to Scientific Devices on the afternoon before the sales call with the unit in the back seat. That evening, I held a training session for the SD crew, and Bill and I had a chance to chat and get to know each other. I liked Bill immediately. He had emigrated from Scotland many years earlier, and had been at SD for several years. Somehow he had retained his soothing Scottish accent. Bill came across as an easygoing guy, but he had a hard edge that indicated he was all about business. I liked that.

The next morning, Bill and I made our way over to Concord Data Systems. On the way we decided I would perform the demonstration. Bill insisted his role was not to be the technical expert but to close the sale. At any rate, he had just gotten his first training session the night before and wasn't yet qualified to demonstrate the product.

At Concord, a technician named Bruce ushered us into a lab. We sat the TAS 1010 on a table and began hooking up cables. When we finished, Bruce convened a group of engineers and technicians. Most were young, and all except one were white. They were a scruffy bunch, dressed in jeans and T-shirts. One guy with a beard and a biker jacket looked like a misfit even among this group of misfits.

With everyone assembled around the lab bench, Bill introduced me as the president of TAS. He said I was an ex-Bell Labs engineer, and that my company had developed a new telephone network simulator that was better than anything else out there—better than Bradley and much better than AEA.

The Concord guys apparently knew and respected Bill, because they paid close attention. Each time Bill referred to me, the guys would look at me as if sizing me up. I was full of confidence because I had already sold our product to demanding Bell Labs engineers. I was sure Concord would be an easy sale.

I started the demo by revealing more of my background: Bell Labs MTS, developer of computer-controlled test equipment and advanced data communications devices. I explained that I had seen the need for a product that was technically sound like a Bradley Box but more accurate. I said we had already sold units to key people at Bell Labs, and that the Bell Labs transmission impairments guru, Jim Ingle, had blessed our product. Then I launched my demo. I showed each function of the TAS 1010, and demonstrated how each telephone network impairment affected their modem's performance.

Most watched and listened intently, but the guy in the biker outfit kept interrupting with questions. How did we know we were doing the impairment functions correctly? Why should he buy our product when he didn't have any problems with the Bradley Box they already owned? Why should he buy a product from a company that hadn't been around very long? Each question sounded skeptical if not downright dismissive. Each time I explained one of the TAS 1010's advantages, he said it wasn't important. He seemed like a redneck biker just trying to give me a hard time. After several minutes of verbal sparring, I felt patches of sweat forming in the armpits of my shirt. I was thinking *this asshole is giving me a hard time because I'm black!*

I wasn't going to stand there and take any more of The Biker's abuse, so I blurted out a question for him.

"Why did you ask me to come here?"

I don't know what I expected him to say. I could imagine him saying, "Well, we need a product just like yours, but I'll be damned if we're going to buy it from a nigger!" I knew he wouldn't say that, but I just *knew* that was what he wanted to say.

What he actually said was worse, because it seemed to doom our sales prospects at Concord.

The Biker glared at me and said, "I don't know *why* you're here—we ordered a simulator from AEA a couple of weeks ago."

I wanted to punch him. My hands were clenched into clammy fists. I was ready to curse the whole group and take my equipment and go home.

"You already ordered a simulator from AEA?"

The Biker said, "Yep."

I felt angry and defeated. I felt like an idiot.

At that moment, I heard Bill Graham's soothing Scottish voice repeat the question I had just asked: "So you ordered the AEA product?"

The Biker again said, "Yep."

Bill paused for a moment. I thought he might be feeling embarrassed for bringing me on such a bogus sales call. Then Bill refocused on The Biker and said coolly, "When are you expecting to get it?"

The Biker said, "I don't know—they say they're all backed up. It'll probably be another month or so."

I saw a mischievous twinkle in Bill's eye. He looked straight at The Biker and asked, "If TAS promises to deliver in two weeks, will you cancel your AEA order and go with us?"

The Biker didn't hesitate. He said, "Yeah, I probably would."

Bill turned to me and said, "Can TAS deliver to Concord within two weeks?"

"Sure," I said. "If we get an order today, we can have the product here next week."

Bill and I walked out of Concord with their commitment to order a TAS 1010.

I was dumbfounded. Just a few minutes earlier, I'd have bet my company that we wouldn't get an order, but Bill had pulled victory from the jaws of defeat. And he was so *cool*. He didn't react to my spat with The Biker, and he didn't panic when The Biker said he'd already ordered our competitor's product. Bill just calmly analyzed the situation and identified the one thing that would get us the sale—quick delivery.

During the long drive home from Boston I reflected on the day's events. I had been sure The Biker was giving me a hard time because I was black. I was sure race was an issue—I could feel it in The Biker's belligerent questions. There was no way to know that for sure, but clearly race was not the *deciding* issue. Bill proved that by getting the order. He viewed my spat with The Biker as just another obstacle to the sale. Maybe The Biker was objecting to my race, or maybe he didn't like arrogant Bell Labs engineers, or maybe he was an asshole to every vendor. Whatever The Biker's objection, Bill proved it was only an objection, not a defeat. Like an objection to price, color, or a certain feature, the objection could be overcome by showing a larger benefit. That's what Bill did, and his performance was classic.

I couldn't help thinking about the little speech Pete McLaughlin made the first time I met the Scientific Devices guys—the one about reps being like whores. "I don't have to like you, and you don't have to like me. Just get in, get out, and get paid."

37. THE MISSING ONE

AS WE ENTERED 1985, STEVE, CHARLES, AND I FELT WE LACKED one key person. We had taken our company through its first year, but we were engineers, not marketers. To Bell Labs engineers, calling someone a marketer or a salesman was akin to calling him a lightweight or even a charlatan. We realized, of course, that no company could survive without moving product off the shelf. I had tried to attract a marketing expert while we were still headquartered in my basement, but didn't succeed. That left no choice but to add "marketing director" and "sales manager" to my portfolio of duties. Steve and Charles helped by advising me, critiquing my decisions, and analyzing the competition. Hard work, technical excellence, and dogged determination got us by that first year, but we needed to do more than get by. We believed we needed someone who could find customers, establish our product in the marketplace, and set our product direction.

Family, friends, and colleagues kept telling us that a company run solely by engineers was a company headed for disaster. That's why I was so receptive when one of my Bell Labs buddies called unexpectedly and proposed a fix. One of my buddy's young coworkers, Geoff Harris, was unhappy at Bell Labs and was considering a career change. My friend thought Geoff's skills were a good fit for TAS. When I saw Geoff's résumé, I jumped at the chance to meet him. He had a bachelor's degree in engineering from a small Southern school, Bell Labs work experience, and a Harvard MBA in marketing. Like the three TAS founders, Geoff was young, ambitious, highly educated, and black. Those attributes suggested Geoff could easily be TAS's missing "fourth founder."

I found Geoff personable and articulate. He seemed to share our goals and motivations, and was intrigued by the idea of joining a team of young black men striving to build their own high-tech company. He identified not just with our business but with our *cause*—the struggle to prove our talents in a field dominated by white men. Steve and Charles met Geoff soon after I did, and they, too, were impressed. We all believed he was a little green, but that with experience he could emerge as a real star. In February 1985 we hired Geoff as vice president of sales and marketing.

Geoff jumped right into learning about our products and operations. He read and absorbed our brochures and other marketing materials. He itched to meet our sales representatives and visit potential customers. Soon after Geoff started, I sat down with him and outlined five key job responsibilities. I expected him to: 1) visit customers with our sales reps to boost sales; 2) train the reps to sell our products on their own; 3) identify new customers and new product applications; 4) produce marketing materials such as brochures and application notes; and 5) identify and research new product opportunities.

Geoff readily accepted objective #1, and in no time we were on the road visiting potential customers. At that time, the reps were not capable of demonstrating our product, so I would make the sales presentation and the rep would—we all hoped—close the sale and get the order. Geoff was eager to take over the presentations, so I gave him that opportunity.

I had no concerns about putting Geoff in front of sales reps or customers. He was always presentable—he typically wore a nice suit, fresh white shirt, and stylish tie. He appeared professional and articulate. I was proud to have him represent our company; however, during Geoff's first few presentations I noted some concerns. Geoff usually did a fine job making the "canned" product presentation, but when it was time to demonstrate the product, he often appeared nervous. If a customer asked a question, he would get defensive. At first, I chalked his behavior up to inexperience and lack of product knowledge, but it wasn't his inability to address every question that most concerned me, it was the defensiveness and anxiety he displayed to customers.

Geoff's strange demeanor was amplified when something went wrong during a demonstration. It might have been a flaw in our equipment or setup, or operator error, or a problem with the customer's equipment. When a glitch occurred, an engineer in the audience was sure to pounce and claim that something was wrong with our product. The correct response was to downplay the problem, or take note of it and promise to get back to the customer later. Unfortunately, Geoff's typical response to a customer challenge was to redirect it to me.

"Yeah, that doesn't look right," Geoff would say. "What's the problem, Dave? I don't think it should work that way, do you? Looks like something might be wrong with our box!"

I would have to contain my fury for fear of losing the customer. Closing the sale seemed of secondary importance to Geoff. His first priority seemed to be making *himself* look good, and showing that any demo problem didn't result from his lack of knowledge or preparation, but was instead a problem with the "box." I realized Geoff was trying to preserve his personal credibility, but his behavior was maddening—more important, it probably cost us some sales.

My confidence in Geoff receded, first to hope and then to doubt. I hoped Geoff would get up to speed, develop a savvier sales approach, and live up to my other expectations. I hoped the rough edges would become smooth, and that Geoff would become the key team member we needed. Despite those hopes, as the weeks passed I began to doubt Geoff was, after all, the right person for the job.

Geoff's performance wasn't all bad. He exhibited tremendous energy, and was always champing at the bit to visit customers. He often said, "We aren't going to sell anything by sitting around the office, Dave." He was right, of course. My engineer DNA guided me toward introspection and over-preparation. Geoff forced me to get out on the road, and that was good.

The summer of 1985 brought an entirely new imperative to my business efforts and the growing struggle with Geoff. On June 22 I was sitting on my backyard deck with John Borum and his wife Vera. I was grilling steaks, and Diane was in the kitchen preparing sautéed string beans. A relaxing evening with good friends, delicious food, and good wine was in store when Diane emerged and announced: "I think my water broke."

Diane and I immediately left our guests and our uneaten steaks and dashed to Riverview Hospital in Red Bank. At 10:00 p.m. the next night, after twenty-three laborious hours, Stacy Sheng Tarver was born. She seemed like a perfect little package, a gift from God. The scale recorded her weight at seven pounds seven ounces, which I took as a sign that she would lead a charmed life. That night, TAS became much more than an experiment or a way to prove my worth or an extension of the civil rights struggle—it was the means to provide for my precious daughter.

Diane took an eight-week maternity leave, and for the first time she had to step away from her high-powered research career. That alone was enough to cause tension and anxiety, but she also faced the challenges of motherhood largely on her own. I had to keep driving our business. Our competitive position was so tenuous that I was afraid to slow down, even for a few days.

A few weeks after Stacy's birth, Geoff and I were scheduled for two days of sales calls in New England and Florida. After a full day in the office, and before Geoff arrived at the house to pick me up, Diane and I argued about some household chore I had left undone. She didn't want me to escape on yet another road trip. When Geoff pulled up the driveway, Diane followed me onto the back porch carrying a writhing, crying baby Stacy. As I ducked into Geoff's car, Diane cried out, "Get a job!"—as if my work at TAS were an expensive and time-consuming hobby I should no longer indulge. I felt terrible, but I had to go.

Geoff and I reached the Boston area around 2:00 a.m. We were both exhausted, but I noticed that besides being tired, Geoff seemed quite agitated. It was as though the travel stress had changed his personality. He wisecracked about any number of TAS matters, suggesting our product wasn't as good as AEA's and that I was making bad strategic decisions. It was a weird, perplexing transformation. I told myself, *you just don't know people until you see how they react to stress.* I noted, half seriously, that our interview process should have included a late-night sleep-deprived road trip.

We got up early, met with our sales reps, and visited a few customers. That afternoon, we drove back to Newark Airport and boarded a flight to Florida. The next morning we did customer presentations all day in Tampa, and that night we drove to Melbourne to make a presentation at a hotel there. By the time we arrived, I was physically, mentally, and emotionally spent. I didn't want to be there—I wanted to be home with my wife and my newborn daughter. I was annoyed, because if Geoff had been up to speed, I wouldn't have needed to do so much traveling. That's when I realized things weren't working the way I'd intended, and probably wouldn't get better. Geoff wasn't reducing my workload, he was increasing it.

Back in New Jersey, I decided to discuss the situation with Geoff and try to reduce the scope of his job. I thought if I could get him to be effective at the functions we most needed—selling to existing customers and finding new ones—he would provide enough value to justify his employment. I figured I could continue to handle marketing and advertising, and that if Geoff didn't have to worry about those functions, he would be more effective. I arranged to meet him on a Saturday afternoon to discuss the situation and present my plan. I chose Saturday for privacy—our administrative assistant and lab tech didn't work weekends.

Geoff arrived at the appointed time. Steve was working in the lab, but otherwise the office was empty. As we sat down in the conference room, I sensed that our meeting would not go well. Geoff appeared jumpy, and I was nervous too. I hoped I wouldn't have to fire him, because I had never fired anyone, and because I didn't know how he would react.

I told Geoff I appreciated his enthusiasm and his desire to get out to see customers, and that I was pleased our customers and sales reps respected him. "On the other hand," I said, "you don't yet have enough knowledge about our product or our customers, and you need to concentrate on sales and technical support until you've gained enough experience to handle your other job responsibilities."

Geoff waited for me to finish, and then said, "You're trying to say I can't handle the job? We've lost sales because there have been problems with the product! You guys don't know the first thing about marketing. If you did, you'd be looking into some of the new products I've been suggesting. This is not my fault! I'm not going to accept that this is my fault!"

He jumped up from the table and dashed from the room, slamming the door so hard it sounded like an explosion. I got up in time to see Geoff storm out of the building. Steve ran in from the lab.

"What's going on, man?" Steve said. He looked worried.

"That was Geoff slamming the door. Our talk didn't go too well."

I must have looked pretty shaken, because Steve said, "You okay? I thought you guys might have been going to blows up here."

"Yeah, I'm all right," I said. I hadn't noticed until then that my hands were trembling. "I guess we'll have to find another sales and marketing cat, though."

I spent the weekend reflecting and agonizing over the situation. The more I thought about it, the more I realized I had placed Geoff in an impossible situation. He didn't know our product well enough to smooth over its idiosyncrasies during demonstrations. He hadn't grown up with it, hadn't painstakingly designed and debugged each of its circuits. It wasn't his "baby"—to him it was just a "box." Geoff didn't know our customers the way Steve and Charles and I did. We understood them, because despite cultural differences, they were nerdy engineers, just like us. Geoff was young and inexperienced, but I had assumed a Harvard MBA would immediately understand our business and give us sage advice. I was wrong. An effective marketing manager would need to know our product, our marketplace, our customers, and our competition. He would need to understand our operations and our production capabilities. I knew only one person who fit that description—me.

TAS was operating in an extremely technical field, selling an exceedingly complex product to highly sophisticated engineers. I concluded that no magic marketing expert was going to tell me how to succeed in that venture. Anyone who could do that was probably creating his own multimillion-dollar company. I decided to stop looking outside for an "expert." The fun, the challenge, and the excruciating difficulty of running TAS was that, do or die, it was my baby.

First thing Monday I met with Geoff again. We were both much calmer, probably because we were both emotionally drained. We agreed our association wasn't working, and that Geoff would go. It was a sad day, because I liked Geoff. He was an energetic young man who believed in our company, and he had been proud to be one of four black men on a trailblazing business journey.

38. Southern Surprise

"HEY DAVE! OVER HERE!"

I scanned the crowded concourse at Atlanta's Hartsfield International Airport and saw Charlie Nixon hustling toward me. Charlie was president of Commtec Associates, our new Southeast sales rep firm. A lanky guy with a slim, ruddy-red face, wearing faded jeans, a big belt buckle, and a plaid cotton shirt—Charlie looked more like a country music star than a sales rep. He had recently left a career at AT&T to start his own firm. Charlie was eager to ramp up his business, and I was eager to sell more equipment in the South.

It was late summer, more than a year since we'd shipped our first product and more than a month since Geoff's departure. I was once again alone on the road wearing my sales and marketing hat. For the most part, sales were still coming one by one, and each seemed to require monumental effort. Steve, Charles, and I were paying our bills and paying ourselves modest salaries, but we were by no means comfortable. Even at our established major customer, AT&T, each sale required diligence, follow-up, and large chunks of time. Our customer list was growing, but no client approached the size or stature of Ma Bell.

I was concerned about our ability to grow the company. That was why I'd been so hopeful a few weeks earlier when I received a call from Charlie. He wanted me to come to Atlanta and make a sales presentation to Hayes Microcomputer Products Inc. Hayes, just a few years old, was already one of the world's largest modem manufacturers. The company had achieved that distinction by focusing on the new and exploding personal computer market.

"Welcome to Atlanta," Charlie said. "You ready to go get some business?" He sounded as if we were about to go to a bar and try to pick up women.

Charlie's excitement was infectious, and my reply was enthusiastic. "Oh yeah, I'm ready!"

We made our way to baggage claim. I had packed a TAS 1010 in our standard white shipping box and checked it as baggage. I was relieved to see it emerge, undamaged, on the carousel. We headed to the parking lot, where I carefully placed the box on the back seat of Charlie's car.

We drove from the airport, south of the city, to Hayes headquarters in Norcross, north of Atlanta. I expected Hayes to be situated in a small industrial park and to have perhaps a hundred employees. To my amazement, we pulled up to a large, modern building on a verdant campus. The workforce must have numbered in the hundreds. I asked myself, *could something as simple and basic as a modem support a company this large?*

Apparently it could. After Charlie and I signed in at the reception desk, the engineer who was Charlie's contact came out to greet us.

"Hello, fellas," our host said. "I'm Neil Hodges, and I'll be escorting you into the building today."

Neil was younger than I expected, and though he had a bit of a Southern drawl, it was less severe than Charlie's.

"I'm Charlie Nixon—I spoke to you on the phone. This is Dave Tarver, president of TAS." Charlie was on his native Southern turf, and he seemed perfectly comfortable.

"Oh, boy, you brought out the big dog!" Neil said with a snicker.

"Well, Neil, I decided to stop messin' around and just go straight to the top!"

Neil and Charlie laughed. I managed an uneasy smile. I was in foreign territory, and didn't feel as if I could relax and be myself.

"I'm glad to be here, Neil. This is an impressive place."

I sounded stiff and formal. It was the best I could manage.

"Yes, Mr. Tarver. Hayes is just a few years old, and already it's the leading maker of modems for the PC industry. That's why we have such tight security in this building. We're very concerned about our technology falling into our competitors' hands. A lot of companies out there would love to be in Hayes's position right now."

"I see."

Neil's presentation struck me as cocky, but I wasn't about to argue. Bell Labs management regarded modems as old technology. Bell Labs engineers weren't bragging about building analog modems in 1985. Most had moved on to newer projects involving high-speed digital communications. Hayes seemed to be doing pretty well in a business Bell Labs and AT&T had all but abandoned.

Neil walked Charlie and me through the security doors and into the bowels of the Hayes building. We were headed to the conference room where I would be making my sales presentation. As we walked, Neil explained how Hayes got its start and reiterated the company's position in the marketplace—not *one* of the leading modem manufacturers, but "*the* leading maker of modems for the PC industry." Neil also told us Hayes not only made the world's best modems but also produced a savvy communications software package called SmartCom, and that together those products were a real PC industry juggernaut. Heading to the conference room I took note of the very expensive, very modern office environment. The rich carpeting, glass-enclosed conference rooms, and sleek engineering workstations befitted a dynamic, successful company. Just seeing the place made me hopeful and excited. Hayes really seemed to have great potential. *If only we can get their business*, I thought.

Neil got Charlie and me settled in the conference room and went to round up the engineers for my presentation. Charlie and I started setting up the equipment.

"So, Dave, you think we can do some business with these boys?" Charlie asked.

"Man, I sure hope so. This place is a lot larger than I expected, and all for making modems!"

"How 'bout that!" Charlie's response revealed more than a bit of Southern pride.

After a few minutes, the Hayes "boys" began filing into the conference room. I wasn't surprised that all of them were white. I managed to make eye contact with several as they made their way to their seats, but most avoided me or spoke to their peers in hushed tones. I didn't know if they had expected a black presenter, and I was primed for any signs of hostility. I was glad Charlie was there to "balance out" our presentation. *When in Rome...*, I thought.

Just then, a big surprise: a former Bell Labs colleague entered the room. Don Pierce—Dr. Donald Pierce—had been an engineer in the Data Communication Laboratory at Bell Labs when I worked there. Our paths had hardly ever crossed, but I knew he managed something called the Data Bank, the maintenance facility for our modem prototypes. It wasn't a glamorous job, and I always wondered why someone with a PhD would be assigned to it. To me, the Data Bank was a glorified modem repair shop.

As Dr. Pierce entered the room, we made eye contact and he immediately came over. He greeted me as if I were a long-lost friend. "Welcome to Hayes," he said. "I'm heading up modem research here, and we're anxious to hear how your telephone network simulator might help us."

Heading up modem research, I thought. *That's a big step up from "modem repair supervisor."* Dr. Pierce's position at Hayes seemed to better fit his education. He may have been underutilized at Bell Labs, but at Hayes he was The Man. Relieved to have an old colleague among the group, I relaxed just a bit.

"Well, Dr. Pierce, I think our simulator can be a big help here. The modem group at the Labs really likes it, and it's starting to catch on at other companies, too."

I knew my offhand reference to Bell Labs would get him. "The Labs" was still viewed as the mecca for all things related to telecommunications, and if our product was good enough for them, it was good enough for Hayes.

The demo went well, but there were some bumps. Our host, Neil Hodges, kept piping in with questions that seemed intended to show how smart he was rather than to elicit information. A few other attendees seemed to ignore the presentation and avoided eye contact with me. Overall, though, the Hayes engineers seemed impressed by our product, and toward the end of the presentation Don Pierce asked how soon they could get a "demo unit." At that point, I knew we were near a sale.

After the presentation, Neil Hodges and a dour-looking guy named Raymond Grant buttonholed me for several minutes with detailed technical questions. They could not, of course, stump me. I had designed the product, and I knew the function of every component. Neil posed questions in a good-natured way and seemed satisfied with my answers. Raymond was stone-faced and unsmiling. He had fewer questions, but his seemed laced with skepticism. Worse, when I answered one of Raymond's questions, his face would turn beet red, as though he was embarrassed or angry that I might know something he didn't. For a moment, I pictured Raymond wearing a Ku Klux Klan robe. The image seemed very believable.

Charlie Nixon sensed the growing tension, and he stepped in to rescue the sale: "We're gonna ship a demo unit to you boys from the factory so y'all can play with it all you want. I'm sure if you have some suggestions on how we can improve the product, the engineers at the factory will be glad to listen. Y'all are the experts in the modem business, and we just wanna git y'all a product you can work with. Ain't that right, Dave?"

I was right at the brink of confrontation with Neil and Raymond, but Charlie's question forced me to back down. "That's right, Charlie," I said. I was thinking *Das rat, missa Chahlay*. I didn't mind doing a little shuckin' and jivin' if that's what it took to get an order. If Charlie wanted these guys to think someone back at the factory, perhaps a *white* someone, knew more about our product than I did, that was okay. I just wanted the sale, and I wanted Hayes as a customer. We finished packing up as fast as we could and got the hell out. We both felt there was nothing to be gained by standing around talking to Neil and Raymond.

Later that night, we did make a "sale," but the "customer" wasn't Hayes. Back at the hotel, we left the TAS 1010, packed in its big white box, on Charlie's back seat. The next morning we discovered someone had smashed the rear window and made off with it. I was appalled. Our hotel seemed to be in a safe, upscale area. I couldn't believe any criminal in his right mind would steal a telephone network simulator. It wasn't exactly something you could sell at a local pawnshop.

At first, Charlie seemed more concerned about the damage to his car than about my stolen equipment, but he turned and asked: "Do you guys have insurance on your demo equipment?"

"Yes, as a matter of fact we do." We had never filed a claim, but I had included demonstration units in our business insurance coverage.

"Well, hell, man, that's just like a sale! You just made your first sale in Georgia! Congratulations—and don't forget my commission!"

I hoped Charlie was right—that our insurance would indeed cover the theft of our demo unit—and I hoped he was kidding about getting a commission on the "sale."

A few weeks later, Hayes placed an order for not one, but two units. I was ecstatic. Also, our insurance claim for the stolen TAS 1010 resulted in full payment. Not a bad result for our first foray into Georgia.

A few more weeks went by, and things seemed to be going well at Hayes. The engineers had a few questions about our product, but there were no significant problems. Then Neil Hodges called and dropped a bombshell:

"Hello, Dave, this is Neil Hodges."

"Hey, Neil. How are things going there?"

"Pretty well, Dave, pretty well. Raymond Grant and I—you remember Raymond, don't you?—Raymond and I are working on a new test system for the Hayes production floor. Every modem Hayes makes will be tested on our system before it leaves the factory. We're thinking of designing your TAS 1010 into our system. If all goes well, we would need somewhere between thirty and fifty of your boxes. I guess that would represent a pretty sizable order for you fellas."

I nearly dropped the phone. I had been happy to sell *two* systems to Hayes. In my wildest dreams I never imagined that we would sell them thirty.

"Yes, that would be great, Neil," I said cautiously. "What do we have to do to move that along?"

"Well, Raymond and I have to finish evaluating your system. Then we'll give our recommendation to management, and they'll make the final decision. But as far as you're concerned, Raymond and I will be making the decision."

I was getting to know Neil, and he struck me as someone who had a high opinion of himself and his importance to Hayes. Still, he seemed like a decent guy who appreciated our product. Raymond, on the other hand, was a wild card. I couldn't get that first mind's-eye image of him—beet-red face framed by a Klan hood—out of my mind. I didn't think I could count on Raymond to support our cause. I just hoped he was a minor player who couldn't block Neil's decision to buy from us.

I called Neil and Raymond every few days for several weeks to see if they were having any problems designing the TAS 1010 into their modem test system. I put my coworkers on notice: "If Neil and Raymond call, notify me immediately so I can personally respond to their technical questions."

Whenever I spoke to Raymond, he was all business. There was no kidding around, or, as with Neil, self-aggrandizement. Raymond would ask a technical question about our system, and when I answered, he would thank me and hang up. That was it. I couldn't tell if he was a supporter or a detractor. I hoped his questions weren't intended to gather ammunition to shoot down our proposal.

After a few tense weeks, Neil Hodges finally called with his decision. "Dave," he said, "you'll be happy to know that we've decided to use your channel simulator in our production test system."

He was right—I *was* happy. Overcome with joy, in fact. I wanted to end the phone call immediately, before I could say something that would screw up the deal. That was rule #1 in sales: "When you've made the sale, shut up!"

Then I remembered another sales maxim: "Always thank the customer."

"Neil, that's great," I said. "Thank you very much."

"Well, Dave, there is one more thing."

Somehow I knew there would be a catch. "What's that, Neil?"

"I made the recommendation to management, but this is such an important purchase for Hayes that they want to come and visit your factory. If our proposal goes through, Hayes will be spending a lot of money with TAS, and we need to be sure we'll get good support and that you guys will be around. Our managers visit every major supplier. They even paid a visit to Hewlett-Packard not too long ago."

That sounded like trouble. It was one thing for Hayes management to visit a place like HP, at that time the largest test equipment vendor in the world; it was quite another for them to visit TAS, a tiny five-person company in the basement of the Red Bank Mini Mall. I began to dread the Hayes "opportunity." I wished Neil and Raymond would just buy our simulators one at a time so the purchases wouldn't draw top management's attention. Now senior managers, perhaps including Dennis Hayes himself, would focus on the possibility of spending millions for the products of a tiny, black-owned company.

Later, I learned that three top Hayes managers would make the TAS inspection tour. The visit could not be avoided. If I refused, we would lose our biggest-ever sales opportunity. All we could do was put the best possible face on the situation. I thought about holding the meeting at a hotel, but that wouldn't fly. The Hayes managers weren't just coming to meet TAS management, they wanted to see our facilities. We would have to host them at the Red Bank Mini-Mall and hope for the best.

A few days before the Hayes visit, I got a call from Raymond Grant.

"Hello, Raymond," I said cautiously. I was deeply suspicious of his motives, and I couldn't shake my initial impression of him.

"Hey, David. I know Hayes management is going to visit TAS in a couple of days. I want to give you some information that might help you."

I didn't know how to take that. Beyond answering narrow technical questions about our system, I had never had a conversation with Raymond. Was he genuinely trying to help, or was he trying to sabotage our opportunity? Okay, I thought. No harm in listening.

"Oh, thanks.... Uh, let me get a pencil so I can take some notes."

"Okay, but you can't mention that I called. I'm just trying to help you guys out because I think you have a better product than your competitor. I've been the one programming your simulator for the past few weeks, and I find it's much easier to use."

Raymond's furtive tone did nothing to dampen my suspicions, but his message sounded promising.

"Of course, Raymond. I won't mention our conversation to anyone. I appreciate the help and the feedback."

"Okay, so listen to this. The guys you're going to be seeing are vice presidents. These guys don't know anything about the technical aspects of your product. They only care about two things: how your product can save money for Hayes, and whether your company will still be around in a few years. You've got to be sure to deal with both of those issues. Neil and I have already dealt with the technical issues, so you don't have to worry about that."

"Well, okay, Raymond. I think we can handle that. I'll be sure to stress our answers to both of those issues during my presentation."

"Okay, good luck."

With that, Raymond hung up the phone, and I sat dumbfounded and confused. Had my initial impression about him been wrong? Was he sincerely trying to help us? I reflected on Raymond's motive and convinced myself that he was sincere. Raymond's job was to write the test automation software that would control our simulator, and ours *was* much easier to control than AEA's. If Hayes bought our product, it would make Raymond's job much easier, so his actions made sense. I still didn't know whether he was a racist or not, but it seemed that his racial attitudes, whatever they were, didn't influence his business decisions.

I took Raymond's message to heart as I crafted my presentation. In a moment of inspired clarity, I realized Hayes would be crazy *not* to buy our simulator. My analysis showed that, even though each TAS 1010 (plus accessories) cost $20,000, the cost per modem tested was just thirty-eight cents. That negligible amount was the cost of our unit spread over the huge number of modems tested during our product's five-year life span. On the other hand, each defective modem Hayes shipped cost hundreds of dollars to replace. The analysis clearly showed that our product would save Hayes lots of money.

The second issue—whether TAS was a viable company with staying power—seemed much more difficult. Our Mini-Mall location did not shout out: "Rock-solid business here!" Nor did our staff size—just five full-time employees, including the founders. Then it hit me:

Hayes had no choice. We were just as viable as our main competitor, and we had a better product. It wasn't as if Hayes could get a telephone network simulator from a more established company. The only alternative—designing and building their own simulator—was prohibitively expensive and would drain precious resources from their main business, modem development. Instead of being defensive about our company's size, I decided to point out the education and experience of the three TAS founders and to stress our determination to succeed and grow. I also planned to stress our substantial AT&T business, and to insinuate that if our products were good enough for Ma Bell, they were good enough for Hayes. Those were plausible arguments, and I hoped they were winning arguments. We didn't have to be the best possible choice; we just had to be the best *available* one.

On the morning of the Hayes visit, I sat in my office nervously awaiting our guests. Just as my anxiety peaked, I heard the doorbell. I heard our office assistant Maureen's effervescent greeting, and heard the Hayes managers respond: "Well, hello there, young lady... Yes, we're glad to be here... Yes, we had a very nice flight... This is a nice place you have here." They sounded like down-to-earth folks, and they seemed to be getting along with Maureen just fine. It was show time. I gathered my courage and walked out to the foyer.

"Hello! Welcome to TAS."

The three Hayes managers, who had clustered around Maureen, turned and looked at me as if I were crashing their party.

"I'm David Tarver, President of TAS. I'm glad you all could make the trip here today."

I didn't know if the Hayes guys were expecting the president of TAS to be a black man, but I noticed no signs of shock.

"I'm Mr. White," one of the Hayes visitors said, "and this is Mr. Jones, and this is Mr. Rodriguez."

I shook hands with each man and ushered them into the tiny and drab space that served as our conference room, then went to the lab to get Steve and Charles. We had decided they would sit in on my presentation and join the Hayes folks and me for lunch.

Sitting around the conference table, we began with small talk. Charles told our guests he was a graduate of Morehouse College and

Georgia Tech, hoping his Georgia connection would enhance our prospects. Steve informed the Hayes guys he had graduated from the University of South Carolina, again hoping the Dixie connection would resonate. I remained mostly silent; in fact, for a while I felt myself floating above the scene.

There we were, three black engineers sitting at our conference table with three senior managers from one of the world's fastest growing technology businesses—a Deep South company. I took note of our cheesy conference table. I noted the harsh fluorescent light and the cheap nylon pile carpet. I noted the chintzy plastic chairs, and the dime-store whiteboard on which I was about to make my presentation. Everything about the scene was improbable, but it was happening, because we had developed something Hayes needed and no one else on earth could provide. At that moment, I was overcome with intense pride. The game was on, and I was ready.

When I sensed the small talk subsiding, I launched my presentation. I mentioned that Bell Labs engineers were using our equipment to perfect their modem designs, and felt ours was the only equipment that could meet their needs. Then I hit our guests with what they had come to hear, according to Raymond Grant: the cost justification. They followed intently as I walked them through the analysis, and by the time I finished, they were all nodding in agreement. They too concluded that it would cost Hayes more money if they didn't buy our simulator, and that they would be crazy not to invest in our equipment. I couldn't have imagined a better result. It was the best sales presentation I'd ever given. I didn't even have to deal with the issue of our Mini-Mall location—the Hayes guys were apparently sold. It was time for lunch.

Steve, Charles, and I walked our guests to a nearby restaurant. On the way, I grew nervous again. I was fine dealing with the Hayes guys in a formal context, but I was afraid to engage in small talk or other unscripted conversation. My presentation had gone beautifully, and I didn't want to screw it up. I was worried about revealing how little we had in common with these white, Southern execs, afraid we'd kill our chances of doing business.

If Steve and Charles shared my nervousness, they didn't show it.

They chatted easily with our visitors about sports, Georgia universities, living "up North" versus "down South," and other subjects. As I listened, I recalled the many times I'd gone to lunch with Steve and Charles and had grown impatient as they engaged in similar banter. I'd always felt we needed to focus on our work; now I was happy to see them talking so easily with the Hayes guys about nothing in particular. By the time our food arrived, a casual observer might have mistaken us for old college buddies.

I sensed that things were going extremely well, but we still hadn't closed the deal. I couldn't come right out and ask our guests for their business, or what they thought about our meager offices and our tiny staff. I was afraid of the answers I might get, so I just ate and occasionally smiled. I decided to wait until we got back to the office to try to wring a commitment from the Hayes execs. No point in pushing too hard—things were going very well.

At that moment, I noticed that the playful conversation between Steve and Charles and the Hayes guys had ceased. The next thing I knew, Steve looked intently at our guests and said: "So, now that you guys have seen our company, what'd y'all think? Does it bother you that we're located in a little mini-mall?"

I thought I would die. *This is why you shouldn't have brought Steve and Charles to lunch,* I told myself. Trying not to look at Steve, I looked plaintively at the Hayes managers, hoping they wouldn't be taken aback.

Mr. White spoke up. "Let me tell you something about Hayes. When Dennis Hayes started the company, he was building modems on his dining-room table. When he moved the business out of his house, he rented a retail storefront, not much different from your place. What you guys are doing reminds me of the early days of Hayes."

I couldn't believe what I was hearing. Mr. White seemed to be indicating that, instead of being a liability, our humble headquarters was an asset.

Emboldened, Steve asked another question: "So, do you guys have any questions or concerns about what you've seen?"

Mr. White spoke up again: "No, everything looks good. It looks like you have a great little company here, and Neil Hodges and Raymond Grant really like your product. I think we can move forward."

Steve's direct approach worked—we had closed the sale.

On the walk back to the office I was positively giddy, but I still just wanted the day to end. It looked like we had the sale, so I didn't want to do or say anything to mess it up. I led a brief tour of our little factory, and then our guests were off. As they departed, each made a point of saying goodbye to Maureen. I stood at the front door and watched as they walked past the tanning salon and up the stairs. I noticed for the first time that the shirt under my sport coat was soaking wet. I stood frozen for a few more minutes, nervous and excited. It had clearly been the best day in the history of our company. Still, all we had was a sort of promise from the Hayes guys, a verbal statement of intent. We didn't have the order yet.

Neil Hodges called a few days later, and said, "Well, Dave, I guess you guys did it. Our test system has been approved by the powers that be here at Hayes. They approved our first order for five of your boxes, and there'll be a lot more after that."

I could hardly contain my joy. Five TAS 1010s on one order! As soon as I got off the phone, I ran back to the lab to tell Steve and Charles the good news: "We're going to get the business from Hayes! The first order is on the way!"

"That's good," Steve said. "But we don't *have* the order yet." Charles added, "Yeah, and we still have to build the units after we get it."

Their dour responses didn't bother me. I could tell they were as happy and hopeful as I was. They both had a way of keeping my feet planted firmly on the ground. Sometimes it was really irritating, but on this occasion I didn't mind.

Over the next few months, Hayes ordered ten more TAS 1010s. Charlie Nixon had to hire a salesman to cover the Atlanta area and service Hayes. I made several trips there to make sure things were going smoothly. I got to know Raymond Grant a little better. I found he was a hardworking, no-nonsense guy. He never mentioned the surreptitious help he'd given me, and I never brought it up. After several months, I still had no clue how he felt about black people in general or me in particular. I was dying to know.

I finally got a chance to find out. I was visiting Hayes to make a presentation to the engineering group and provide support to Neil and Raymond. Charlie's new salesman invited them to join us for lunch. I had never socialized with Neil and Raymond—all our previous interactions had occurred in a Hayes office, conference room, or lab. I was uneasy about lunching with them, but we already had their business, so I thought it couldn't do any harm.

A few minutes after we were seated at the restaurant, Charlie's new salesman excused himself to make a phone call, and Neil left to use the men's room. I found myself alone and face-to-face with Raymond Grant, the fellow I had initially viewed as a Southern racist but who had given me sound advice about winning the business from Hayes. I felt the need to say something.

"Raymond, I want to thank you for the advice you gave me before the inspection visit. It turned out to be very helpful." My tone was low, almost a whisper, because no one was supposed to know that Raymond had given me any advice, and there were likely lots of Hayes engineers in the restaurant.

Raymond's response stunned me. He looked at me earnestly and said, "I just wanted to be sure you guys got a fair shot. You have a good product. Sometimes people see the color of someone's skin and can't get past that."

I could hardly believe what I was hearing. A few months earlier, I had seen Raymond's red face and his serious, unsmiling demeanor and assumed that he didn't like me and didn't like black people. I had even imagined him as a member of the Klan. As I sat listening to him, I was overcome with shame.

"Well, I really appreciate what you did, Raymond," I muttered.

"What I did wouldn't have mattered if you guys didn't have the best product," Raymond said. Then he cracked the faintest of smiles. I knew he was telling the truth, but Raymond's motives seemed to go beyond cold business and technical calculations. I felt as if he wanted us to succeed.

In the ensuing months, our Hayes business exploded. In less than a year, Hayes became our largest customer. Raymond Grant, Neil Hodges, and the Hayes managers hadn't let the fact that we were a tiny, black-owned company obscure their judgment. Perhaps that kind of wisdom is what made Hayes so phenomenally successful.

39. Double-Cross

IN 1984, OUR FIRST YEAR IN BUSINESS, TAS sales revenues fell just short of $100,000. In 1985 we registered eightfold growth, reaching almost $800,000. Late that year we abandoned the Red Bank Mini-Mall for a 3,000-square-foot building on Birch Avenue, one of two "industrial" streets in Little Silver. We hired two new employees—an assembler and a production supervisor. TAS was starting to look and feel like a "real" company as we entered our third year, and I was looking for ways to expand further. Meanwhile, several of our sales representatives were asking the same question: "Are you guys going to be at Interface?"

Interface '86, at the Georgia World Congress Center in Atlanta, was billed as the biggest telecommunications tradeshow of the year. I had seen magazine ads pitching Interface, but hadn't considered exhibiting there. I was consumed by the day-to-day challenges of running TAS. Exhibiting our products at a tradeshow seemed like a huge distraction.

Our sales reps saw things differently. Everybody who was anybody in telecommunications was going to be at Interface '86, they insisted, and an up-and-coming company like TAS should be there too. The reps told of small companies like ours striking it rich by landing big orders at a tradeshow. Even if we didn't get a big order, they reasoned, we would present our products to dozens of potential customers in just a few days.

Slowly, the reps won me over. They convinced me that participation at Interface would signal to the industry that TAS had "arrived." I imagined people flocking to our booth and poring over our equipment. I imagined working side by side with our sales reps, building camaraderie while greeting new customers and booking a flood of new orders.

If I needed more convincing, Charlie Nixon provided it: "You know, those AEA boys are going to be at Interface," he drawled. Charlie had cut right to the heart of the matter. I knew that if our competitive nemesis was going to be at Interface '86, then we had to be there too.

I presented the Interface suggestion to Steve and Charles, who were reluctant at first. They realized that any time spent at a tradeshow would be time away from their design and production duties. Steve, Charles, and I comprised 100 percent of our product development team, and none of us could afford to waste time, but the promise of more business and fear of AEA led us to agree that we would go to Atlanta. We decided to take along two members of our tiny staff: Maureen, our administrative assistant, would serve as exhibit-booth hostess, and Tom, our production supervisor, would set up and maintain our equipment and displays.

Preparing for the show was anything but glamorous. Registration paperwork needed to be done, and an exhibit had to be assembled, but the most challenging task was getting the display, the equipment, and the five of us to Atlanta. I feared our products would get lost or damaged if we shipped them commercially, and our shoestring travel budget would not accommodate five airline tickets. I solved the problem by renting an old Winnebago motor home to carry our equipment. Steve and I planned to take turns driving. Charles and Tom would follow in a rented car. Maureen would fly to Atlanta and meet us at the show.

We crammed the equipment and exhibit materials into the Winnebago and, one March evening, set out on the fifteen-hour drive. It wasn't pleasant. The Winnebago was musty and uncomfortable, and drove like a storage shed on wheels. After several hours, the unpleasant environment started to weigh on my disposition. I staved off despair only by talking with Steve about our plans and dreams, and by telling myself everything would be fine once we got to Atlanta.

We drove all night and arrived around noon. We headed straight for the Georgia World Congress Center, unloaded, then headed to our rooms at the adjacent Omni Hotel to relax and freshen up. The Interface '86 venue was remarkable, and the Omni was stunning. We had arrived at our first tradeshow, and I felt as if we had entered the Promised Land.

Inside the convention center I was impressed by the vast exhibit hall and the ocean of carpet that stretched from one end to the other. Hundreds of people scurried back and forth, some erecting displays, others pushing equipment carts. After we found our assigned space and set up our display, I walked around the hall. TAS looked like small potatoes compared with most companies at the show. Motorola and AT&T and Hewlett-Packard had exhibits many times the size of ours, tended by throngs of well-dressed staffers. Even many small companies had more impressive exhibits than ours. Still, I didn't feel discouraged. We were in the midst of the action at the telecommunications industry's most prominent tradeshow. Most of our customers were exhibitors, and potential new customers seemed to abound. Our display was up and our equipment was intact and working. I looked forward to landing some of those big orders the sales reps had talked about.

My excitement began to fade when Interface kicked off the next day. I suddenly realized our exhibit booth was in the hinterlands of the convention center, on an aisle assigned to minor players and first-timers. The vast majority of traffic flowed past the major players' exhibits and avoided our area altogether. Most people who trickled by our exhibit didn't understand our product, and those who did weren't interested. Visitors briefly poked their noses into our booth seeking free Frisbees or stress-reliever balls. Seeing none, they scurried away.

Some of our sales reps came by to help, but there was precious little for them to do. Maureen was like flypaper to the reps, so we had no problem attracting them to our booth. Steve and I tried to fill the idle time by training them to use and demonstrate our equipment, but their attention span was short. Between flirting with Maureen and running off to see the exhibits of their other "principals," they got very little done.

On the second day, three men wearing expensive suits stopped at our booth. Maureen and I were on duty, but the visitors simply nodded at her and walked straight over to me. The apparent group leader, an intense fellow with steely blue eyes, identified himself as Hans and introduced his associates. All three looked as if they could have been from East German intelligence, and indeed Hans's business card contained the name of a mysterious-sounding German company. I didn't know what these guys were up to, but they made me nervous. Hans and his associates seemed extremely interested in our equipment, but I didn't feel comfortable giving them much information. We engaged in a verbal joust for several minutes, with Hans asking pointed questions about our simulator and me giving oblique answers. Then it occurred to me that I should ask *Hans* some questions.

"How do you plan to use our equipment?" I asked. I tried to remove any trace of cynicism from my voice, but I am sure some slipped through.

Hans fixed his steely gaze on me and replied, "We don't intend to use your equipment. We are the sales representatives for AEA. I believe they are your competitor in this area."

I was furious. How could Hans think I would demonstrate our equipment and give him ammunition to compete with us? Why hadn't he introduced himself as an AEA rep?

"Well, I appreciate you stopping by," I said sarcastically. As far as I was concerned, the "demo" was over.

"Thank you for the demonstration," Hans replied. "You are welcome to stop by the AEA booth. We will give you a complete demonstration of our equipment. We are not afraid to show it to you."

Hans and his pals turned and left, talking with each other as they sauntered away. They seemed cool and confident. My heart was racing and my eyes were watering.

I thought the visit from Hans would be the low point of the Interface show, but I was wrong. That came the next day after a lunch meeting with Mel Mayo, vice president of Phoenix Microsystems.

Mel's company supplied an instrument called a data analyzer, which was a critical accessory to our telephone channel simulator; critical because it filled a void I'd discovered during our early days in business. Sometime in 1985, I was demonstrating our product at a Boston area company called Infinet. The head of R&D there was Andy Salazar, a former Bell Labs engineer. After I demoed the TAS 1010 to his engineers, Andy and I retreated to his office and he told me: "You guys have a great simulator, but what I really need is something that will automatically test my modem." I chewed on that awhile and realized that Andy was right—we would have a much more salable product if we could, in addition to simulating telephone network conditions, inject data into the "sending" modem, collect it at the "receiving" modem, and record the errors our simulator caused. I settled on the Phoenix box to perform the analysis—send data, receive data, count errors—and devised a software package called TASKIT to automate the testing process. The Phoenix data analyzer and TASKIT software allowed us to offer our customers a complete, automatic modem test system. Andy Salazar was one of the first customers, and many more followed.

The purpose of the lunch with Mel was to agree on an annual purchase contract for the Phoenix data analyzers. My interest was to ensure a steady supply of the Phoenix box at the lowest possible price. Mel's was to secure our commitment to buy the largest possible quantity of his company's product.

We had a great lunch. Mel expressed appreciation for our business, and we quickly agreed on purchase contract terms. I had warm feelings for Mel, because from the beginning of our business relationship, I sensed that he was trying to help us. We shared the profit motive, but in his genteel, almost fatherly way, Mel signaled that our relationship was about more than dollars and cents. We shook hands on our deal and walked back to the convention center together. When we got there, Mel headed toward the Phoenix booth, and I went to tell Steve and Charles the good news.

On the way to our booth, I spotted Bill Graham, the rep who had intervened to save me from The Biker at Concord Data Systems, rushing toward me. He grabbed my arm, cast furtive glances left and right, and pulled a flier from his pocket. In a hushed but urgent voice, he said, "Look at this!"

At first I didn't realize what Bill had shoved under my nose. The flier said "Phoenix Microsystems" at the top, and when I scanned the page, I saw a picture of a test instrument. The title over it read, "Phoenix 5100 Telephone Network Simulator."

"They must have seen how well you guys were doing and decided to get in on the action," Bill said.

I was so upset I could hardly breathe. Mel's company was planning to compete with us! My face was hot and my heart was pounding. Sweat poured down the sides of my torso.

"I can't believe Phoenix would do this. I just had lunch with Mel Mayo, and he didn't say anything about this." I heard myself sounding wounded and betrayed.

"I know—that's why I wanted you to see this," Bill said in his most sympathetic voice.

Betrayed was exactly how I felt. Phoenix didn't have to compete with us—they were a successful company with multiple products and rapidly growing sales much greater than ours. Why would they use precious engineering resources to come after our business? It just didn't make sense.

I tried to suck it up and look confident in front of Bill. The last thing I needed was for a key sales rep to think I was frightened. Truth is, I was.

"I'm going over to the Phoenix booth to see Mel about this. Do you mind if I borrow the flier?"

"By all means, take it." Bill said. "Just don't tell the Phoenix guys who gave it to you."

"No problem, Bill. Thanks."

I turned and flew across the industrial-grade carpeting, through the sea of exhibits with rented hostesses and cheap giveaway items, until I reached the Phoenix booth. Mel was just returning. I hurried over and took him by the arm.

"Mel, I want to ask you about something." My tone was urgent, angry, hurt.

"Sure, Dave. What is it?"

We moved a few steps away from the booth, and I showed him

the flier.

"What is this?"

Mel glanced at the flier and looked up at me. "Oh, this… this is nothing, Dave. It's just something some of our engineers wanted to do. It's not a competitor to what you guys are doing. Your system is much more complete. Our guys saw a need to provide something much less expensive—an entry-level product. Besides, it's not even out yet."

I wasn't buying it. "Mel, this data sheet says "Telephone Network Simulator. You know that's exactly what we make. We're buying lots of product from you, and now you guys are competing with us. Why didn't you mention this?"

"Dave, I don't think this little thing is going to affect your business one bit. We don't even expect to sell many. This is just something some of our customers were asking for, and our guys thought we could do it. We have a right to do that, don't we?"

Mel's tone was no longer warm and congenial. It was stiff and disingenuous. He was making no sense. Why would Phoenix develop a product if they didn't expect to sell many? I didn't know what bothered me more—that Phoenix would decide to compete with us, or that Mel thought I was stupid enough to believe him.

"Well, Mel, you guys do what you have to do, and we'll do what we have to do." With that not-so-subtle threat, I turned and walked back to the TAS booth.

Steve and Maureen were waiting. Someone had given Steve a copy of the Phoenix flier, and he hurried over to show it to me.

"Yeah, I know," I said. I just got the flier from Bill Graham, and I went over to talk to Mel Mayo about it. I don't know why these guys would do this."

"I don't know either, but I don't put nothin' past the good-ol' boys," Steve said.

Just when I thought things couldn't look any bleaker, Charlie Nixon dropped by. He, too, had learned about the new Phoenix product. One by one, several other reps arrived. All had long faces and wanted to talk about our new competition—what it meant, how we planned to respond—but the look on my face told them to back off. They tried to look sympathetic as they gathered in a corner and waited to hear what I would say. I felt as if a close relative had just died, and my reps were at the wake offering awkward condolences.

I tried my best to look and sound unconcerned. "No, this really isn't a serious competitor to our product... No, I don't know why Phoenix would do this... This just shows that our market has lots of potential." I don't know whether the sales reps were convinced. I certainly wasn't.

One by one, the reps eased away as quietly and somberly as they had arrived, leaving Steve, Maureen, and me in the booth. For the next several hours, Steve and I assessed the situation. We agreed that Phoenix's entry into our business was a big problem. They were larger and more successful, albeit in a different market niche. They had more experience with modems and modem manufacturers. In fact, the founders of Phoenix had *come* from a modem manufacturer. TAS and Phoenix had many of the same sales reps and distributors. If Phoenix played hardball, they could force some of our reps to drop our products, because most made more money selling Phoenix products than ours. The biggest problem was that Phoenix's simulator was priced at one-third the cost of ours.

Since those crazy days in the basement of my home we had harbored one "big fear." What if a "real" company decided to compete with us? Not some rinky-dink outfit like AEA, but a bona fide player such as Hewlett-Packard or Tektronix. We feared we could quickly lose our business and our livelihood. Phoenix wasn't a major player, but they were a lot bigger and more successful than AEA. They were large enough to trigger the "big fear."

After wallowing in that fear for a while, Steve and I began to claw our way out. Phoenix Microsystems was not Hewlett-Packard. Phoenix didn't have unlimited resources, and couldn't possibly focus on our market as intensely as we could. Modem test systems were our only business, and we had a big head start. Phoenix was simply copying our product in an attempt to latch onto our success. We had some big advantages: Our product was more complete and more sophisticated; Steve, Charles, and I were highly capable engineers totally focused on one product line. The only way Phoenix could catch us was if we slowed down, and we weren't about to do that. By the end of the afternoon, Steve and I started to feel a little better, and we made two key decisions: we would develop our own data analyzer and end our dependence on Phoenix, and we would work like dogs to prevent those Phoenix yokels from beating us at our own game.

The end of the week finally arrived, and with it the end of Interface '86. The tradeshow my reps had billed as glamorous, exciting, and profitable turned out to be the most agonizing week in my business life. We didn't get one dollar of new business. Instead, we exposed ourselves to our main competitor and got double-crossed by a key supplier.

Before Interface, I saw TAS as a graceful schooner sailing glassy waters, a strong and predictable wind in its sails. After the tradeshow, TAS seemed like a fragile dinghy riding strong currents toward a troubling storm. The before/after difference was partly mere perspective, partly reality. Yes, our sales had increased eightfold in just one year, but we were still very small potatoes in a very large stew. Yes, we were learning to outfox our key competitor, AEA, at some major accounts, but we were about to face a larger, better-funded competitor—Phoenix. I knew I couldn't bury my head in the sand and ignore the troubling new perspective I'd gained at Interface, but I couldn't quell the overriding feeling that I never wanted to attend another tradeshow.

On Friday, we tore down our exhibit and loaded the Winnebago for the long drive back to New Jersey. Schooner or dinghy, rough seas or calm, TAS was moving forward, and we were determined to reach our destination.

40. Nuts, Misfits, and Rejects

THE TAS WORK ENVIRONMENT was, to put it mildly, intense. Steve, Charles, and I immersed ourselves in designing and adding features to our products as we simultaneously managed all other aspects of the business. We lived in constant fear of being overtaken by a larger, deep-pocketed competitor, so we worked long hours to stay ahead. We encouraged our few employees to work hard too, and to focus on business objectives rather than the number of hours they clocked. We wanted them to feel they had a stake in TAS, and we assured them they would reap their fair share of the rewards if the company succeeded.

Sometimes our fervor didn't sit well with the employees. Where Steve, Charles, and I saw an almost sacred mission, our employees merely saw a job. They didn't understand why we wanted them to work extraordinarily hard, and we didn't understand why they wouldn't. One April morning, barely two weeks after we returned, wounded and discouraged, from Interface '86, that untenable situation came to a head.

I arrived at the office and found two envelopes on my chair, both addressed simply to "David." I opened the top one and discovered it was from my manufacturing manager, Tom. Tom had only recently received his Industrial Engineering degree, and TAS was his first full-time job. I liked Tom and thought his academic preparation and youthful energy gave him great potential. In a matter of months, though, it became obvious that Tom wasn't working hard enough to master his duties. He was young and carefree, still a college kid at heart. As I read Tom's message, I was shocked to find it was a resignation letter. He had some unpleasant things to say about the company and about me. I was surprised because Tom had never before said any of those things to me.

"TAS is run like a dictatorship. I fail to see why we ever have discussions on any 'open' issues, because whatever David says is what goes. This puts David in a very comfortable position, but it puts everyone else in a very uncomfortable position... Without happy employees TAS will go nowhere, and I do not know of any happy employees... I'm sure other people will be following my lead."

More than a bit stunned, I opened the second letter. This one was from Maureen, our very first employee, and she too was resigning. Maureen's letter was much more diplomatic, but reading between the lines, I could tell she was unhappy.

"As of Friday, April 4, I am handing in my one-week notice... I have accepted a position with another company, which I feel better allows me to reach the career goals I am seeking... I would like to thank you for hiring me in June 1984 and for all the learning opportunities I had at TAS."

The two letters clearly indicated that Tom and Maureen had coordinated their resignations, and their abrupt exodus signaled their disdain for the company. I felt kicked in the stomach. Tom and Maureen had worked long hours on our Interface '86 excursion, but it hadn't occurred to me that they resented the assignment. I had no answers for their letters, only questions. I knew we were all working hard, but were things really that bad? Why hadn't Tom and Maureen expressed their grievances to my face? Why did Tom have such uncomplimentary things to say?

I looked outside my office window and saw that Tom's car was gone. Apparently he had dropped off his resignation letter and left. I sat at my desk in paralyzed silence for several minutes, and then walked back to the lab to give Steve and Charles the news. I caught a glimpse of Maureen at her desk, but I was in no mood to talk to her. I asked my cofounders to join me in the conference room, and showed them the letters. The coordinated exodus surprised them as it had me, but I wanted to know what they thought it all meant. We were losing two of our four employees, and Tom was implying we would lose more. I was afraid that if Steve and Charles blamed me, our company would fall apart.

Steve spoke up first. "Tom is a bum. He's just a big kid, and he wasn't able to do the work. He was the ringleader, and Maureen is so young and impressionable that she's following him out the door. This is stupid!"

I was relieved. Steve and I saw eye to eye. Charles was more conciliatory.

"Maybe I should have a talk with Tom to see what his problem is. We have a lot of work to do, and we can't afford to lose him."

Besides his engineering duties, Charles oversaw our manufacturing operations. He was worried that Tom's departure would increase his own heavy workload.

Steve responded in his usual direct way: "Charles, we're doing all the work anyway. It's taking us more time to train that bum than to just do the work ourselves. I say if he and Maureen want to leave, good riddance!"

I felt the same way, but I knew we couldn't grow the company if we couldn't keep good people. If Steve and Charles and I continued to do all the jobs there would be no one to design new products, and our business would grind to a halt. I didn't want to try to keep Tom and Maureen, but we needed to hire replacements who would stick.

"Well, if you want to talk with Tom, that's fine, Charles," I said. "My gut feeling is that he's gone and we need to hire new people. We're making money, and we'll be making even more with those two gone. We need to take this opportunity to go out and find better people—people who can do the job and aren't afraid to work hard."

I wasn't in the mood for a long discussion. What I said sounded like a decree, but I felt too wounded to put it any other way. Our company was growing—succeeding—and I wasn't about to let two entry-level employees spoil it. We had to move on.

We replaced Tom and Maureen, but that certainly didn't spell the end of our personnel problems.

LaVerne, our new office manager/bookkeeper, provided the next surprise. LaVerne was a middle-aged woman who had responded to our HELP WANTED ad in the *Asbury Park Press*. Her previous job was office manager for a bakery that specialized in cheese blintzes. LaVerne was overweight but energetic, with solid experience and a no-nonsense attitude. I had high hopes she would fit right in.

LaVerne did whatever office work we needed, from answering phones to bookkeeping. She interviewed and hired an office assistant to replace Maureen. Whenever LaVerne thought we might lose money, such as when a supplier didn't give us an appropriate discount or a customer didn't pay on time, she pointed it out right away. She seemed to treat the company's money as she would her own. After a few months, though, LaVerne changed. She started showing up at work late or not at all. Her bookkeeping got sloppy.

One morning LaVerne was absent and the receptionist brought the day's stack of incoming mail straight to me. One item was an American Express credit card statement that included something called a "Management Report." I had arranged Amex cards for Steve, Charles, and me so we could more easily handle business-related expenses. The account statements always went to LaVerne, so I had never seen one. The management report included a list of the previous month's charges, and I spotted some unusual ones. One was from a place called Fashion Bug. I assumed the charge was a mistake, but then I saw a few others I didn't recognize. I flipped to the next page and saw photocopied receipts for the questionable charges, and there, to my surprise, was my name. Someone had forged my signature!

Steve was in an adjacent office, so I called out to him, "Hey, Steve, come look at this!" I pointed out the bogus charges. Steve fished his own Amex statement from his inbox and was amazed to find that his report also contained bogus charges. He flipped to the photocopied receipts and found someone had forged his signature, too. LaVerne's job description made her the logical culprit, but we couldn't be sure. We figured she had been intercepting the management reports. If LaVerne hadn't been out of the office that day, we might never have discovered the fraud.

When LaVerne returned the next day, Steve and I summoned her to the conference room and confronted her. At first she denied any knowledge of the forgeries, but when we showed her the reports and the charge receipts she started sobbing and blubbering uncontrollably. She said, "I love you guys. I would never do anything to hurt you. I just needed to get a few things. I was going to pay you back, I swear!"

We considered pressing charges, but in the end we just accepted LaVerne's resignation. We never saw or heard from her again.

That was still not the end of our personnel woes. We replaced the departed Tom with Bob, a fifty-something guy who had a wealth of manufacturing experience. Like LaVerne, Bob responded to a newspaper ad and performed very well in his interview. Bob was a down-to-earth Brooklyn native who seemed ready to roll up his sleeves and give hands-on attention to our manufacturing operations. Steve, Charles, and I agreed that with Bob in charge of manufacturing, we wouldn't miss Tom one bit.

Bob was a solid performer. He immediately took over assembly of our products and helped plan and execute our parts orders. He recruited and hired a young woman named Alice to help with assembly and shipping and receiving. That hire, however, proved to be Bob's undoing.

Alice was ex-military, and like Bob, she was a no-nonsense, hardworking employee. She didn't have much in the way of civilian skills or experience, but she was a willing worker. Late one afternoon I was alerted to an incident involving Alice and Bob. Alice claimed Bob had cornered her on the receiving dock and placed his hands on her breasts.

That was very bad news. Bob was a capable employee, and we badly needed his work and expertise, but I knew that if Alice's allegations were true, Bob would have to go. We couldn't have a sexual predator supervising our manufacturing operations. I summoned Bob. To my amazement, he didn't deny the incident. He simply tried to pass it off as innocent horseplay. When I told him Alice didn't see it that way, and that it was unacceptable in any case, an evil smirk creased his face, and he clammed up. That was the end of Bob's tenure at TAS.

Our personnel difficulties were a major distraction but did not throw us off course. Steve, Charles, and I did all the product designs, and we just kept churning them out. If an employee left, one of us stepped in and filled the role until we found a replacement. Soon though, there was too much design work for the three of us, and we needed to hire some engineers. If we thought our personnel problems were going to be confined to non-engineer employees, we were sorely mistaken.

We recruited engineers in two ways: newspaper ads and head-hunter agencies. We sought engineers who were like us: top graduates from leading universities who had product development experience. Unfortunately, we soon found that engineers like us were not seeking a company like ours. Engineers like us wanted to do what we had done—work for a prestigious, stable outfit like Bell Labs. That became evident very quickly when the résumés came in. We saw hardly any blue-chip candidates. Most had dubious education backgrounds, or were working for a small company we hadn't heard of (perhaps a company like ours), or had a telltale job history that revealed brief stays at a series of companies.

Occasionally we got lucky and spotted a solid candidate. One was a young Asian engineer named Jenny, who had recently graduated from a top-tier engineering school. She had solid grades, and in her employment interview she came across as very knowledgeable. We hired Jenny as our first full-time software development engineer. I was excited because up to that point *I'd* been the software development engineer. Jenny's hiring would give me much more time to run the company.

After just a few days, Jenny had a good understanding of our software, and I was confident she would be working independently in a few weeks. Jenny had the potential to be a great hire. With her position filled, I focused on recruiting other key people. Our efforts to properly staff the company were gaining traction.

Two weeks after Jenny started, I stopped by her office to review her progress. She seemed edgy and distracted, but I chalked that up to new-employee jitters. As I stood to leave, she handed me a letter. I opened it on the spot, and to my considerable surprise and chagrin found myself reading another resignation letter. Jenny was quitting a job she had started just two weeks earlier! Incredulous, I asked her why. She looked straight at me and calmly said she had accepted a job offer from Bell Communications Research (BellCore). She said they wanted her to start right away, so she would not be giving us the customary two weeks' notice—she would leave at the end of the week. She said two weeks' notice probably wouldn't help anyway, since she was just getting started at TAS and hadn't done any real work.

Angry and disappointed and hurt, I resisted the urge to strangle Jenny. I couldn't believe she would do this, and I resented her cavalier attitude. Jenny was heaping disrespect upon me and my company and everything we were trying to build. In effect, she was saying: I was willing to accept a job at your crappy little company as long as nothing else was available, but now that I have an offer from BellCore, a *real* company, how could you possibly expect me to work here?

I spent a few days reflecting on Jenny's departure, and realized she was in some ways like Steve and Charles and me. She was a top engineer from a solid school, and after her college career she wanted to work for a prestigious company, just as we had done. The big difference was that none of us had accepted a key position at a small company and then totally disrespected that company by resigning two weeks later. I understood Jenny's motives, but I thought what she did was completely unethical and inexcusable.

Jenny's sudden resignation cut deep. I wondered if TAS would ever be able to hire and keep good employees. I wondered if we would we be able to attract solid people, or if we would forever be consigned the nuts, misfits, and rejects of corporate America. I wondered, too, if we were having trouble hiring and retaining people because we, the company principals, were black. Tom, Maureen, LaVerne, and Bob all were white. Jenny was Asian American. Whatever the reasons for our personnel problems, I knew that if our track record in hiring and retaining employees didn't change, our company was doomed.

After each employee departure, I reviewed the situation with Steve and Charles. I dreaded those sessions because I was embarrassed. People seemed to be leaving left and right, and I thought that reflected badly on my management style and judgment. Every time I sat with Steve and Charles, the same questions arose: Why did the person leave? Was there something wrong with us, or was the employee the problem? Why would someone work for TAS if they could make more money at a larger, more established company? Why would a talented young white or Asian engineer work for a small, black-owned company? Was I being unreasonable when I asked people to work harder for us than they would elsewhere?

I was convinced we wouldn't succeed unless we had talented, highly motivated employees who worked extremely hard. We couldn't staff every conceivable role with a different person, so each TAS employee had to do the work of two or three people. Steve agreed, but he felt our personnel problems were merely symptoms of a larger issue: a general decline in the American work ethic. Steve was willing to work incredibly hard to achieve success, and he couldn't understand why others would not. Charles, on the other hand, didn't think people should necessarily be willing to work harder at TAS than elsewhere. He seemed to think there was something wrong with our little company, and that to be successful we would have to structure ourselves more like a "real" company.

I disagreed with Charles's point of view, and resented it, but I did accept that we couldn't just hire people at random, tell them what we needed from them, and expect it to happen. I decided we weren't being selective *enough* in our hiring, and that we didn't have a clear enough picture of the kind of person we wanted. I had a series of discussions with Steve and Charles to review our hiring and retention strategies. Our talks centered on one question: why were we ourselves working at TAS? I believed that if we could get in touch with our own motivations, we might begin to understand what would motivate someone else.

We spent hours hashing it out, but ultimately our own reasons for working at TAS came down to three things: we loved the work, we wanted greater freedom and impact than we could have at a large company like Bell Labs, and we wanted to make more money than we could at a large corporation. We decided to position ourselves, and sell ourselves, as exactly what we were: a small, close-knit team working to be the very best in a niche market. We wouldn't try to compete with the salaries and reputations of companies like Bell Labs and BellCore.

We would sell our unique attributes: a chance to get in on the ground floor of a great little company destined to do big things; a chance to design great products that people would actually use and appreciate; and a chance, through profit-sharing and stock options, to share the financial rewards that would follow success.

Our new personnel vision didn't come a moment too soon. Six of our first eighteen hires quit, but we just kept going. Like a basketball star who keeps missing but keeps shooting anyway, we kept hiring replacements. I even found three immigrant engineers working in my favorite Chinese restaurant, helped them secure U.S. work visas, and hired them. Our persistence kept the company growing. During our first year in Little Silver, sales more than doubled. The next year sales nearly doubled again, and by the end of 1987 our annual revenues exceeded $3 million. Once we enunciated a clear vision, some of our new hires started to stick. We managed to hire and retain a manufacturing supervisor, operations director, office manager, finance director, sales manager, software engineer, and more. Step by step, we built a company that attracted more than just nuts, misfits, and rejects. TAS was gradually becoming a company that talented people actually *wanted* to work for.

41. California Showdown

I HAD SEEN A SHOWDOWN COMING FOR WEEKS, and it was finally happening. Rockwell Semiconductor Products in Irvine, California, was a hugely important account, and we were doing zero business there. Our main competitor, AEA, had sold several systems to Rockwell before we arrived on the scene. Rockwell engineers were on the verge of purchasing many more modem test systems, and were about to choose the company that would supply them. The winner stood to get millions of dollars in new business. The loser would suffer a serious marketplace defeat.

I felt fortunate that the showdown was even happening. For months, we had been trying to break into Rockwell, to no avail. Rockwell designed and manufactured the integrated circuit "chipset" that comprised the brains of a modem. The company's chipset business was poised to explode because manufacturers were starting to imbed modems in all kinds of products, such as computers and point-of-sale terminals and even vending machines. If we secured Rockwell's business, we would sell modem test systems not only to Rockwell but also to its many customers.

Our sales rep in Southern California, Mondy Lariz, arranged the Rockwell showdown. I had hired Mondy after a series of disastrous California sales reps—people who talked a good game at first and then did nothing to improve our sales. Some of those rep firms seemed to think they were too big and important to give TAS much attention.

Some were, in the final analysis, simply incompetent. Mondy had his own small firm, and he seemed to be doing well. Most important, Mondy seemed like a good guy—a personable, technically astute person—and we got along well. I liked his earnest, direct way of speaking and the fact that he was young and hungry. I appreciated his Latino heritage because I thought it would help him relate to my situation—a black man trying to make it in an industry dominated by white men.

For several months leading up to the showdown, Mondy had tried to sell our products to the engineers at Rockwell. He had been rebuffed repeatedly, but refused to give up. After considerable investigation, Mondy learned that a couple of engineers in a key department had already purchased products from AEA, and that those engineers were blocking consideration of our product. The department head was a senior engineer named Frank Weldon, and his assistant was a young Asian, Mike Chen. According to Mondy, Frank wasn't interested in hearing about our product, and was using Mike to spread good propaganda about AEA and bad propaganda about TAS. Mondy didn't know the reason for Frank's refusal to consider our product. I had my suspicions.

When Mondy related the situation to me, I thought *here we go again*. From our first day in business, every new sale was an ordeal. Most potential customers we visited already owned at least one AEA box, and though nearly all had problems with it, they were reluctant to consider our product. That was true even at Bell Labs, where people knew me, and where we had Jim Ingle vouching for us. We were a new, small company facing an entrenched competitor. Our company founder/president/chief salesman was black. I didn't know how much each of those factors influenced potential customers, but they usually resulted in a protracted struggle.

I had accompanied Mondy on several Rockwell sales calls and had met most key engineering staff members. I remembered seeing Frank Weldon at one of my presentations and noted that he seemed dour and uninterested. He had a distant look in his eyes the entire time, and when I finished he didn't ask a single question. I remembered Mike Chen, too. The thing that struck me about Mike was that, even though he was a young guy, he didn't seem the least bit friendly or open to discussion. He was atypical, because most of the young Rockwell engineers I had encountered were outgoing and sociable. Not Mike. He seemed to be following Frank's lead in giving us the brush-off.

Our sales prospects at Rockwell looked pretty bleak, and after months of repeated rejection, I held out little hope of getting business there. After each unsuccessful sales visit to California, I would return to New Jersey with my tail between my legs, but Mondy just kept pounding away at the Rockwell guys. He spoke not only to Frank and Mike but also to any other engineer who would listen. Finally, one day, he called with good news.

"David," Mondy began. " I got us in!"

"You mean you got an order from Rockwell?" I knew Mondy had been working hard, but I couldn't believe he'd actually gotten Rockwell to buy something from us.

"Well, no, but they agreed to evaluate your product against the AEA and write up a comparison."

"I see," I said slowly. "Who's going to be doing the comparison?"

"Frank Weldon."

I was incredulous. "Frank Weldon! That guy doesn't like us, Mondy. He and Mike Chen are using AEA, and they barely want to talk to us when we're there. You expect them to do a fair evaluation? I don't."

"What do you want me to do, Dave? I got us a shot. If we're going to get into Rockwell, we have to get past Weldon. He's the guy responsible for evaluating test equipment."

Mondy sounded simultaneously hurt and pissed off, and I immediately understood why. He had put a lot of effort and time into forcing a technical shootout. Rockwell had been disinclined to do business with TAS, but Mondy kept insisting that we had a better product. He pretty much challenged, cajoled, and shamed the Rockwell engineers into comparing our product with AEA. Mondy was right: this was our best shot—perhaps our *only* shot—at getting into Rockwell. I had no choice but to go along and make the best of it. If I didn't, I might not only lose Rockwell—I might lose Mondy, too.

"Okay, Mondy," I said reluctantly. You're right, but we have to do something to make sure it's not all left up to Frank and Mike. If they make the final decision, we're screwed." Though I didn't say it to Mondy, I thought we were screwed anyway.

Mondy was silent for a few seconds, and then said, "Okay, Dave. I'll make sure Frank and Mike give us the written results of their evaluation, and I'll try to arrange a meeting for them to present their results to all of engineering."

"That's a good idea," I said. I tried my best to sound encouraging, and I did think there was a small chance that Mondy's approach might work. I was convinced that if Frank and Mike did an honest evaluation, and if they didn't pigeonhole the results, we would win. I just didn't trust them to be honest. I knew we had a better product than AEA—we had proved that on numerous occasions. The only way we would lose to AEA was if Frank and Mike were to distort or bury the results of their evaluation.

True to his word, Mondy secured the promise of a written evaluation from Frank Weldon, and to my surprise, he even arranged a meeting with the Rockwell engineering staff to review the results. Just knowing about the impending review meeting made me more confident and determined. I couldn't wait to hear what Frank Weldon and Mike Chen would say.

Frank and Mike spent a few weeks performing their evaluation, and then the day of the big review finally arrived; the day we would learn the results of the shootout; the day we would find out who would get Rockwell's future business.

I was seated on one side of the large table in the engineering director's conference room. There was nothing especially fancy about the room. The hard, tightly woven nylon carpeting; the long table with the smooth, synthetic finish; the chairs with their padded bottoms and spring-loaded backrests; the large whiteboard on the wall at one end of the room; all of that was standard high-tech company stuff. The conference room was at the front of the building, and the Southern California sun was streaming in through translucent curtains. The building's air-conditioning system kept the room cool—a little too cool for me.

I looked around the table at the Rockwell engineers. They represented several departments: research and development, system test, field support, and manufacturing. All of the attendees were white men except for one Asian, Mike Chen. Mike's boss, Frank Weldon, was seated next to him. I stared at Mike and then at Frank. Like jurors who were about to render a guilty verdict, both refused to make eye contact with me. Their unsmiling faces betrayed unease, as though they didn't want to be there. *This isn't going to go well,* I thought.

The agenda for the meeting was straightforward: Mike Chen was to present his report on the comparison between our product and AEA's and then open the floor for questions and discussion. At the end of the meeting, the Rockwell engineers would decide which product was best suited to their needs. The stakes couldn't have been higher.

The last person to enter the room was Bill Benson, Director of Engineering. Mondy had introduced me to Benson once as we passed in the hallway, but I hadn't expected him to be at the meeting. I glanced over at Mondy, who had positioned himself near Frank and Mike. Mondy lowered his chin and gave me a look as if to say, *see, I told you I would set this thing up right!*

Benson took his seat, and it was time for the meeting to begin. Mike Chen passed copies of his evaluation report around the table and then got up to make his presentation on the overhead projector. I glanced at Mondy again—show time!

I wasn't even listening to Mike as he began his presentation. I was busy flipping through the report to find his conclusions. On the last page, I found a table that summarized his test results:

	AEA	**TAS**	**Comment**
Phase Jitter Level Accuracy	+/– 5%	+/– 0.5%	AEA is more accurate
Phase Jitter Frequency Accuracy	+/– 10%	+/–0.1%	AEA is better
Nonlinear Distortion Accuracy	+/– 1 dB	+/– 1 dB	Both satisfactory
Amplitude Modulation Frequency Accuracy	+/– 10%	+/– 0.1%	AEA is better

My eyes were immediately drawn to the rightmost column of the table—the one headed "Comment." Each entry in that column indicated that the AEA product was equal to or better than ours. I could hardly contain my outrage. Had Mike falsified the evaluation results? I knew for a fact that we had a more precise, more accurate product than AEA. I wondered how Mike could come up with a different conclusion, unless he'd cheated.

Then I looked at the other columns to see the numbers Mike had based his comments on, and I saw the most bizarre thing I had seen in all my years in the engineering profession. Mike's numbers, the results of his measurements, clearly indicated that our product was better. Mike had not falsified his measurements; he had falsified the conclusions. Somewhere between his objective measurements and his subjective comments, Mike had introduced a blatant bias in favor of AEA. When I saw that bias presented in black and white, I was astounded. I couldn't wait for the chance to pounce and expose the non sequitur in Mike's sham evaluation. As it turned out, I didn't have to.

Mike finished his presentation and then opened the floor for questions. For someone who had just uttered a series of blatant lies, he was astonishingly cool. Frank Weldon was cool, too. I was about to expose them both when one of the engineering department heads, Bob Simpson, spoke up.

"On your Phase Jitter result, you state that AEA is better, even though it looks like the TAS product is ten times more accurate. How do you conclude that AEA is better?"

Mike glanced briefly at Frank, and then said, "In the real world, phase jitter is not so accurate; it's all over the map. That's why we concluded that the AEA is better—it's more like real-world conditions."

Bob Simpson's face scrunched in amazement. "That's hogwash," he said. When I'm doing my tests, I need my system to be accurate so I can get repeatable results."

Mike Chen just shrugged and called on the next person. One by one, led by Bob Simpson's bold example, the engineers in the room picked apart the report produced by Chen and Weldon. By the end of the meeting Mike and Frank had not only discredited themselves, but had also confirmed the superiority of the TAS product. They had shown themselves to be able technicians, because their measurements were correct, but dishonest judges, because their conclusions didn't agree with their observations. Mondy and I didn't have to say a word!

Mike and Frank were shamed by their colleagues, and they deserved it. They had violated one of the most solemn principles in the engineering profession—that one must always be honest and objective about technical facts. They presented conclusions that were directly opposite the facts in order to satisfy their bias in favor of AEA, and they'd been exposed.

In the car on the way back to Mondy's office, he and I discussed what had just transpired. Why would Frank and Mike so blatantly slant their conclusions toward AEA? Their behavior was irrational because it went against their own and their company's best interests. Were they motivated by racism, or were they simply more comfortable and familiar with the AEA product? Mondy said that in all his years as a manufacturer's rep he had never seen anything like it.

I smiled. Mondy hadn't expected Frank and Mike to be biased against us, but I had. For me, their behavior wasn't an aberration. I had encountered people like them throughout my career, and the result was usually bad. The difference this time was that we were able to expose their bias to a wider audience that included fair, open-minded people.

My smile grew into an ear-to-ear grin. I had never expected such a clear victory at Rockwell, such a total repudiation of two very unfair people. Mondy and I were silent for the remainder of the trip to his office. It was a beautiful, sunny Southern California afternoon, the end of a perfect day, and there was nothing more to say. We rolled the windows down and let the warm breeze dance over us, and I started singing a song by The Persuasions, which my friend Reggie Barnett had introduced me to when we were in high school:

> *Buffalo Soldier, Buffalo Soldier*
>
> *Will you survive, in this new land?*
>
> *Buffalo Soldier*
>
> *Tell me when will they call you a man?*

42. International

THOUGH OUR APPEARANCE AT THE INTERFACE '86 tradeshow in Atlanta seemed disastrous at the time, it yielded some surprising benefits. The steely-eyed Hans, who'd strolled up to our booth looking like an East German spy, turned out to be a nice guy. Though he was the U.S. distributor for our competitor AEA, he later provided useful market intelligence. Moreover, the two associates who accompanied him at the show turned out to be good guys, too. They were Gianni Migliorini and Giordano Bruni, principals of a successful Italian distribution company who later expressed interest in carrying our products. Those small victories caused me to violate my pledge to never attend another tradeshow, and in early 1987 we exhibited at a Washington, D.C., show called CommNet. That show introduced me to two more potential distributors—one from England and the other from France. In June of that year, those introductions led to my first trip to Europe.

I boarded a plane at Kennedy Airport for the overnight flight to London, planning to first visit the U.K. distributor, Ian Farr, and then hop over to France. When I'd first met Ian at CommNet he struck me as abrasive, condescending, and goofy. He looked and acted more like a comedian than a businessman. I learned, however, that Ian was an effective salesman with a small but rapidly growing company, Phoenix Datacom Limited. Ian's company wasn't affiliated with Phoenix Microsystems, the company that double-crossed us in Atlanta, but he did distribute their products. He promised not to sell their telephone network simulator, though, so I signed him up.

I was full of confidence as I crossed the Atlantic. TAS's good reputation was spreading, customers were starting to seek us out, and our business was exploding. In 1986, TAS had blown past the million-dollar milestone on the way to sales of $1.7 million. Besides England, France, and Italy, two distributors in Japan had expressed interest in our products. It seemed like the right time to "go international." One reason was simply to gain access to new markets; another was that AEA already had a foothold in several countries. I feared that leaving the rest of the world to them would put us at a serious disadvantage.

The morning after leaving New York, my flight descended through thick fog and landed at Heathrow Airport. Despite the approach of summer, the skies were overcast and the outside air was brisk. Walking through the arrival terminal, I was struck by the cold, gray airport hallways. After I retrieved my bags and proceeded to customs, I noted that all of the agents were of Indian descent, and none displayed a trace of friendliness. It seemed I had landed in a dank, miserable, unwelcoming country.

I felt both hope and anxiety as I scanned the arrival concourse looking for Ian. I hoped my trip would tap a new source of growth for TAS, but I was anxious because expanding into Europe would require an awful lot of work. I was already spread thin just trying to keep our U.S. business growing.

After a few moments, I spotted Ian's round, red face bobbing toward me.

"Hey, David!" I was surprised by his hearty greeting—he seemed genuinely happy to see me. "So you decided to visit our little country. That's mighty big of you."

Ian was being his playful, comedic self, so I went along.

"Yes, Ian, I had lots of time to spare, so I decided to visit England for a quick holiday."

Ian grabbed one of my bags and ushered me to a sleek new Jaguar sedan. He looked proud and aristocratic as we drove from the airport. "It's a beautiful car, and it purrs like a kitten, but it's powerful, you see?" He gave the accelerator a pump.

I felt like royalty cruising the outskirts of London in Ian's new

Jag. I had no idea where we were, or where Ian's town of Aylesbury was situated relative to London. I only knew we were headed to Ian's office to start doing business, and I was ready. After an hour or so of Ian's small talk and my drifting between sleep and awake, we arrived.

Aylesbury was one of those towns that managed to look gray even on a sunny day, and absolutely dismal on a cloudy day. We arrived on a cloudy day. Ian took me first to my hotel, the Bell Inn, to drop my things and get breakfast. The inn looked as if it dated to medieval times. Some might have thought it quaint, but to me it was old and spooky.

After breakfast we made the short trip to Ian's office. The first order of business was a meeting with Ian and his sales staff. I followed him as he strutted into his office, sat in his plush leather desk chair, and offered me a seat on the sofa. After thirty minutes of talking to his secretary, checking messages, and returning calls, Ian escorted me into a decent-sized conference room. I was surprised to see the TAS equipment already prepared for demonstration, and two of Ian's staff members waiting.

"David, you remember John Nice, don't you?" Ian asked.

I turned to John Nice [pronounced "Neece"] and shook his hand. John was in his late thirties or early forties, medium height, slightly pudgy, with thinning black hair. His goatee gave him a mischievous appearance.

"Yes, of course. Good to see you again, John."

It *was* good to see John. I remembered the enthusiasm he had shown for our equipment when we first met. John had impressed me then as that rare salesperson who could understand our equipment's technical details *and* effectively communicate its benefits. I looked forward to seeing whether my first impressions were accurate.

"John has been working with your kit quite a bit," Ian said. "He has lots of questions for you."

"Good," I said. "That's why I'm here."

Ian turned to his other staff member. "Now this fellow is Mr. David Crow. David is one of our top salesmen, after me-self of course. He's going to be selling your kit."

David Crow was young, striking, confident. I was even more impressed when Ian told me David had been a member of the Royal Air Force.

We got right down to business. John Nice started with a practice presentation of our equipment, during which he peppered me with questions. I was amazed at how proficient John had become after working with our equipment for just a few months. His presentation was every bit as good as my own. In fact, in some ways it was better.

After John finished, David Crow asked me several questions about our sales strategy. He asked good questions, and I could see that he was more focused on making money than John was. John was the professor; David was the salesman.

By the end of the meeting, I was champing at the bit to see some actual customers. Ian had arranged visits to two key modem manufacturers. We piled the equipment into the trunk of Ian's Jag and headed to our first appointment. John Nice, Ian, and I made the trip. David Crow had scheduled other sales calls.

We drove for about an hour to the site of our first potential customer—one of England's largest modem manufacturers. After signing in, we were shown to a large conference room. A few minutes later, Ian's contact entered. He had dark skin and short, close-cropped hair. He was clearly of African descent, but when he opened his mouth, he sounded every bit like a dyed-in-the-wool Englishman. He greeted Ian and John with hearty handshakes, and then Ian introduced me.

I was surprised and happy to meet a fellow black engineer in England, but if our host was especially happy to meet me, he didn't show it. He was cordial, but offered no knowing smile or secret handshake.

When the pleasantries ended, John began his demonstration. After a few minutes, he and the customer engaged in a lively exchange, and Ian threw in an occasional quip. Each man seemed to speak a different English dialect. They understood each other, but I couldn't follow the discussion. I sat back and marveled at the scene.

After perhaps twenty minutes of banter, Ian turned to me and said, "Well, I guess we'll be off now." We said our good-byes and made our way, with equipment, back to Ian's car.

Once under way, Ian said, "I think that went rather well, don't you, John?"

"Smashing," crowed John from the back seat.

"What did you think, Dave?" Ian asked.

I hadn't understood enough of what was discussed to offer an opinion. The customer was clearly interested in our equipment, and John seemed to address his questions thoroughly. I was still marveling at what I had just seen. For the first time, I had witnessed a demonstration of our equipment in another country, carried out by someone other than me, in a language (or at least a dialect) I barely understood. I had an odd feeling of pride. My company had reached another milestone.

"Ian, I have to say—that was the first time I've ever encountered an English-speaking black person I couldn't understand." Ian and I both laughed.

We reached Aylesbury near the end of the workday. Ian took me to his home and introduced me to his wife and kids, then dropped me down the road at the Bell Inn. Exhausted, I flopped onto the bed and immediately fell asleep.

I woke up a few hours later. The room was cold and dark, and for a moment I didn't remember where I was. I sat up and looked around. I had never felt so far from home, so detached from the world. I felt the urge to call home. I looked at my watch—only 9:00 p.m. I looked around for a phone but couldn't find one.

I needed to get out of there. I bolted from my room and exited the Bell Inn. I walked in the direction of a red phone booth I remembered seeing down the road. I heard a car roaring toward me from behind and jumped out of its path just before it sped past. I had forgotten that the English drove on the "wrong" side and was nearly run down. Heart racing, I ran to the booth and pulled the door closed. Try as I might, I couldn't figure out how to place a call to the U.S.

Still shaking from my near-death experience, I walked farther down the road to Ian Farr's house. Once there, I peered into a window next to the front door. To my surprise, Ian's wife was on the other side peering out. She looked startled and hurried away, and a minute later Ian appeared at the door.

"Come in, Dave. What brings you out so late? You nearly scared me dear wife to death!"

"Oh, sorry, Ian. I wasn't sure this was your house. I went to the phone booth to call home just now and was nearly run over. I just can't get used to you guys driving on the wrong side of the road."

Ian laughed. "Well, you nearly learned the hard way. Anyway, come in and use the phone here. It's quite simple, really."

After the call, I suddenly felt exhausted again. I thanked Ian for his hospitality, apologized to his wife, and walked back to the Bell Inn. I turned on the TV and crawled into bed. I was asleep soon after my head hit the pillow.

A few hours later, I awoke again. It was the middle of the night, and an old horror movie was on TV. Jet lag had me in the right place at the wrong time. I sat up in my spooky room and waited for morning.

Next morning I grabbed continental breakfast before Ian arrived. The day's agenda consisted of another customer visit followed by a meeting with Ian to iron out our distribution agreement.

The customer meeting went well. John Nice presented our product to several engineers while Ian and I watched. John's presentation was outstanding, and the customers seemed quite interested. This time I was able to follow the discussion more easily, though the result was the same—we packed up and left without knowing the customer's intent.

By the time we got back to Ian's office, I was ready to finish my business and leave miserable little Aylesbury. It was time to talk turkey. Ian and I haggled over several issues: his prices for our products (too high); whether he would purchase his own demonstration equipment (he didn't want to); his advertising budget (too low). By the end of the discussion, I wasn't sure Ian Farr was the right distributor for TAS.

On the other hand, Ian was doing a great job for his other suppliers, and he had promised not to carry any product that competed with ours. I was convinced John Nice and David Crow would do an excellent job presenting and supporting our products. We badly needed a U.K. distributor, and Ian was my only live prospect. I decided, without enthusiasm, to go with him.

The next morning, I was on a plane to Paris for a meeting with the other potential distributor, MB Electronique.

I was even more apprehensive about visiting France. For one thing, it was my first visit to a country where the dominant language was not English. For another, I had often heard that the French didn't particularly care for Americans, and could be real snobs. Ian didn't dampen my concerns. The last thing he said as he dropped me off at Heathrow was, "Mind the Frogs!" That was his humorous way of telling me to beware of the French.

As the plane approached Charles De Gaulle Airport, my apprehension gradually turned to hope and excitement. Unfortunately, the excitement didn't last long. At customs, an agent looked over my passport and asked for my visa. I didn't have one. The agent summoned a supervisor, and I was ushered into a small office. The supervisor said that without a visa I would have to return immediately to the U.K. After thirty minutes of explaining and begging, and after the supervisor spoke by phone with my MB contact, Guy Sereys, to verify my reason for visiting France, I was granted a twenty-four-hour visa. I had exactly one day to finish my business and get out.

One of Guy's subordinates, Jean-Michel Hure, met me outside customs. Jean-Michel was tall and lean, and was wearing a nice business suit, white shirt, and tie. He gave me a cordial but formal greeting, and then apologized for my difficulties. "But you know, you must have a visa for entering France," he said.

I wanted to say, *no shit!* Instead, I just said, "Yes, I know that now."

We boarded Jean-Michel's plebeian Renault and drove to the MB Electronique offices. It was a jolting switch from Ian Farr's Jaguar. Jean-Michel chain-smoked and drove intently. We shared occasional bits of conversation.

Guy Sereys greeted me at MB Electronique as if I were an old friend. Guy was fortyish, medium height and build, and sharply dressed, verging on elegant. He looked and acted more like a movie actor or politician than a sales manager.

Guy took me on a brief but impressive tour. MB was much larger than Ian Farr's company, and served as French distributor for some big-name U.S. electronics manufacturers. The tour erased the sour taste of my arrival in France. I began to feel optimistic.

Guy, Jean-Michel, and I spent the afternoon discussing a possible business relationship. The conversation centered on the usual issues, but one loomed largest: MB sold AEA products, and Guy seemed reluctant to drop them. That was a big problem. I was worried that TAS might not be very important to MB Electronique.

After our business meeting, Guy and Jean-Michel treated me to an extravagant dinner in Paris. I found the wine extraordinary, but was unfamiliar with most of the food. As each dish arrived, I asked Guy to identify it. My requests seemed to annoy him. When I asked Guy about one particularly unusual dish, he wiggled a finger beside each of his temples and said, "escargot!" I picked around the little mollusks before pushing them aside.

Despite my culinary naïveté, I enjoyed socializing with Guy and Jean-Michel. We talked about our families, music, politics, and how we'd arrived at our respective positions. Amid laughter and good food and wine, I warmed to my French colleagues. France didn't seem so bad after all. At one point, I felt as if I needed to pinch myself. I was having a business dinner in the heart of Paris, representing the company I had created on the strength of my own imagination and ingenuity. I felt happy and proud.

Driving through Paris after dinner, the sights were dazzling. I gawked at the glittering, magnificent Eiffel Tower, but I read and mispronounced the name of all the other sites we passed. As we turned onto a wide boulevard, a huge marble arch loomed ahead. I looked to my right and saw a sign that read CHAMPS ELYSÉES. I turned to Guy and asked, "What does that sign mean? Does it have something to do with boxing?"

Guy emitted a clearly audible sigh and said, "Shounz-ahlee-zay! Shounz-ahlee-zay! This is the boulevard we are traveling on. And in front of us is the *Arc de Triomphe*."

To Guy and Jean-Michel, I must have seemed like *un vrai plouc* (a real hick). I know I felt like one.

After dropping Jean-Michel at the office, Guy proceeded to the hotel he had reserved for me in a little town outside Paris. It was well past midnight. The lobby was empty but for a young male clerk behind a rustic check-in desk. Guy and the clerk conversed in French that grew progressively louder and more agitated. After several minutes Guy turned to me and said, "He says they have no more rooms."

I was incredulous. I was extremely tired. I just wanted to shower and crawl into bed.

"Didn't you guarantee the room for late arrival?" I asked.

"Yes, of course I did do that, but the clerk says that when it became so late, he had to give the room to someone else!"

"Well, let's just go to another hotel," I said.

"There is no other hotel nearby," Guy said.

"Then I guess the clerk will just have to find me a room. I do have a reservation, don't I?"

Guy turned back to the desk clerk and they resumed their heated conversation. I didn't know what they were saying, and I didn't care. I just wanted my room.

I saw a phone booth across the lobby, and while Guy and the clerk argued, I took advantage of the five-hour time difference to check on Diane and Stacy and to see how things were going at the office. When I rejoined Guy and the clerk they were still going at it. Finally, Guy turned and said, "The clerk says you may stay in his room."

"Fine," I said. I assumed the problem was solved, and that the clerk would show me to his room at the hotel. I thought it might even be nicer than the standard guest room.

The clerk came from behind the desk and walked out the front door. Guy followed him, and I followed Guy. Then Guy got in his car, and I got in beside him. The clerk sat in the back seat. I was puzzled but said nothing.

We drove a few minutes in the chilly night air until we reached a high-rise brick building that looked like an apartment complex. It certainly didn't look like a hotel. We went inside and climbed a few flights of stairs, then walked down a dimly lit hallway to a nondescript door. The clerk used his key, and we all stepped inside.

I was immediately hit by a sickening smell that seemed like a mixture of extreme body odor and rotting food. Black-light posters hung on the wall, and soiled clothes were spread across the floor. I surveyed the scene for less than a minute, then went downstairs and waited beside Guy's car. Guy and the clerk arrived a minute later, and I said to Guy, "I will sleep in the hotel lobby or I'll sleep in the street, but I will not sleep in there."

"Okay," Guy said. He looked at me as if I were a petulant child. "Let's go, then."

Guy drove us back to the hotel. He and the clerk exchanged a few brief words in French before the clerk left the car. Guy then pulled away from the hotel, and I wondered what our next stop would be.

A few minutes later we arrived at a nice-looking apartment building, much smaller than the clerk's and in a much nicer neighborhood. "Here we are," Guy said. "This is my home."

Guy and I collected my bags and walked up a few stairs to his apartment. Inside, he called out a greeting and a petite, attractive young woman with short blond hair came running. Guy introduced the woman by her first name—he didn't say whether she was his wife or just a friend—and told her I would be spending the night. The woman prepared the living-room pullout sofa and I bedded down. It was nearly 2:00 a.m. Angry, confused, and tired, I fell asleep immediately.

Guy woke me early and we had a simple breakfast of croissants and coffee. I didn't see his companion. Guy said nothing about the previous night, nor did I. I felt I had seen enough to know I didn't want MB Electronique to be my distributor in France. I didn't even know if I wanted to *have* a distributor in France. I just wanted to get the hell out of there.

After breakfast, Guy took me to see a potential customer, but the visit was a total dud. The customer already owned an AEA simulator and didn't seem the least bit interested in ours.

Guy and I reached the airport with time to spare on my twenty-four-hour visa. I flew back to Heathrow and then boarded a plane for the U.S. On the return trip across the Atlantic I wondered, given the lack of sales in England and the disaster in France, if TAS would ever find success on the international stage. In truth, the seeds of that success had already been planted, a very long way from either Aylesbury or Paris.

43. WORLD CLASS

IN EARLY 1987, WHEN I FIRST MET IAN FARR AND GUY SEREYS at our CommNet show exhibit, the TAS customer roster included major players such as AT&T, Hayes, Rockwell, and Motorola. We couldn't relax, though, because we were fearful of the competition and short on staff. I worked around the clock, consumed by my sales and marketing and software development chores. One morning I was at my desk immersed in thought when the receptionist's voice jolted me back to reality.

"David, someone for you on line one. I can barely understand him. Sounds like his name is YAM-OTO or something. I think he's calling from Japan."

Slightly annoyed, I picked up the phone.

"Hello, this is David Tarver."

"My name Mineo Yamamoto, from Toyo, Japan." The voice was delicate, the English heavily accented and choppy, but I understood the caller quite well. "Is this sales director?"

"Yeah, uh, yes." I was president, marketing director, sales director, chief software developer, personnel director, and sometimes janitor. No need to get into all that, though. The caller wanted the sales director, so that's the hat I put on.

"I am interested to purchase one unit TAS 1010 for my customer in Japan. Can you ship such unit within two weeks?"

Wow! This guy wasn't wasting time, but getting an order like that out of the blue sounded too good to be true. I had never heard of "Yamma-moto" or "Toyo Japan" before, so I was skeptical.

"Yes, we can sell you a TAS 1010, but first I need more information." I spoke slowly to make sure the caller could understand, and because I wanted to fudge the fact that we might not be able to ship within two weeks. "Now, please tell me what company you are representing."

"My company Toyo Corporation Japan. We are distributor for electronic instrument product. I visiting customer right now, and customer wishes to purchase your product, but customer has no wish to purchase product directly from supplier. Customer wishes to purchase product from Toyo Corporation."

There was something disarming about this guy, and he seemed earnestly interested in our product. But what if this "customer" had an ulterior motive? What if this "Toyo" company was out to clone our bread-and-butter product? Could that be why this Yamamoto fellow would order our product without even seeing it? Was that why he wanted such quick delivery? I remembered how Japanese companies like Yamaha and Roland had taken over the music synthesizer business, rendering my first business idea stillborn. I was not about to let that happen to TAS. We were just starting to get serious traction in the marketplace. I was not about to help anyone copy our product, least of all the Japanese.

I cut to the heart of the matter. "Why does your customer want our product?" I asked.

"Customer is designing modem and fax product. Customer see your product at AT&T and decide to purchase."

The story was plausible, but I still thought it could be a ruse.

"Who is the customer?"

"Toyo's customer is Sharp Corporation. Sharp is very good Toyo customer. Sharp wishes to purchase your product from Toyo. Please confirm you can ship product in two weeks."

"Yes—okay, I will confirm," I replied. I was still suspicious, but didn't want to let my paranoia ruin a sale—at least not at that point.

"Please send your quote by fax immediately so I may present to customer. I must provide Toyo's quote to Sharp in the morning. You may fax quote to this number."

Mr. Yamamoto slowly recited his fax number, which I recorded on my desk blotter. Then he signed off, and I was left to contemplate our newest customer's motives.

Immediately after hanging up, I sought out Steve and Charles to see what they thought about shipping our product to Japan; whether they were equally concerned about someone copying our TAS 1010. We sat at our conference table. I related the high points of the call and my concerns, and then waited for their reactions.

Charles understood my concerns, but he also understood the importance of a potential new revenue source. He seemed as unsure as I was about how to proceed. Steve, on the other hand, was resolute.

"Go ahead and sell it to 'em, man. Anybody would have a hard time copying our product without schematics and source code. Besides, if they wanted to copy it, they could just buy it through a company in this country and we'd never even know about it. Go ahead and sell it—just don't give them any design information along with it."

Steve was right. If someone wanted to copy our product, they wouldn't have to come to us. We had shipped products to overseas trading companies with U.S. addresses, and in those cases we had no idea where the product ended up. It would be dumb to forgo a sale because of my paranoia. I had no intention of providing design information. We hadn't provided that to any customer, though some had requested it. Steve, Charles, and I all agreed: we would sell the product to Toyo, but under no circumstances would we give them design information.

I prepared a sales quote and faxed it to Mr. Yamamoto. The next day, I received a fax order for one TAS 1010 Channel Simulator from Toyo Corporation. It was our first order from Japan. A few weeks later, we shipped the product. As it left the back door of our little factory, I had my fingers crossed, hoping the sale would result in a new source of business for TAS rather than a new source of competition.

A few weeks later, Mr. Yamamoto called again. He seemed very upbeat.

"We have received your product TAS 1010 and delivered to Toyo's customer Sharp Corporation."

That was good news. I had worried that our product would be damaged in transit to Japan, but apparently everything was all right.

"That's good news, Mr. Yamamoto," I said. "I hope we can do much more business in the future."

"Yes, Tarver-san," Yamamoto replied. "You will soon receive a call from Mr. Bill Berkman. He is Toyo's man in U.S. He will discuss with you Toyo's possibility for future business with TAS products."

Toyo's man in U.S.? That sounded mildly mysterious. What could a U.S. rep possibly have to discuss that Mr. Yamamoto couldn't say for himself? And what was the "possibility for future business with TAS products?"

A few days later, Bill Berkman called and got right to the point.

"I represent Toyo in their dealings with companies here in the U.S. Mr. Yamamoto explained to me that Toyo is interested in selling your company's products in Japan. He asked me to call you to provide some more information about Toyo and share some of their possible concerns."

"That sounds good, Mr. Berkman." I did my best to sound as confident and formal as he did. "We've already shipped our product to Toyo, and we'd be glad to talk about doing more business in Japan."

Berkman said, "Well, if you want to do business with Toyo, there are a number of things about the company and about doing business in Japan that you should understand. I can help you with that."

"That's fine," I said. I was all ears.

"First of all," Berkman began, "you must understand that Toyo is a large and respected company, probably the largest distributor of specialized test and measurement equipment in Japan. When you see their product catalog, you'll see that it resembles the Hewlett-Packard catalog."

I was immediately impressed and excited. At TAS, we tried to produce our products and literature to HP standards. When we had an important product decision to make, we would often ask ourselves, "How would HP do it?" If Japanese customers regarded Toyo's catalog as comparable to HP's, then having our product in Toyo's catalog would be a real coup.

Berkman continued. "Toyo is not just a trading company. They don't just buy your product and resell it. Toyo is a full-service distributor. They fully represent your product to Japanese customers. If they feel your product has potential, they will make Japanese-language brochures, they will include it in their catalog, and they will produce advertising in Japanese trade magazines. Toyo can establish your product as a leader with Japanese companies, but you as a manufacturer must do your part."

Uh-oh, I thought. I was waiting for this.

Berkman went on. "Service is of the utmost importance to Japanese customers. If a company like Toyo doesn't provide excellent service, they lose face with the customer. All business in Japan is based on honor and respect, and once a company loses either, they are no longer able to do business in Japan. This is very different from the business ethic in the U.S., where a company can provide poor service but can get by because they have a leading-edge product or a low price. In Japan, honor and respect count for much more than product features and cost. Do you understand?"

"I think so," I said. I had never thought of honor and respect as foreign business concepts. I thought every company should provide great service and should honor and respect its customers. That's certainly what we were trying to do at TAS. I felt Berkman was trying to say something more, so I just listened.

"Service is of the utmost importance," he said. "If you want to be a Toyo supplier, you must do what you say. You must deliver your product on time, and you must deliver the features you promise. You must provide great service to Toyo, and give them the tools to provide great service to their customers. TAS must never cause Toyo to lose face in front of the customer—that would be the worst possible thing."

I was getting a little irritated by Berkman's lecture, but tried my best not to let it show. I wanted him to know we would provide great service to Toyo, not just because Japanese customers expected it but because that was how we did business everywhere.

I said, "Mr. Berkman, you can be sure that we will provide great service to Toyo. We are a new company and we have to give great service to everyone just to survive."

"Yes, well, I understand that, but you must understand that the Japanese market is different. You cannot simply visit customers or call them. If Toyo Corporation represents your product in Japan, they will be the face of TAS in Japan, and they will be the ones who provide service to Japanese customers."

"I understand that."

"So you must understand that sometimes Toyo's customers will need to have their products repaired in a timely way. They can't wait for a product to be shipped back to the U.S. for repair. They will rely on Toyo to do repairs in Japan."

Aha! I saw where Berkman was going.

"So you may have to provide schematic diagrams and software listings and other design information to Toyo so they can provide local service to Japanese customers. This is very important to Toyo."

So that is what the long speech was about. Give schematic diagrams and software information to Toyo so they could repair, modify, and maybe even *copy* our products! I wanted to tell Bill Berkman flat-out that that would never happen. It was the one thing Steve, Charles, and I had agreed upon. It was just too risky to provide design information to anyone outside our company. Our survival depended on our technical expertise, and we were not going to give it away, even if it meant losing some sales.

I came up with a diplomatic answer, one that would not immediately foreclose the possibility of future sales to Toyo:

"Mr. Berkman, I'm sure that if Toyo represents us in Japan, we will find ways to provide the kind of service Japanese customers require. We are committed to doing that."

My answer seemed to satisfy him for the moment. He moved to his next agenda item.

"Mr. Yamamoto and I would like to visit the TAS factory so we can meet you and your staff, and so we can talk about concluding a distribution agreement. Is that all right with you?"

"Sure," I said. "We would be glad to have you and Mr. Yamamoto visit us, and we are very interested in having Toyo represent us in Japan."

Berkman's call left me exhilarated but also apprehensive. Our little company was important enough for someone to come all the way from Japan to talk about selling our products. I would finally meet the mysterious Mr. Yamamoto who had ordered our product sight unseen. On the other hand, I still didn't trust Toyo's motives.

A few weeks later, Messrs. Yamamoto and Berkman arrived at our Little Silver "factory." Our building was neat and tidy, but it didn't look like the home of a world-class electronics manufacturer. That didn't seem to bother either of our visitors. Both smiled broadly as I greeted them in our small foyer. Mr. Yamamoto bowed deeply when Bill Berkman introduced him. I bowed stiffly and uncomfortably in return. Berkman grabbed my hand and shook it heartily. I sensed enthusiasm in both men, and an eagerness to get down to business.

Mr. Yamamoto fit my preconceived image quite well. He was of medium height and build with straight black hair and a friendly, almost comedic face. Bill Berkman was much larger, a tall white man with a protruding midsection, a full head of brownish hair, and a handlebar mustache. Mr. Yamamoto seemed friendly and even funny despite a formal veneer. Berkman was less formal, with a booming voice and confident manner. He was gregarious, almost overbearing. We retreated to the conference room to get acquainted. Mr. Yamamoto presented me with a small gift. I was embarrassed because I had no gift for him. We then exchanged information on our respective companies. Yamamoto and Berkman wanted to know more about our U.S. customers, and I gave them a partial list. Mr. Yamamoto told me Toyo was a supplier to most of Japan's major electronics manufacturers, and maintained a solid reputation and good relationships with all. He handed me a folder containing extensive information about the Japanese communication test equipment market, and his plan for marketing and selling TAS products there.

Yamamoto's presentation was the best I had ever seen. Our U.S. sales reps, with much closer proximity to our products and customers, had never produced a plan remotely approaching his. It was top-notch—as good as anything we had produced internally.

Afterward, I gave Yamamoto and Berkman a tour, beginning with a chance to talk with Steve and Charles. I took our guests to the manufacturing floor so they could see our operations firsthand. Contrary to my concerns prior to the earlier Hayes inspection tour, I had no worries about Yamamoto and Berkman seeing how small our company was. By now TAS had a close-knit group of skilled, dedicated people, and I was confident they would impress our guests.

When we returned to the conference room to wrap up our discussions, Mr. Yamamoto surprised me by ordering another TAS 1010. He said Toyo wanted it to back up the unit they'd sold to Sharp, and to use for sales demonstrations. The order represented a big commitment on Toyo's part.

I was convinced we wanted Toyo to represent TAS in Japan. I presented Mr. Yamamoto a distribution agreement for his superiors to sign, and we agreed in principle that Toyo would be our distributor. Mr. Yamamoto had one more request.

"I invite you to Japan to provide TAS products seminar for Japanese customers. It is important for Japanese customer to meet product supplier and have such technical information as you can provide. Will you visit Japan later this year and provide seminar for Japanese customer? If you can provide such seminar many more sales are possible."

Yamamoto had me with that last line. He seemed to know just what would motivate me to say yes.

"Yes, Mr. Yamamoto. I will visit Japan later this year."

I didn't know all that would be involved in such a trip, and certainly didn't know how I would be received. I sensed Mr. Yamamoto was earnestly trying to establish our products in Japan, and I began to trust him. After his terrific marketing presentation, the additional TAS 1010 order, and the promise of more sales in Japan, there was no way I could refuse.

A few months later, in November, I was on a plane bound for Tokyo. I had been to Asia only once before, on a vacation tour of China with Diane the previous year. This time I was alone on a United Airlines nonstop flight from Newark to Tokyo's Narita Airport. I left in the afternoon, and with the twelve-hour flight and the fourteen-hour time difference, I landed the following afternoon, Tokyo time. As soon as I stepped off the plane, I began forming impressions. Many of the airport signs were in English as well as Japanese, including a huge, colorful poster hung from the arrival concourse ceiling. The poster depicted active Japanese people, and the caption said: COMPUTERS AND COMMUNICATIONS. I smiled. The poster made it seem as if all of Japan was excited about the market TAS was serving.

I breezed through immigration and was pleasantly surprised to find my luggage waiting for me at baggage claim. I encountered no lines at customs. The agent looked at my documents, glanced at my bags, and waved me through. I was officially in Japan. I immediately spotted Mr. Yamamoto quick-stepping his way toward me, holding a hand-printed sign that said MR. DAVID TARVER. He smiled broadly as he approached.

"*Ohayo gozaimasu,* Tarver-san!" Mr. Yamamoto exclaimed. "Welcome to Japan!"

"I'm happy to be here, Mr. Yamamoto," I said. I *was* happy and excited, and couldn't wait to see Japan.

Mr. Yamamoto walked me to his car and we headed for Tokyo. For a little more than an hour I took in the scenery—an orderly mixture of industrial buildings, office towers, and rice fields. The architecture in the commercial areas we passed was quite simple, and though there were colors, the overall impression Japan presented was shades of gray.

After a little over an hour, we arrived in Tokyo. The traffic was much more intense there, but we finally made our way to the Palace Hotel. It didn't look new or particularly fancy, but its main feature, Mr. Yamamoto explained, was that it was situated across the street from the Imperial Palace. I looked across the street, and sure enough, there was a gray brick wall and a wide moat surrounding the stately grounds.

As we entered the hotel lobby, two of Mr. Yamamoto's colleagues rushed over to us. Apparently their assignment was to greet their American guest. Mr. Yamamoto said they were sales engineers who would help me prepare and present the seminar, and that we would meet briefly in the lobby to discuss the arrangements. Jet lag had left me very sleepy, and the last thing I wanted was a business meeting. I smiled, however, and shook hands with my new associates. I visited my room just long enough to put down my bags and splash water on my face, and then I returned to the lobby. The meeting was more form than substance, and after thirty minutes or so my hosts left me standing alone in the lobby.

I was so tired I couldn't see straight, but I was also desperately hungry, and wandered the hotel lobby looking for a place to eat. Seeing nothing, I went down one level and found a restaurant called The Humming. Most dishes pictured on the menu looked strange to me. Raw meat and fish, sometimes plopped atop a clump of rice. Bowls of what looked like noodle soup, each topped with different meats and vegetables. Some items looked fried, but bore no resemblance to the fried chicken I was accustomed to. My gaze settled on two items: a tuna salad sandwich with potato chips, and spaghetti Bolognese. The tuna sandwich seemed like a safe bet, so I ordered one.

After eating I went up to my room and fell on the bed, and within a few minutes was fast asleep. I awoke at 4:00 a.m. and couldn't get back to sleep. I called home to let Diane know I had arrived safely. I tossed and turned until 7:00 a.m., then fell asleep again.

When the nightstand phone rang, the clock read 9:00 a.m. I grabbed the receiver and heard Mr. Yamamoto's voice.

"This is Mineo Yamamoto of Toyo Corporation. Is this Tarver-san?"

Mr. Yamamoto was downstairs ready to go, and I was still in bed! If he hadn't called, my jet-lagged body clock probably would have let me sleep until noon.

"Yes, Mr. Yamamoto—I'll be right down."

Embarrassed and near terrified, I jumped out of bed, opened my suitcase, grabbed some clothes, and ran into the bathroom to brush my teeth. In fifteen minutes I was downstairs greeting Mr. Yamamoto. When he saw me step from the elevator, he bowed slightly and smiled. I think he knew I was sleeping when he called, but he didn't seem to mind.

"Tarver-san, Ohayo gozaimasu! Did you sleep well?"

"Good morning, Mr. Yamamoto. Yes, I slept very well; perhaps *too* well."

We set out on the short walk to Toyo's offices. As we walked, Mr.

Yamamoto again pointed out the Imperial Palace grounds. I couldn't see the actual palace, but the grounds were impressive.

A few minutes later we arrived at Toyo. The building was several stories high and was located in a busy commercial area. Toyo was no fly-by-night outfit. The building's size alone communicated that they were a substantial company.

Mr. Yamamoto took me up several floors to his department. We stepped off the elevator into a large room buzzing with activity. The room was filled with desks that were not separated by walls or even partitions. Each was stacked high with papers, brochures, and notebooks. Most staffers were men. Everyone seemed to be talking on the phone or to colleagues in the room. The scene was organized chaos.

Mr. Yamamoto took me around the room and introduced me to several staffers, including the sales engineers I'd met the previous evening at the hotel. Everyone seemed friendly and happy to meet me. It was hard to get beyond hello because most of the staff didn't speak English nearly as well as Mr. Yamamoto, and my Japanese was nonexistent. Despite that barrier, I felt welcome. I didn't detect a whiff of animosity or indifference.

We moved to a conference room to prepare for the seminar. A few staffers joined us. Mr. Yamamoto began by reviewing the agenda for my stay in Japan. He passed out a memo, on Toyo letterhead, with each agenda item spelled out in Japanese and in English:

> prepare seminar
>
> rehearse seminar
>
> present seminar
>
> review Toyo's plan to sell TAS products
>
> review TAS product plan
>
> visit Atsugi service center
>
> visit Toyo's customer Sharp

None of the items bothered me, but I hadn't expected to rehearse the seminar. I had never found it necessary to rehearse for the many U.S. seminars I'd conducted. When I questioned the rehearsal, Mr. Yamamoto said it was necessary because he would have to translate everything I presented. I hadn't thought about that; it made sense.

I had prepared slides and product demonstrations before leaving the U.S. I went slowly through each slide and demo for my host and his sales engineers. Mr. Yamamoto would often stop me and ask for clarification, then turn to the other engineers and translate. A long discussion in Japanese often ensued, and I would sit and wait. Finally, Mr. Yamamoto would turn to me and say, "Sorry, Tarver-san—go ahead."

We worked hard all morning, and as noon approached I felt intense hunger. Right on cue, the group's secretary, Miss Kobayashi, came in to find out what we wanted for lunch. I chose sandwiches and orange soda. Mr. Yamamoto and the Toyo guys chose Japanese-style box lunches. My sandwiches were the best I had tasted in a long time. The box lunches looked good too—much better than the Japanese food I had seen in The Humming.

After lunch, we continued through my seminar slides. Mr. Yamamoto continued to ask questions and take copious notes. By midafternoon I was dead tired, but we kept going. Around 4:00 p.m. Mr. Yamamoto called for a break. He and his crew went back to their office. I stayed in the conference room and slept. Mr. Yamamoto returned a little after 5:00 and walked me back to the Palace Hotel. He offered to take me to dinner, but I needed to sleep. We agreed to meet in the lobby at 9:00 a.m. I went to my room, plopped on the bed, and immediately dozed off.

When I awoke, the clock read 9:00 p.m. I looked around my room, and for the first time realized how spartan it was. The room was small and the furnishings were basic, but it was neat and clean. The hotel was far from new, but everything seemed well maintained. My suitcase was on the floor at the foot of the bed, and most of my clothes were still in it. I hadn't had time to unpack.

I was hungry again, so I went back down to The Humming for dinner. This time I opted for the spaghetti bolognese, and it wasn't bad. Back upstairs I unpacked, called home to check on Diane and Stacy, then checked to see what was on TV. Two channels carried English-language news and one was showing an old English-language movie. I was wide-awake, so I watched the movie and reflected on the day's events. I was starting to feel nervous about the seminar, because I worried about how the Japanese engineers would respond to me. The companies slated to attend the seminar were the crown jewels of the Japanese electronics industry, and I was just a guy from a little startup company in New Jersey. Who was I to instruct these guys on anything? Why should they receive me as some kind of expert? Full of anxiety, I drifted off to sleep.

I awoke again at 4:00 a.m., but this time I had the presence of mind to set the alarm for 7:00. I tossed and turned, half asleep, until it sounded. After a leisurely shower, I got dressed and went down to The Humming for breakfast. Nothing looked appetizing—the eggs weren't nearly done enough and the sausage was downright scary. I wasn't familiar enough with the Japanese-style breakfast to try it, so I made the safe choice—raisin bran, toast, and plum jam.

Determined to get there first, I made sure to be in the lobby by 8:55 a.m., but Mr. Yamamoto had already arrived.

When we got to Toyo's facility, Mr. Yamamoto took me straight to the seminar room. The desks for the participants were in place, as were the podium and the overhead projector. A seminar booklet was placed on each desk. On the periphery of the room, the Toyo guys had arranged a few demo stations with TAS equipment and illustrative placards. Mr. Yamamoto handed me one of the booklets to inspect, and at that moment I realized he had been working most of the night to prepare them. The booklets looked professional—better than anything we had produced for our own U.S. seminars. Each contained all of my slides and support materials, and the information was presented in both English and Japanese. Mr. Yamamoto and his colleagues had translated and reproduced my materials during the night, and had set up the seminar room—all while I slept at the Palace Hotel. I was overwhelmed with respect for Mr. Yamamoto and his team.

I smiled broadly and said, "Mr. Yamamoto, you have done an excellent job. Everything looks great!"

Mr. Yamamoto returned the smile, and he looked proud. He bowed and said, "*Arigato*—thank you for your help with Toyo's seminar preparations."

The main agenda item was to rehearse the seminar from start to finish. I presented each slide, and Mr. Yamamoto translated. The Toyo sales engineers played the role of seminar attendees. Occasionally one asked a question, and a discussion in Japanese ensued. We went on this way well into the afternoon, pausing only for lunch. My presentation went smoothly, and I felt much more confident afterward. We spent a couple more hours rehearsing the equipment demonstrations, and then sat down for a final review.

Mr. Yamamoto presented a list of the names and company affiliations of the people who would be attending the seminar. I didn't recognize a single person, but I did recognize the companies: Sony, Matsushita, Sharp, Ricoh, NEC—the list was a *Who's Who* in the consumer electronics industry. What qualified me to teach modem testing techniques to these experts? I was just a guy from Flint, president of a tiny electronics company I started in my basement. Now, standing in the heart of Tokyo, I was expected to instruct the leaders of the global electronics industry.

I didn't know if Mr. Yamamoto could detect my growing fear, but if so he didn't show it. Maybe he didn't fully appreciate the difference between these Japanese giants and my struggling company. Maybe he saw me as some kind of expert that I was not. Maybe he was blind to the potential for racial discrimination on the part of his countrymen. How would he feel if the Japanese engineers laughed me out of the room or gave me the silent treatment? I had to raise the issue.

"Mr. Yamamoto," I began slowly, "I have a concern. The engineers coming to Toyo's seminar are from very large companies. My company is small and has not been in business very long. I am worried about how the Japanese engineers will respond to me, and whether they will accept me."

The Toyo sales engineers were still at the conference table, and all were silent. I was sure most didn't understand what I'd said, but I wasn't addressing them. I wanted Mr. Yamamoto's opinion. I hadn't explicitly raised the issue of race—I referred only to issues of company size and experience. Still, I hoped Mr. Yamamoto would read between the lines when I asked how the Japanese would *respond* to me and whether they would *accept* me.

I studied Mr. Yamamoto's face, looking for that change I had often seen in whites when they were unexpectedly confronted with the issue of race. I had seen it many times—mouth closed tightly and twisted, eyes averted to avoid contact. I didn't see that on Mr. Yamamoto's face. He looked straight at me and said: "Japanese engineers know about Bell Labs. Japanese engineers will buy TAS product if they know you are from Bell Labs and Bell Labs engineers use TAS product. In your presentation you must say you are from Bell Labs. Also, you must say TAS product conforms to testing standards."

I completely understood what Mr. Yamamoto was saying: that to Japanese engineers, my pedigree was more important than my persona. I hadn't realized the extent to which Japanese engineers revered Bell Labs, but apparently the Bell Labs history of the TAS founders would be a huge factor in how they viewed my company. I would have no problem stressing that connection, because Bell Labs was where I obtained the knowledge and experience to start TAS. I'd have no problem stressing that our product conformed to industry standards, because I'd participated in the industry committees that developed those standards. If emphasizing my pedigree was all it would take to convince the Japanese engineers to buy from TAS, we were home free. Still, I wasn't entirely convinced. I was sure Mr. Yamamoto was missing an important angle—the racial angle—because it was something he hadn't experienced. I had lots of experience with white U.S. engineers turning sour when they realized TAS's president was a black man. To them, it didn't matter that I was from Bell Labs, or that I had helped write the testing standards, or even that our product was the best.

It was late evening before I made it back to the Palace Hotel. It had been an exhausting day. Mr. Yamamoto and his Toyo colleagues had done a marvelous job preparing for the seminar. As I lay down, I replayed my conversation with Mr. Yamamoto over and over. I convinced myself that the only thing that could keep our seminar from being a success was if the Japanese engineers refused to accept me—a black man, former Bell Labs manager, president of TAS—as an authority on modem testing.

The next morning, I made sure to get up early. I ate a hearty breakfast, then walked the few blocks to the Toyo building. It was a brisk November morning, and the cool air filled me with energy and excitement. It was show time, and I was ready to see, once and for all, how engineers from Japan's elite electronics manufacturers would respond to me. My apprehension melted, replaced by searing anticipation. I was convinced that on that very day I was going to learn something important about my Japanese colleagues, and was reminded why I had started TAS. I wanted to build a product and a successful company, for sure, but also I wanted to learn more about other people and other cultures.

Upon arriving at the Toyo building, I went straight to the seminar room. Mr. Yamamoto and his subordinates were making final preparations. Mr. Yamamoto flashed a big smile and said, "Hi, Tarver-san!" His enthusiasm was palpable and infectious. The Toyo guys seemed in a great mood. They were all nattily dressed, too, each sporting a dark suit, white shirt, and colorful tie. As I made my way toward the front, each Toyo staffer smiled, nodded, and shook my hand. I felt like a real star, on top of the world. I stood at the speaker's podium and surveyed the room. This was the most professional-looking seminar setup I had ever been part of.

Attendees began arriving around 9:30, and signed in at a table near the door. Miss Kobayashi greeted them, checked them off the invitation list, and gave them nametags. As the attendees filtered in, two things struck me. First was the serious professional demeanor of our guests. Each was dressed to the nines in a nice suit, crisp white shirt, and tie. As each guest approached the reception table, he bowed and presented his business card before accepting a nametag from Miss Kobayashi. The whole procedure was beyond formal, almost solemn. The second thing was how different each of our attendees looked. I had always assumed that one Japanese person looked pretty much like another, but the facial features of our guests differed markedly. In fact, the appearance of the Japanese engineers varied as much as I'd ever seen in a roomful of people. All Japanese people definitely did not look alike.

I stood near the reception table with Mr. Yamamoto. As the guests finished signing in, Mr. Yamamoto introduced some of them to me. He made a point of introducing an engineer from Sharp, Toyo's first TAS customer. He then introduced me to the representatives of Rockwell Japan and Texas Instruments Japan, because both were key TAS customers in the U.S. I exchanged business cards with those gentlemen in the formal Japanese fashion, and with Mr. Yamamoto's translation, engaged in brief small talk.

Mr. Yamamoto went to the podium and introduced me. He spoke Japanese, but occasionally inserted an English phrase. I heard him say "Bell Labs" more than a few times. I surveyed the audience as I moved to the podium. Thirty or so engineers, all but one of them male, were seated in front of me. All looked serious and studious. Each had the seminar booklet open. Some were already taking notes. I thought, *let the show begin!*

"Ohayo gozaimasu!" I said with a smile, and then bowed to the audience. I was not without a showman's instinct. I knew only a few Japanese phrases, but was determined to use them all.

"Ohayo gozaimasu," the audience responded.

"My name is David Tarver, and I come to you from Telecom Analysis Systems, also called T-A-S, in New Jersey, in the USA. Before I started TAS, I was a member of technical staff at Bell Labs."

I paused so Mr. Yamamoto could translate. It sounded as if he was repeating his introduction, but that was okay—apparently one could not say "Bell Labs" too many times.

Launching into the seminar material, I spoke slowly and clearly, and paused at regular intervals so Mr. Yamamoto could translate. I tried to be engaging and entertaining. I used colorful markers to emphasize points on the slides, and wove in references to U.S. and Japanese telecommunications standards. In short, I gave a masterful presentation. The Japanese engineers paid rapt attention—they seemed to watch my every move, and they scribbled frequently in their seminar books. I became more and more aware, however, that no one was asking questions or making comments. I tried not to let the lack of feedback bother me—I chalked it up to the language barrier. I figured I'd wait until the end of the presentation and then invite the audience to ask questions.

I finished presenting all the slides, and then, with the assistance of the Toyo sales engineers, I demonstrated our equipment by testing a pair of modems on the spot. Again, the attendees seemed enchanted, but no one asked a question. Finally it was time to wrap up. I made an explicit appeal to the audience for feedback.

"I would be glad to answer any questions you have about anything you heard in this seminar."

Wait for translation. Nothing.

"I would also like to know what you think about the seminar. Please feel free to give me any feedback you have."

I was in uncharted territory. We hadn't rehearsed asking for feedback. Mr. Yamamoto seemed surprised by my solicitation, but he went ahead with the translation.

Again no response.

"Thank you for coming, Arigato gozaimasu," I said sheepishly, and then stepped away from the podium.

Mr. Yamamoto took over and did his own wrap-up. He spoke entirely in Japanese, but I could tell he was thanking the attendees profusely for coming. Then he had his sales engineers pass out a document to them. It was several pages long and written in Japanese. While the attendees perused the document, Mr. Yamamoto told me he had just passed out a questionnaire regarding the seminar. He said customers might prefer to respond in writing. I wasn't buying it. I concluded that my presentation had bombed, and that the attendees were probably not interested in TAS products. The emotional high that began with my morning walk to the Toyo offices, and peaked as I began the presentation, was gone. Far from winning in Japan, I figured I had just laid an egg.

After all the guests had left, I sat down with the Toyo guys in the seminar room to review our day's work. Mr. Yamamoto asked what I thought of the seminar.

"I thought we did a great job," I said. "I'm so glad we rehearsed the seminar, because everything went smoothly. This seminar was better than anything we have presented in the U.S."

Mr. Yamamoto just nodded. "*Domo arigato*," he said.

I went on. "I was disappointed that the clients didn't have any questions. Even at the end, when I encouraged them to give feedback, they said nothing."

I must have looked downtrodden. Mr. Yamamoto remained his usual upbeat self.

"What did you think of the seminar?" I asked.

"Oh, it was very good. Japanese engineers often do not ask questions in seminar setting. Engineer from Sony does not wish to ask question in front of engineer from Sharp or other competitor. Engineer may be concerned that his question will give information to competitor. That is why we give questionnaire to attendee, so he may write a question or comment."

"I see," I said, still unconvinced. Could Japanese engineers be that different from American engineers? In our U.S. seminars, we hosted clients from competing companies all the time, but no one had a problem speaking up with questions or comments.

At that moment, Miss Kobayashi entered carrying a stack of papers for Mr. Yamamoto. He thumbed through them, looked up, and said, "Questionnaire responses."

I wasn't sure I wanted to hear what the Japanese engineers thought. I had a feeling I was right and Mr. Yamamoto was wrong; that the silence of the attendees meant they were rejecting me and my company. I tried to tell myself that the seminar was a learning experience, and that my purpose in starting TAS was to gain such experiences, but that was not comforting. I really did want to win in Japan, but if the key companies wouldn't buy our equipment, we were doomed to lose.

Mr. Yamamoto started tabulating the questionnaire responses on a piece of paper. "Thirty-three persons attended seminar," he said. "Twenty persons say they wish to purchase TAS modem test system in near future."

I couldn't believe what I was hearing. Almost two-thirds of the people who came said they planned to buy our equipment! That was a much better success rate than we'd ever achieved in the U.S. Apparently, Mr. Yamamoto was right. The fact that I was black and ran a small startup company with two other black guys was not a deciding factor for the Japanese engineers. The deciding factors were that we came from Bell Labs, produced an excellent product, and had a great Japanese distributor—Toyo.

Later that night, I had dinner with Mr. Fukumoto and Mr. Yumoto. Fukumoto was Mr. Yamamoto's department head, and Yumoto was director of the entire organization. The dinner meeting had been arranged soon after my arrival, but I wasn't sure of its purpose. I wondered why Mr. Yamamoto's superiors were interested in taking me to dinner, and why Mr. Yamamoto wasn't joining us.

We traveled to the restaurant by car. As we got under way, Mr. Yumoto said, "We are going to have *Shabu Shabu*. Do you know what is this?"

"No," I said. I was hoping it was nothing exotic, and especially nothing raw.

At the restaurant we were seated at a table with a large pot at its center. The pot sat on a burner and was filled with boiling broth. A waitress arrived with a tray of raw red meat.

Uh-oh, I thought. *Not raw meat—not now!*

Mr. Yumoto picked up his chopsticks and grabbed one of the pieces of meat. It was the leanest-looking beef I'd ever seen. He dipped the meat into the broth and swished it around for a few seconds. When he withdrew it from the cauldron, it was completely cooked. Next, Yumoto dipped the meat into one of the sauces that had been strategically placed on the table. Finally, he lifted the meat to his mouth and ate it.

"You see? Shabu Shabu!" he exclaimed. Mr. Yumoto beckoned me to follow suit, and I did, glad that I had some experience using chopsticks.

The beef was delicious, and the sauces complemented it perfectly. I was relieved at not having to eat raw meat. Gradually, while enjoying the tasty beef and sauces and succulent vegetables and warm tea, I began to relax from the stresses of the momentous day.

Mr. Yumoto, Mr. Fukumoto, and I shared small talk about our families. Then we each recounted how we'd become engineers and arrived at our respective positions. I told them how happy I was with the results of the seminar and how impressed I was with Mr. Yamamoto and the other Toyo staff members. My body was practically glowing with warm feelings. Here on the other side of the world I found people who were willing to work hard for me without regard to my skin color, who just wanted to be the best at their craft, who appreciated the hard work required to start a technology company. I felt connected to my hosts, and I wanted to work hard to make them succeed, to justify and reward their trust and support.

In that moment of warmth and connectedness, I admitted to Fukumoto and Yumoto that I had been concerned about how the Japanese engineers would receive me. Mr. Yumoto looked up from his plate of Shabu Shabu and vegetables. His face was glowing beet red, but it was filled with sincere expression.

"You know, we are minorities too," he said. He didn't elaborate, he just resumed eating his meal and drinking his tea. I glanced at Mr. Fukumoto. He had a wry, disarming look on his face, a look that said he was deferring to Mr. Yumoto.

I didn't know what to say, but I didn't feel I needed to say anything. I thought I knew what Mr. Yumoto meant. I thought he was saying that despite all the business success Japanese people had achieved, they too were struggling for respect.

WITH TOYO STAFF MEMBERS DURING FIRST VISIT TO JAPAN. MR. YAMAMOTO IS SECOND FROM LEFT.

I spent a few more days in Japan preparing Mr. Yamamoto and his team to sell and service TAS products. On the flight home, I felt tremendous satisfaction. Our fledgling company had established a foothold in Japan, one of the toughest electronic equipment markets in the world.

In the following months the orders began to flow from Japan. TAS products were a big hit there. In fact, during some months our shipments to Japan exceeded our U.S. shipments. I couldn't have imagined a better result.

Japanese engineers believed we made the world's best modem test system. I already knew that. For the first time, though, I believed that achievement would be enough to ensure our global success. Steve, Charles, and I, three young black Americans, had built ourselves a world-class company.

44. Mandela is Free, Bill

MY TRIP TO LONDON AND PARIS was a scouting mission, our first direct contact with a market outside North America. The Tokyo seminar four months later marked the moment TAS truly arrived on the world stage. Between those two journeys, I said *no* to an international sale. The dollars were not especially large, but the principles were huge.

"Are you crazy?" Bill Graham asked, as what had begun as an innocuous business lunch began to heat up. "David, you don't want to give up a thirty-thousand-dollar sale because of some loony romantic idea you have about South Africa. Take the business! You'll be helping yourself far more than you're hurting any poor black South Africans, and I guarantee you won't be helping Mr. Mandela one bit by refusing to sell there. Besides, if they can't get the equipment from you, they'll just buy it from AEA. All you're doing is helping your main competitor!"

It was late summer 1987. I had flown up to Boston to join Bill for a day of sales calls at Motorola, and would fly back to New Jersey that afternoon. Ian Farr had called from England a few days earlier to say he had a South African customer who wanted to buy some TAS equipment. I told Ian I wasn't interested in doing business in South Africa, and Ian was upset with me. When I related this to Bill, our genial conversation turned into a heated argument.

"I don't care if it doesn't help," I said. "I'm not going to sell our equipment there. Who knows, they might use it to help suppress black folks. It's just the principle of the thing. I won't do it!"

I liked Bill Graham a lot. He was one of the best salesmen I'd ever known. I had learned an important sales lesson from Bill that day he turned a lost sale at Concord Data Systems into a key win. Since those early days, with Bill's help, we had established business with several leading Boston-area telecommunications equipment companies. I didn't know if it was his Scottish accent or his sales acumen or his easy manner, but Bill had a way with customers, and I had grown to trust him. I was surprised when Bill adamantly stated that I should ignore the brutal apartheid regime and sell our equipment in South Africa. His attitude and demeanor seemed out of character.

"Well, suit yourself," Bill said. "But I think it's a stupid decision—absolutely stupid."

Bill went on to argue that blacks had it pretty good in South Africa, far better than before the Europeans arrived. He claimed that the rivalry and brutality among South African tribes was far worse than anything the Boers had perpetrated against the blacks.

I knew little of South African history or politics, but I knew what I saw on TV and read in the newspapers—that white South Africans believed themselves superior to "coloreds," and that anyone who was not white was not deemed a full citizen. Having grown up during the American civil rights movement, having visited the South as a kid when bathrooms and drinking fountains were designated WHITE and COLORED, I couldn't accept what was happening in South Africa.

Bill and I argued long after our table was cleared, until I suddenly realized I was in danger of missing my flight back to New Jersey. I ended our argument with one final taunt: "I won't sell equipment to South Africa until Mandela is free!"

I didn't know much about Nelson Mandela, but invoking his name felt like something cool to say, and I knew it would get under Bill's skin. He didn't say one more word on the subject. We just jumped into his car and headed for Logan Airport. Bill focused on the traffic while I worried about missing my flight. He dropped me at the terminal and I dashed to the departure gate. Too late. We had argued so vehemently that I missed my flight—but I had made my point. That made it worth waiting for the next shuttle.

Earlier in the life of TAS, it might not have been so easy for me to pass up a sure sale, but business was good. Sales were on track to double in 1987, as they had in 1986. We hired a new employee or two every month, and our customer list was growing rapidly. We were busily developing new products: a smaller, less expensive telephone network simulator, a data analyzer to replace the Phoenix product, and improvements and additions to our bread-and-butter TAS 1010. Steve, Charles, and I could barely keep a handle on it all. In three years, the company had grown from three guys in my basement to eighteen people in a small industrial building in Little Silver. Things were so hectic we decided to take a "time-out" to assess where our business was heading.

In October, Steve and I and two newly hired managers retreated with our wives to Mount Airy Lodge in the Pocono Mountains of eastern Pennsylvania. Charles and Marlene didn't make the trip because of urgent family business. The new managers were Jim Scott, administrative and finance director, and Rick Cathers, manufacturing manager. I hired my former Bell Labs colleague, Dr. Clifton Smith, to facilitate the meeting. Cliff had left Bell Labs and started his own management-consulting firm. I felt sure he was the person we needed to help solidify our team and create sound management practices.

I thought I was taking my team to a rustic but modern resort where we could work hard by day and relax and party with our wives by night. I was surprised to find that Mount Airy Lodge seemed stuck in its 1950s heyday, with pink heart-shaped tubs in the rooms, and photos of retired or dead entertainers lining the hallways.

Despite the dated venue, the retreat was highly productive. Clifton Smith first helped us define our company's mission. Although that seemed like a reasonable thing to do, we had never given it much thought. When we launched TAS, our mission was to survive, finish our prototype, and build and sell our first production units. After we started growing, our mission became simply to keep growing. With Clifton's guidance, we decided TAS's mission was to design and manufacture high-quality electronic tools for developers and manufacturers of communications equipment. That statement encompassed what we were already doing and provided guidance for our future activities.

We next turned to our vision statement. We hadn't spent much time thinking about that, either, other than to recall that Steve, Charles, and I all planned to become millionaires. We half-jokingly tried that vision on Dr. Smith, but he rejected it. He said every entrepreneur wanted to be a millionaire, and that our vision needed to cover more than the desire to strike it rich. After much wrangling, we came up with a series of vision statements such as "we will be a world-class provider of telecommunications test equipment," and "we will provide testing tools for all major communication technologies." We crafted ambitious statements because Clifton encouraged us to "think big."

The next—and most important—subject Clifton helped us tackle was teamwork. Our management team generally worked well, but we did hit rough patches, especially when one team member usurped another's authority or didn't adequately consider his input. For example, I sometimes overruled design decisions Steve made, even though hardware design was his responsibility. Charles sometimes changed a manufacturing procedure Rick had devised, even though Rick was the manufacturing manager. I probably stepped on more toes than anyone, but I certainly wasn't the only culprit. At times our conflicts led to quiet frustration, and other times to vociferous arguments. Clifton calmly and carefully walked us through the roles, responsibilities, and authority of each team member. The structured, dispassionate discussion resulted in agreements among the managers that removed the major sources of conflict.

The most helpful aspect of our discussion was Dr. Smith's definition of teamwork. He said the best teams operate by "sharing uncertainty." He said many people, especially engineers, approach a discussion willing to share only what they already know. Engineers are accustomed to solving a problem and then giving the answer. Steve, Charles, and I often operated that way, which frustrated our growing management team. We found it difficult to say, "Here's a problem I'm considering, and I'm not sure what the solution is. What do you think?" Clifton helped us understand that our team would be much more powerful if we could explore problems together rather than argue over preconceived solutions.

We spent nearly a week at Mount Airy. Many things we worked out were mere common sense, but common sense had often gotten lost in the heat of battle. It was helpful to have a detached third party walk us through our issues. I suppose that, like marriage counseling, that kind of intervention works only when the parties want it to. In the case of the Graphics Workstation project team at Bell Labs, it didn't work. In the case of our TAS team, it did.

We ended 1987 with 80 percent growth. We were expanding so quickly that we again needed a larger facility. In June 1988 we moved into a new industrial park building in Eatontown, ten minutes' drive from Little Silver. We were the major tenant, and the sign out front gave us top billing. For the first time, we had a nicer building than some of our customers. Steve, Charles, and I each had a large, comfortable front office. We finally looked like a bona fide high-tech company.

In keeping with our new mission, we defined even more new products. We devised network simulators for rapidly emerging digital communication networks, and began developing a replacement for the original TAS 1010 Channel Simulator. The new projects required hiring more engineers and staff. Fortunately, our sales kept growing, which allowed us to keep investing in the company. We expanded our distribution network to Italy, Spain, Hong Kong, and Taiwan. By the end of 1988, we were celebrating another record year and another major milestone—more than $5 million in sales.

On paper, Steve, Charles, and I had reached our financial goal—we were millionaires. We didn't feel like millionaires, though, because our net worth was tied up in the company. Most profits went to increasing inventory, hiring new employees, and paying down business loans. We were making money at a record pace—but we were spending it at a record pace, too.

Rapid growth brought other problems. Some new employees turned out to be complete duds, and motivating others to work hard proved difficult. We needed to get new products out the door to keep growing, but new products were slow in coming. It was becoming more and more difficult to increase sales. Several large customers provided good repeat business, but we still had to fight for every new customer.

On May 6, 1989, a momentous event occurred: Diane gave birth to our second child, Aaron Michael. Aaron's birth came at a difficult time in our relationship. While I had focused on realizing my business dreams, Diane and I had become emotionally estranged. Aaron's birth made me realize how detached I was. Even when I was at home with Diane and Stacy and Aaron, my mind was consumed with the challenges and rewards of managing and expanding a multimillion-dollar business. It was a selfish obsession, but I couldn't let go.

In 1989 TAS set another record, as sales exceeded $6 million. That represented our smallest year-to-year growth—about 25 percent—since we started the company, but by normal measures it was still impressive.

In June 1990 I received a call from an old friend and mentor, Darwin Davis, a senior vice president at the Equitable Life Assurance Society (Equitable Insurance) in New York City. It was at Darwin's home in Flint, twenty-one years earlier, that I'd met my first serious girlfriend, Gay Carlton. I was a tenth-grader then, and Darwin was an executive in Equitable's Detroit office. Darwin moved quickly up the ladder, first in Detroit and then at the New York headquarters. By the time I started TAS, he was among the leading black corporate executives in the U.S.

Darwin invited me to his Manhattan office for lunch several times, and on those occasions we talked about my business concerns, accomplishments, and plans. Darwin also invited Diane and me to several annual Equitable Black Achievement Awards events, where I was privileged to meet African American leaders in industry, government, entertainment, and sports. Darwin was generous and helpful, and I got the feeling he was proud of me because, like him, I was a kid from Flint who was making it in the business world.

On that June day in 1990, Darwin was calling to invite me to a New York event that was beyond anything I could have imagined. Nelson Mandela, recently released from prison in South Africa, was making his first post-release visit to the U.S., and New York mayor David Dinkins was hosting a reception for him at Windows on the World, atop the World Trade Center. When I heard Darwin's invitation, I had to put the phone down for a moment and collect myself. I couldn't believe I was going to be in the same room with, and perhaps meet, Nelson Mandela. I took a deep breath, and then told Darwin I would be honored to attend.

A few weeks later, I was at Windows on the World amid a crush of dignitaries. Attendees came from the top echelons of New York society: politicians, business leaders, socialites... and me, president of an up-and-coming technology company, who happened to be a friend of Darwin Davis. I was seated at a long table with several other guests. I didn't recognize any of them, but based upon their demeanor and conversation, I surmised that they were corporate executives. I looked toward the dais and immediately recognized Mayor Dinkins, and then I saw the guest of honor—Nelson Mandela. Nothing in his dignified manner indicated that he had just spent twenty-five years in prison. Everything about the man, from his regal bearing to the deference shown him by everyone in the room, indicated he was a revered figure and a statesman.

When Mandela spoke to us that morning, I was amazed by his gentle tone. His speech was not fiery or political, but merely a request to the New York business community and the country at large to support South Africa's nascent post-apartheid democracy. It was a simple message, but judging from the response in the room, it was effective.

After the speeches, many people made their way to the dais to shake hands and be photographed with Mandela. Two suddenly giddy businessmen at my table debated going up for a few minutes, and off they went. After snapping their pictures, they returned to the table and encouraged others to go.

I remained in my seat. I was too much in awe of Mandela to go up and pose, and besides, it seemed a cheesy thing to do. I was sure 99 percent of those in the room—including me—had done nothing to help secure Mandela's release or to alleviate the decades of suffering endured by South African blacks. Who were we to think we should rush up and share a recorded image with the Great One?

After the picture taking and handshaking were complete, the event ended. The attendees were ushered out of the ballroom and instructed to stand on either side of the hallway. Then the dignitaries marched slowly out of the restaurant, between the columns of guests. I stood at the front of the crowd on one side of the hallway, and was sure Mandela was going to pass right in front of me. A few moments later, Mayor Dinkins emerged, with Nelson Mandela right behind him. I wanted to do something to acknowledge Mr. Mandela, but didn't know what to do. As Mayor Dinkins passed, he looked at me and smiled. A second later, Nelson Mandela was in front of me. He walked slowly, purposefully, looking at everyone but focusing on no one. He looked at me, but I don't think he *saw* me. I wanted badly to extend my hand to him, but I froze. I didn't feel worthy to shake his hand.

On the drive back to New Jersey, I was in a sort of daze, caught in a world between reality and dreams. As I floated through that world, an image appeared before me: Bill Graham and I were arguing about selling some TAS equipment to South Africa, and I said we wouldn't sell anything to South Africa "until Mandela is free!"

I felt myself smiling from ear to ear, and said aloud: "Mandela is free, Bill."

45. Hitting the Wall

THE FIRST ENGINEER SCOWLED AT ME. "I need to send my *own* files through your system. Your canned files don't give me enough flexibility!"

I jotted a note and replied with a question: "Do you want to be able to send any file, or are there specific files you're interested in?"

"I need to be able to send *any* file. Why should that be so difficult to do?"

"Okay, I'll see what we can do," I said.

The second engineer didn't seem as angry—merely disappointed. She said, "Your Belgium dial tone is wrong. I can't finish certifying my modem for Europe until you correct it."

I jotted another note and said, "I'll make sure the guys at the factory get right on it."

The next engineer barely let me finish before jumping in: "I need your software to show a pass/fail summary for every test. Your report gives too much information."

Once again, I jotted. "I'll see what we can do about that," I said.

It was late afternoon on January 16, 1991. I was sitting in a conference room at Rockwell Semiconductor Products in Newport Beach, California, listening to engineers complain about our products. Mondy Lariz, the sales rep who had played such an important role in landing the Rockwell account, liked to call these meetings gripe sessions. We held one every six months or so following our first sale to Rockwell in

1986. The main purpose was to let customers blow off steam and to head off any problems that might cause us to lose business.

I had opened the meeting by thanking the attendees for their interest in TAS products and giving a brief update. Then Mondy invited the Rockwell engineers to comment, and that was when they let me have it. As usual, the engineers pummeled me with problems, complaints, and suggested product improvements. I had intended to respond without appearing angry or argumentative, and succeeded for the most part. Like a hard workout, the meeting was painful while in progress but satisfying afterward. I emerged with a to-do list and a few good product improvement ideas. I knew every problem we solved and every improvement suggestion we incorporated would make Rockwell engineers even more committed to TAS products.

Two engineers escorted Mondy and me from the meeting to the laboratory where most of our equipment was located. The end of the workday was approaching, and we were wrapping up our discussions when a young technician rushed in.

"The war just started!" he blurted.

Someone turned on a radio and we heard the first reports: Operation Desert Storm was under way. The U.S. and its allies were invading to drive Saddam Hussein's Iraqi forces from Kuwait. Though the war had been anticipated for months, I worried about how it would affect world stability, the economy, and, frankly, my business.

Mondy and I listened to the reports for a few minutes, then said our good-byes and stepped into the gorgeous late afternoon. The warm Santa Ana winds and setting sun offered no hint that a war was raging on the other side of the world. Mondy drove me to dinner, where we recapped the day's events. Back at my hotel I stayed up late into the night, watching CNN and worrying about the future of TAS.

I had cause to worry. Although our sales had increased during the previous year, 1990, the gain was very small—only about 5 percent—by far the smallest sales increase we'd achieved since starting the company. It was beginning to look as if our business was hitting the wall.

In the months following that Rockwell gripe session, it became clear that the economy was experiencing serious problems. Gloomy forecasts dominated news reports, and talking heads speculated about recession. My sales reps and distributors reported declining orders, not just for TAS products but also for all their product lines. I tried not to worry about the overall economic climate and to stay focused on what was happening with *our* business, to understand why *our* sales had gone soft. I resolved to address the problems at TAS and let the rest of the world take care of itself.

When I dissected our sales slump, the diagnosis was clear: Sales of our bread-and-butter products remained strong but weren't growing. The prescription was also clear: To survive in the near term, we needed to boost sales; to survive in the long term, we needed to develop new products.

Seeing what needed to be done was easy. The tougher challenge was to do it. All three cofounders were stretched to the absolute limit. Steve was managing development of our modem test systems, the source of most TAS sales and profits. He was meticulous, reliable, and thorough. Our modem test systems were carrying the entire company, so we couldn't afford product-line mistakes, hence we couldn't afford to blur Steve's focus.

Charles was managing development of products to test emerging "digital modems." We had hoped those products would fuel significant new growth, but the results had been disappointing. Even though the products hadn't caught on, or perhaps because of it, Charles wasn't in a position to take on anything else.

I was responsible for marketing, sales, and overall management. I was also managing development of TASKIT, the software that controlled our modem test systems. My broad range of activities gave me top-to-bottom control of the company, but didn't leave me time to take on any more responsibilities.

We were spread so thin it was hard to plot a course toward success. We needed new products, but even if we could identify the right products to develop, we had no one to develop them. If we stopped improving existing products to work on new ones, our sales would fall. We faced a difficult choice: spend lots of money to develop new products and risk going broke quickly, or spend much less, stay with existing products, and risk going broke slowly. It was a frustrating dilemma.

Despite the limp economy and lack of new products, TAS recorded a tiny sales gain in 1991. For all practical purposes, though, our sales had remained the same for three years. Despite our efforts, and despite our fervent desire for fast growth, TAS was treading water.

In 1992, we started sinking.

We had hoped to resume growth as the year began, but in fact during the first few months our sales decreased. I concluded that, notwithstanding the risks, we had to bite the bullet and develop some new products. Creating a product that would save the company would be no easy task, but it was a task we had to accomplish.

Desperate for a solution, I visited my old friend Victor Lawrence at Bell Labs. Eight years had passed since Victor placed a company-saving order for five TAS 1010 Channel Simulators. Twelve had passed since my Bell Labs colleagues had deemed Victor "not ready for promotion." Since that woefully wrong assessment, Victor had risen to Bell Labs executive director and was responsible for managing hundreds of researchers. He had also been designated an AT&T Fellow—the highest honor that could be accorded a Bell Labs scientist. I was honored to speak at a reception honoring Victor's achievement, and to sit on the dais with Ghana's U.N. ambassador.

Victor and I had maintained a good relationship since our days as Bell Labs colleagues, and every six months or so he invited me to lunch. Afterward, we would retreat to his office and Victor would "pull back the curtain" on new and emerging technologies so I could plan TAS's product direction. That was the agenda I was hoping for as I drove to visit my old friend.

As usual, we talked and laughed about the "good-old days" at Bell Labs, and he brought me up to date with news of my former colleagues. By the time we finished lunch and sat down in Victor's plush office, I was bursting with anxiety. I got straight to the point, trying my best to convey urgency without sounding desperate.

"Victor, I need to identify a new product line for TAS. Our current products are great, but our sales aren't growing. What do you think we should do?"

Victor didn't hesitate.

"Well, Dave, I think you should consider the wireless business. Cellular telephones are already big business, and that business will be growing very rapidly. You should look into this and develop a product for testing cell phones and other wireless equipment."

I told Victor that no one in my company had any expertise in wireless, and that I didn't see how we could enter that arena. Victor didn't flinch. "My dear friend, Dave Goodman, is giving a seminar on wireless communication at Rutgers very soon. I think you should attend. Some of my guys will be going as well. I think you will find it useful and very interesting."

It was impossible to argue with Victor's polite and optimistic message, but I just didn't see how TAS could make the leap from modems to cell phones. The technology difference was enormous. It was like asking a bicycle manufacturer to start building airplanes. I left our meeting disappointed and even more anxious about our company's future. I had great respect for Victor's advice, though, so I attended Dr. Goodman's seminar. It was a real eye opener. I learned there was an urgent need to simulate wireless communications so engineers could design better cell phones. The designers needed an electronic "proving ground" so they could test their cell-phone designs in the lab instead of driving around asking, "Can you hear me now?" I didn't yet know how we were going to do it, but the seminar convinced me that wireless technology presented a huge opportunity for TAS.

Feeling the same excitement I'd felt as I watched Fred and his friend Leroy Nesbit build their first crystal radio, I started jotting product ideas. I remembered how I was later introduced to ham radio by my father's olive-green Morse code trainer. Recalling those days, I realized wireless technology was nothing new and nothing to be afraid of. Just as Victor had suggested, my company needed to be in the wireless business. I returned from the seminar knowing what we needed to do—even if at that moment we had no one to manage development of a wireless product, no engineers to produce the designs, and no experience with the technology.

Steve, Charles, and I spent the next few months researching the market. We decided to create a product we dubbed a "wireless channel emulator" that would provide a convenient way to test cell phones and other wireless equipment. Already stretched thin, Steve was nonetheless determined to oversee development of our new product. He learned what he needed to know from scratch. We assembled a development staff by hiring some new engineers and reassigning others. Several months after my meeting with Victor, we began developing the product we hoped would save our company.

Meanwhile, the atmosphere at TAS continued to deteriorate. Engineers were missing project deadlines, and sales managers were missing targets for new orders. As I walked the hallways, I saw employees becoming sullen and disenchanted. They seemed to be losing faith in the company's viability. Morale was at an all-time low.

I could see clearly where the company needed to go, but we were not on a trajectory to reach our goals. We didn't have a focused, motivated, committed staff. No amount of new project pizzazz would make up for that. We needed to get everyone motivated and pulling in the same direction. I knew that if something didn't change, the company would fail. I longed to reach the employees on a different level—not on the level of projects and schedules, but on a deeper, more spiritual level.

The solution to our problems could not be an either/or proposition. We needed to change the spirit of the company *and* develop new products. Without a change in spirit, it didn't matter what products we might try to develop, because we would never get them done. Without new products, no amount of company spirit would help, because we'd lack the sales and profits the company needed to survive.

As we entered the summer of 1992, we had solved one huge problem but were facing another. We had identified a new product line to fuel the company's growth, but our employees were in no position to design, build, and sell it. We had a winning game plan, but were fielding a demoralized team.

46. Hard Drive

I ONCE READ THAT FEW COMPANIES MAKE IT to their fifth birthday, and that ten-year-old companies are extremely rare. TAS was eight years old in the summer of 1992, and at the time I wondered if we were nearing the end of the road. Sales were slipping and morale was plummeting. Our plan to enter the wireless market was little more than a dream, a good idea still being researched. I needed to put the company on track and get back to growing our enterprise. We needed to reach that magic tenth birthday, and we needed to do it on an upswing. I didn't know exactly how to achieve that turnaround, but I had an excellent idea where to begin.

I reached out once again to my old friend and adviser Clifton Smith. It had been five years since Dr. Smith made his first formal "house call" on TAS. When he facilitated our 1987 Mount Airy Lodge retreat, the company—with just eighteen employees—was riding high. Now, in mid-1992, we had nearly sixty employees, but we were hardly riding high—we were struggling to survive.

After meeting with Cliff and conferring with Steve and Charles, I decided to convene a team-building meeting to work through the company's morale issues and get everyone pulling in the same direction. In previous years, I limited our team-building participants to the cofounders and two or three other top managers. This time, I decided to widen the scope to include the top-level managers, mid-level managers, and all the engineers. To resolve the issues of stalled sales, manufacturing defects, and slipping product development schedules, we would need the help of all key employees.

We met at a local hotel and spent the whole day discussing the company's status, plans, and problems. The climax of the meeting occurred in the evening, in an exercise Clifton called the Organizational Mirror. Each department took its turn in the center of a large circle formed by the rest of the staff. Clifton then invited and cajoled the staff to share concerns about the department in the hot seat. Staff criticized the sales department for failing to hit our revenue targets. They criticized the product development department for missing deadlines. They criticized manufacturing for the unacceptably high number of defects. Finally, it was the management team's turn to hear the staff's feedback. With some trepidation, I took my seat at the center of the circle, feeling the kind of creeping excitement that could portend either failure or breakthrough.

The first comment came from one of our newer engineers: "Our work hours are much too long. TAS doesn't respect its employees' personal lives."

I expected that comment, because I'd heard that particular employee grumbling in the hallways, and I knew several newer engineers weren't accustomed to working at the torrid pace we had established.

Several employees chimed in. Almost all felt they were working harder than they should. As that line of comment died down, another engineer posed a pair of questions: "Are we choosing the right products to develop? Who does the marketing for the company?"

Those questions were a critique directed straight at me. Everyone knew I managed our marketing efforts and determined what products we would develop.

Then an engineer who had been with us for a few years, a quiet, hardworking, competent guy, shot another dagger: "What about the management team? Sometimes the people who start a company aren't the right ones to grow it."

I felt an intense pressure pushing outward from inside my skull, and could feel the anger coming on. I was thinking that the Organizational Mirror was one of Clifton's worst exercises, and that we should end the meeting and go home. Just then, an engineering technician—a good, solid employee—chimed in with a question: "Has TAS considered bringing in some professional management?"

Another engineer—an older, experienced recent hire—felt the need to pile on. He said the company "may have hit its high-water mark" and might be headed for a fall.

The condescending tone of that last remark really got to me. I tried my best not appear defensive, but it didn't work. On the outside, I felt myself projecting palpable anger. On the inside, I was screaming: *Who do you people think you are? Don't you know we started this company? We built it. We hired you! We didn't have any "professional managers" when we were in my basement, working our asses off into the night. I doubt there are many "professional managers" out there who have started a company from nothing and grown it into a multimillion-dollar business. We are "professional managers," although by "professional managers" you probably mean white managers. Well, if that's what you're thinking, you can just kiss my ass, because we're not going anywhere!*

I didn't say any of that, but I had reached my boiling point. Some key employees were insinuating that the management team was incompetent, and that I was perhaps not the right person to run the company. It didn't matter that most employees were working strictly by the clock instead of doing whatever they could to meet objectives. It didn't matter that they disregarded important product deadlines or failed to think more creatively. To them, the answer to the company's dilemma was to bring in some competent (white?) managers, select different products to work on, and work at a more "normal" (i.e., relaxed) pace.

Through the burning mist of my own anger, I couldn't see any logic in the employees' feedback. If everyone worked shorter hours we would never achieve our objectives. Our engineers had set their own project schedules, so as far as I was concerned they were obligated to meet them, no matter how many hours it took.

I couldn't believe some employees were suggesting that we throw in the towel and develop different products. We were failing to meet deadlines on our *existing* product lines, which were aimed at a market and a set of technologies we already understood. I realized that, as the saying goes, "the grass is always greener on the other side of the fence," but to me, our problem wasn't lack of opportunity, it was lack of effort. We had a clear business plan, and I was certain that if we just executed the plan—if our employees delivered what they were supposed to deliver, on time—we would be just fine.

The insinuation that we might be the wrong people to run the company was especially hurtful. Steve, Charles, and I had built TAS from nothing. Each of us was working twelve to sixteen hours a day. We were working harder than any employee, and certainly harder than any "professional managers" we might hire. I knew I shouldn't take the feedback personally, but I was bitter, hurt, and angry, and nearly ready to throw in the towel.

In the days following the team-building session, I tried and tried to make sense of what I had heard. Steve, Charles, and I met in Steve's office one night and tried to hash it out.

"We just have too many bums," Steve said. "These guys aren't used to working hard. Most of them haven't been part of a successful project, so they don't know what it takes."

"We *are* trying to do an awful lot," said Charles. Sooner or later, something has to give."

"Bullshit, Charles!" Steve said. "We're working harder than all of these bums put together. What about *our* time and *our* families? If we succeed, everyone will be doing better. If we fail, we'll all be up the damn creek."

Try as we might, we couldn't come up with a solution other than working harder and trying to hire some better engineers. I didn't hold out much hope for that solution, because we were already working at our limit, and good engineers were hard to find.

A few days later, Dr. Smith stopped by to review the results of our meeting. He walked slowly into my office and closed the door. His severe expression told me he wasn't bringing any solutions to our problems. Clifton settled into the chair in front of my desk.

"Sooo... what do you think?" he began.

Typical psychologist bullshit, I thought. *He wants me to spill my guts so he can diagnose my "problem."*

I hesitated, then said, "I think the meeting was a disaster. The employees didn't help us solve a single problem. All they did was complain. They're blaming us for the company's problems, and yet they're the ones who leave the building promptly at five o'clock. They're the ones who miss their project deadlines and then go on like it doesn't matter. To tell the truth, Cliff, I'm sick of this bullshit."

I could see Clifton reading the simmering anger in me, but that was okay—I wanted him to see it. I wanted him to know how serious the situation was, so that maybe he would help me find a way out. Cliff's reply was not what I wanted to hear.

"You're going to have to deal with this, Dave. This is how your employees feel. You're the president of this company. You have the authority to make whatever changes you feel you need to make. I suggest you start by seriously considering the feedback you heard the other night."

Just as I suspected, no solutions from Cliff.

"And remember," he said, "every organization is perfectly designed to achieve the results it is achieving."

I had heard that one from Cliff many times, but our "results" at those times were rapid growth and success, so Cliff's maxim was like a hearty pat on the back. This time, it was a swift kick in the ass.

After Clifton left, I was despondent. I'd pinned my hopes on the team-building meeting. I understood my employees' feedback, but it didn't provide any answers. If I were in their position, I might have felt the same way. The company was like a ship that had drifted off course, and the crew's confidence in the captain was rightly shaken.

A few days after Cliff's visit, I spent an evening wandering, alone and dejected, through the Monmouth Mall. As I shuffled past the display window of B. Dalton Bookseller, my attention was drawn to a book cover that featured a picture of Microsoft founder and CEO Bill Gates. The book was *Hard Drive,* a chronicle of Microsoft's origins written by James Wallace and Jim Erickson. We used many Microsoft products at TAS, and I had witnessed from afar that company's phenomenal growth. I rarely spent time reading anything other than technical books, but I took *Hard Drive* home and dove into it. It was a real eye opener.

The book's biggest revelation was the degree of similarity between

Microsoft and TAS. Both companies began as shoestring operations founded by a few technology fanatics. I was surprised to learn that, in its early days, TAS had actually grown faster than Microsoft. However, after TAS reached a few million dollars in sales it started behaving like a mature corporation, and our growth stalled. Microsoft maintained a hard-driving startup culture even as its sales surpassed $100 million. The people at Microsoft were passionate about their products and their mission to put "a PC on every desktop." They hired the most talented people they could find, and those people worked day and night to solve problems and create great products. Because they overwhelmed their problems with phenomenally hard work and a singular focus, Microsoft people succeeded despite their mistakes and missteps.

Bill Gates had been a difficult boss, nerdy and impolitic, driven and difficult to get along with. Many early employees quit because they didn't like the crazy work hours or the manic culture or Gates himself. The Microsoft story confirmed for me that the people who left TAS hadn't done so because the founders had done something wrong or because we were black. *Hard Drive* confirmed that such departures were a natural consequence of our high-growth entrepreneurial culture. I found reassurance that my own driven persona wasn't a defect—quite the contrary, it was the basis for our past accomplishments and a requirement for future success.

I reached a momentous conclusion: TAS had gone soft. We needed to stop functioning like a mature, successful corporation and get back to functioning like a struggling, hungry startup. We needed to be harder, sharper, and more focused. We needed to work more, not less. We needed employees who had the energy and drive and desire to succeed at building something from nothing, not people who wanted the comfort and security of a large corporation. I was convinced that we needed to push the reset button on TAS.

I got together with Steve and Charles and shared the revelations from *Hard Drive*. The Microsoft culture mirrored the way Steve had always approached work, and he was obviously comfortable with it. Charles was less inclined toward the "whatever it takes" approach, but he couldn't argue with the results. Both read *Hard Drive,* then agreed that we needed to refocus the company; that we needed, in engineering parlance, to return to "startup mode."

We knew that merely saying so wouldn't get it done. We decided to spell out what each project, each department, and each person needed to achieve for us to turn the company around. We suspected some employees were unwilling to make the required commitment, and others lacked the needed technical skills. We knew that, as a result, some would leave. We agreed to make sure that any new hires would be compatible with a startup culture, and to prioritize raw talent, energy, and enthusiasm over experience.

We three realized that our new strategy wouldn't work unless we set the example. Each of us felt stretched to the limit, but we each found a way to do more. For Steve, it meant immersing himself in the wireless project and becoming its chief designer. For Charles, it meant working to resolve the logistical and quality problems in manufacturing. For me, it meant literally returning to my basement on nights and weekends to develop software. For all of us, it meant being personally involved in the process of recruiting new employees.

I couldn't effectively execute my part of the new plan if I continued to manage our sales force—that required too much time away from the office. I needed to be on hand to manage the company through tumultuous times. I appointed my marketing director, Nigel Wright, to run the sales force. That was a major change, and not without risk. Sales were our lifeblood, and I had previously tried without success to delegate the leadership role.

Nigel had been with TAS more than a year. He was a tall, handsome, funny, quintessentially British chap. He was a great communicator, well liked by everyone: international distributors, sales reps, customers, and employees. Best of all, Nigel had a solid technical understanding of our products and communicated that understanding clearly and simply. If anyone could handle the sales force, it was Nigel. I had to give it a try.

I called a company-wide meeting to roll out our new strategy. My presentation was clear and straightforward. There was only one slide. It said, "Return to Startup Mode." I called the meeting to order and placed the slide on the projector, then waited a few moments to let the words sink in. Then I summarized the company's problems: missed product development deadlines, manufacturing defects, below-target sales. Next I summarized the prescription for success: stop behaving as if TAS were a mature corporation and start behaving like a startup company. I told the staff about *Hard Drive,* and how Microsoft had behaved like a startup even after its sales reached $100 million. I told them how Microsoft people worked around the clock and slept at the office. I told them that the attitude of Bill Gates and his staff was to win at any cost—to overwhelm mistakes and missteps with incredibly hard work.

I gazed about during my presentation and noticed a range of reactions. A few employees nodded in agreement—they seemed motivated and ready to do whatever I asked. A few others rolled their eyes, pursed their lips, and muttered complaints. The rest just looked stunned. They seemed to be wondering whether the meeting signaled a new beginning or the beginning of the end. I knew how they felt, because I was wondering the same thing.

When I ended the presentation and asked for comments, the room was silent.

The first employee who spoke said, "We're not Microsoft. We're a different company in a different market."

I said, "That's true, but I think the lessons of Microsoft apply to our company. There are more similarities than differences."

Another employee said, "TAS isn't actually a startup, so we can't just say we are."

I replied without hesitation. "Actually, I do think we're a startup. Our problem is that we've been acting like a large, established company, like Bell Labs or IBM. We're too small to do that. We've got to prove ourselves every day. We don't have a cushion to fall back on, and no one is going to give us anything."

Another said, "You can't ask people to give up their lives and their families for work."

I was ready for that familiar sentiment. I replied, "All I want you to do is meet the objectives we set. I'm not asking you to do anything Steve and Charles and I aren't doing. Everybody's life and everybody's family will be better off if we succeed. It won't do any of us any good to fail."

The give-and-take continued. After each of my answers there was a somber silence as the employees pondered the situation and waited for the next person to weigh in.

Finally, Chris, our big, gregarious production supervisor, spoke up.

"Look, I'm not happy about this plan, either. My wife and kids don't like me working around the clock, and I'm tired of this shit, but I'm in this thing too far to quit now. Just tell us what we have to do so we can get back to work."

Chris's comments brought an emphatic end to the discussion. The room again became quiet. The employees looked uncertain and afraid, as if sensing their world was about to change.

I presented copies of *Hard Drive* to all my managers and offered to get a copy to anyone else who wanted to read it. Then I reminded the staff about our bonus and stock option plans and assured them they would share in the company's success.

The employees filed quietly out of the room. During the next few days I observed grumbling in the hallways and hushed conversations in several offices. I could pretty much tell whom we were going to lose. That didn't bother me. Some departures were to be expected, even desired. I'd come to believe that successful companies retained people who fit the culture, and released those who didn't. I was confident about our new strategy and anxious to restart the company. We weren't drifting anymore—we set a clear course toward success. If our plan worked, we would look back and realize that this was the moment we'd turned the company around.

I felt relief and exhilaration in the days after our staff meeting. I had assumed full responsibility for my company's performance and had put a turnaround strategy in place. I had considered my employees' feedback, as Clifton Smith suggested, and had produced a response. My response wasn't a touchy-feely coddling of the employees' short-term wishes, it was a comprehensive strategy that addressed our real problems. I was convinced that if we stuck to our guns, employees who stayed with the company would be much better off in the future.

We began executing our *Hard Drive* strategy in the fall of 1992. After eight years in business, TAS was once again in "startup mode."

47. In the Valley

AS EXPECTED, SEVERAL KEY EMPLOYEES QUIT in the wake of the *Hard Drive* turmoil. No doubt they believed TAS was foundering, and didn't want to go down with the ship. Some left notes, usually angry, on my chair or with my assistant. It was as though leaving a nasty personal message giving me one last tweak would in some small way prove my judgment wrong and theirs correct. I believed the company was foundering, too, but was determined to do all I could to change the culture and lead a turnaround, or at least make the best of a bad situation.

Looking at our results for the fiscal year ended November 30, 1992, one might have thought the naysayers had it right. We recorded an annual sales decline for the first time in TAS history. Sales fell 20 percent. Profits practically vanished. The enterprise we had worked so hard to build seemed to be slipping into oblivion. We were still a multimillion-dollar company, but it felt as if we were dying.

I hit the panic button and decided it was time to consider selling TAS. Knowing nothing about the process, I called an old acquaintance, Pamela Carlton, for help. Pam was the younger sister of my high-school sweetheart, Gay Carlton, and I hadn't seen her since she left high school to attend Williams College. After Williams, Pam had gone on to Yale where she earned a law degree and an MBA. My mentor Darwin Davis had told me Pam was working at Morgan Stanley, and I figured she might be just the person to advise me on selling

TAS.

Pamela and I met for lunch in New York City, and afterward she sent me some books on valuing and selling a company. She also referred me to a high-tech investment-banking firm in San Francisco, Robertson Stephens (RS) and Co. I spoke to Barry Richards at RS, who said a TAS "deal" was too small for his firm, but a friend had a company that might be just right for us. That friend was Peter Hammond, and his company was Clover Valley Group.

Barry assured me Peter was a top-notch investment banker whose company was better positioned to work on "smaller deals." When I called Peter, he sounded as if he already knew about TAS and couldn't wait to work with us.

"Hello, Dave. Barry told me you would be calling. He told me you have an exciting company with an excellent track record."

My first reaction was cynical. *Not exciting enough for Barry to be interested in handling the business,* I thought.

"Barry had some good things to say about your company, too," I said. "I'm interested in exploring a sale or merger of my company, or perhaps even a public offering."

"That's great—that's just the sort of thing we do," Peter said.

Peter told me about himself: military officer, then Stanford MBA, then jobs at large investment banking firms, then out on his own. He had two partners and was looking for a third. He stressed that Clover Valley Group had the expertise of a large firm, but was focused on working with smaller companies like TAS. Peter's pitch sounded good, and I already had a California business trip on my schedule, so I added a side trip to San Francisco to meet with him.

I arrived at the very swanky Pan Pacific Hotel on December 9. I was impressed that Peter had set me up at such a nice place, but when I checked in I realized that I, not Peter, was paying the bill. Had I known that, I would have stayed at my usual hotel—the Holiday Inn.

I met Peter and his partner George Wilson in the lobby the next morning. Both were in their mid-thirties to early forties, and both gave off a distinct Ivy League air. Their khaki pants, Ralph Lauren dress shirts, and carefully styled hair radiated the casual confidence of the upwardly mobile urban financier, California-style.

"Dave—it's great to meet you. We're excited about this opportunity—how about you?

I said I was also excited, and I was. This felt like the first solid step toward selling TAS. I didn't know what the process was going to bring, but the potential seemed huge.

After we exchanged pleasantries, Peter said, "We're going to have breakfast over at the Olympic Club. We can review our agenda there."

The Olympic Club was one of those exclusive urban sports clubs that reek of old money. I surmised it was a hangout for San Francisco's rich and powerful. It was definitely not the sort of place I frequented. A few days earlier, I'd been pleased when my West Coast sales manager and I managed to wolf down a McDonald's breakfast burrito and hash browns before dashing to a morning appointment. Now I was sitting at one of the nicest clubs in San Francisco, dining on a cheese omelet and sourdough toast.

After breakfast, Peter and George took me over to the Robertson Stephens offices. They said Clover Valley's offices were across the bridge in Sausalito and that, for convenience, we would meet in an RS conference room. I started to wonder if Clover Valley actually *had* an office. I also wondered just what the relationship was between RS and Clover Valley Group.

Peter and George shared their professional backgrounds and company history, then invited me to give a presentation about TAS. I took them through the company's origins, the founders, our competitors, our market position, and such. Then I stated the problem: my company was no longer growing.

The two listened intently, and then Peter reframed the problem. Given that we were not growing anymore, he said, we needed to determine our optimum future course. Was the answer to our problem better products, a strategic partner, or more resources? I liked the fact that he was approaching our situation in positive terms, and not simply as a crass discussion about selling the company.

Peter outlined the process Clover Valley Group would execute if we hired them as advisers. First they would perform a "due diligence" analysis of TAS to determine its strengths, weaknesses, opportunities, and external threats. Then they would produce an "offering document" that would introduce the company to potential buyers or partners. Next, they would identify interested parties and market the company to them. Peter's presentation was clear and rational, and I was definitely interested in moving forward.

After Peter's presentation, we broke for lunch at the St. Francis Yacht Club. Like the Olympic Club, the St. Francis was a luxurious place. As we made our way through the dining room toward our table, I thought: *These guys sure do know how to live. I'm breaking my back developing real products, staying at Holiday Inns, and eating at McDonald's. These guys hardly break a sweat, and they don't produce anything, but they live like kings. What's wrong with this picture?*

After lunch we continued our discussions at the yacht club. I expressed concern about disclosing to employees our intent to sell the company, and about sharing our confidential marketing and development plans with potential buyers. Peter assured me he and his partners were experienced with those issues and would handle them appropriately. Then Peter changed the subject to how TAS would pay Clover Valley Group for their services.

First, Peter assured me Clover Valley Group would be happy to serve as advisers to TAS, implying I had met some minimum requirement and he was willing to accept TAS as a client. He reiterated that Clover Valley Group could deliver large-company expertise, but with small-company focus and fees. He said the fee structure would have three components: a monthly retainer, a "success fee" to be paid upon the sale of the company, and expense reimbursements. I bristled, because I saw retainers as a way for lawyers and accountants and other professionals to get paid for doing nothing. I didn't get paid for doing nothing, and I didn't think they should, either. I was also apprehensive about reimbursing Clover Valley Group's expenses, especially in light of the apparently lavish lifestyle the principals enjoyed.

I was disappointed with Peter's compensation schedule, but this was the first real opportunity we had to sell TAS, and Clover Valley Group seemed impressive. I decided that, rather than make a fuss, I would wait to receive a full, formal proposal.

A few days later, Peter faxed the proposal to New Jersey. I blew right through the boilerplate about the Clover Valley Group team and the selling process and went straight to the section labeled COMPENSATION. My attention was immediately drawn to two items: 1) Clover Valley Group required a retainer of $15,000 per month; and 2) they required us to reimburse all their expenses, including charges from lawyers, accountants, and other advisers. They also charged a flat 5 percent commission on the sale—their "success fee." The retainer would be credited toward the success fee, but the expenses would not.

I wasn't sure about the commission, but I was quite sure the retainer and expenses were a bad idea. I certainly didn't like the idea that the Clover Valley guys could live high on the hog, fail to sell our company, and still get paid a few hundred thousand dollars. I would be more comfortable if our advisers were willing to get paid only if we sold the company. The more I read the proposal, the more apprehensive I became. I didn't dismiss Clover Valley Group out of hand, though. I wanted to see if their contract terms were customary, and see how much Peter thought TAS would sell for. After all, if we were able to sell the company for, say, $50 million, we would gladly pay a 5 percent "success fee," and Peter's services would be worth every penny.

The next day, Peter called to see if I had reviewed his proposal.

"So, Dave, what do you think? It's pretty much along the lines of what we discussed when you were here. As I said then, you guys have a really exciting company, and I don't think we'll have a problem rounding up potential buyers."

That gave me the perfect opening.

"Peter, I know you haven't done your valuation yet, but can you give me a rough idea of how much you think TAS will bring?"

Peter hesitated several seconds.

"Well, Dave, it's really hard to say until we do the analysis, but off the cuff I'd say the company should bring anywhere from two to five million."

My disappointment was so profound I didn't know what to say, so I said nothing.

Peter continued: "You see, Dave, your sales haven't been growing, and you indicated that your profit this year will be very small. We would have to convince buyers that that's a temporary situation, and that you can increase profit by cutting R&D and other expenses."

"I see. Well, Peter, give me some more time to go through the proposal in detail and talk to the references you gave me, and I'll get back to you."

On that terse note, we ended the conversation. I was sure Peter sensed my disappointment, but I didn't care. He was probably headed over to the yacht club for cocktails, and I would be at my desk until midnight trying to rescue my company. I forced myself to pick up the proposal and comb through the details again. I noticed some interesting things, such as a provision that Clover Valley Group would be reimbursed for *business-class* air travel. I always traveled coach because I couldn't stand to see my company's resources wasted. I noticed also that Clover Valley Group listed only one transaction in their references that was even remotely similar to ours. I was already negative toward their proposal, but forced myself to call the references. I managed to reach only one.

"They wanted a retainer—I didn't want to pay," the business exec said. "I tried to tell them that charging a retainer wasn't the standard arrangement, and I tried to talk them out of it. They just wouldn't budge. I decided not to use them."

The nerve of those Clover Valley guys! They had referred me to someone who hadn't even used their services. Did they think I wouldn't check? The reference hadn't hired them because the retainer was "not the standard arrangement." I was incredulous. I summed up the situation: Peter gave me a reference who hadn't used Clover Valley's services. Peter would receive $15,000 per month plus expenses whether or not Clover Valley sold TAS. Clover Valley's rough estimated value for TAS—the company built on our blood, sweat, and tears—was a measly two to five million dollars. What had first seemed a solid opportunity was starting to feel like a hustle.

Peter was probably right about one thing, though: Anyone looking to buy our company would be discouraged by the negative sales and profit trend. Heck, that negative trend was the reason I was thinking of selling the company. I wished I had thought of selling during our high growth years, 1985–89.

As the Clover Valley Group opportunity evaporated, I felt trapped. My business was failing, but I couldn't get out. I was thirty-nine, and my dream of retiring at forty to "study anthropology" was looking impossible. Worst of all, there was no time to enjoy life; no time to take a vacation or even spend an afternoon away from the office.

Desperate for a way out, I tried to see the Clover Valley Group proposal in a different light. I could sell TAS at Peter's fire-sale price and extricate myself from the nightmare. I could put at least a million dollars in the bank and go back to work for Bell Labs. It was a tempting thought, but I quickly put it aside. I had promised Steve and Charles more than ten years earlier that we would *all* be millionaires. With their smaller shares, they'd be lucky to walk away with a few hundred thousand dollars. That didn't seem fair, and besides, they were unlikely to go for it.

I didn't want my company to end that way. I didn't want to fail, because everyone who knew me knew TAS was my dream. My identity and self-esteem were inextricably linked to the company.

Soon after the California trip, we gathered all the employees for our annual Christmas luncheon at a nearby hotel. The seasonal celebration was to end with distribution of profit-sharing bonus checks. Given our poor results, the bonuses were tiny, and I wondered how the staff would react. I felt terrible, because I knew everyone was working harder than employees at most other companies. The folks who had stuck with us were good people, and I wanted them to be successful.

We sang Christmas carols and clowned around and had a great time. When it was time to present the checks, I got up and gave a short speech. "I want to thank you all for your hard work. I know this has been a rough year, but things will get better. This isn't some huge company where you can't see the results of your work. Everything we produce comes from the people right here in this room, and you should be proud of that. The bonus this year is very small—maybe enough for a nice lunch at McDonald's." I paused and heard a few uneasy chuckles. "In the future, when the bonuses are much larger, I want you to remember this day. When the company is growing, and our profits are huge again, you'll know that your work made the difference. You'll know that the people in this room produced our success."

I fought back tears. At that moment I didn't care about our financial results. I just wanted everyone in the room to enjoy the moment and come together as a team. I felt that if we could do that, everything else would work out.

My remarks were warmly received. Two assistants began to pass out the bonus checks, and I made my way back to my seat. Numerous employees stopped me. Some wished me a Merry Christmas. Others looked me in the eye and shook my hand. A few gave big, tearful hugs.

I knew right then that I couldn't give up on TAS. Something deep within told me I had to keep pushing forward. The only smart thing, the only fair thing, was to get the company prospering again, and I resolved to do just that—to get the company back on a high-growth track and then look to sell or go public. I wanted to sell from a position of strength, in a way that wouldn't put the company or its employees at risk.

Once I made the decision, I felt relieved and confident. I felt renewed and ready to face the challenge. *Anyone can manage during the good times*, I thought. *It takes a real manager to take a bad situation and turn it around*. One of Steve's favorite sayings reverberated in my head: "If it was easy, everybody would be doin' it."

I have to turn the company around, I thought. *That's why I'm here.*

48. Turnaround

BAD AS IT LOOKED ON PAPER, 1992 was a year of positive transformation for TAS. We entered 1993 with our aggressive *Hard Drive*-inspired strategy in place, new products under development, and a leaner and meaner organization. I was growing more confident that we could turn the company around, but in February my confidence got yet another severe test. Nigel Wright abruptly left the company just a few months after I appointed him sales director. Nigel and I had disagreed over a fundamental issue, the selection of a new eastern region sales manager. The previous guy in that slot was among those who resigned in the wake of my "return to startup mode" edict. Nigel interviewed replacement candidates, and settled on a man with lengthy sales experience but shallow technical background. I was convinced that Nigel's candidate would be like previous failed sales managers—familiar with sales procedures and lingo but incapable of effectively communicating with our savvy engineering customers. Nigel's candidate didn't fit the profile defined by our new strategy. We wanted each new sales manager to have an engineering degree. Nigel's candidate did not. I was determined to stick with our strategy, so I vetoed Nigel's choice.

Nigel felt I was usurping his authority. I felt he was disregarding our new strategy, which was just beginning to take hold. When Nigel left, Steve, Charles, and I engaged in the usual collective introspection and doubt. Was I wrong to veto Nigel's candidate? Wouldn't it have been better to let him have his way than to have him resign? Wasn't this the usual "Dave's way or the highway" approach?

No one suffered from Nigel's departure more than I did. I was back to functioning as president, sales director, marketing director, and software developer. I no longer had the "luxury" of staying in town to run the company by day and develop software by night. I had to go back out on the road with my sales managers. I lugged a computer and CRT monitor on sales trips so I could work on software in my hotel room at night. I was prepared to work around the clock, but I wasn't going to deviate from the *Hard Drive* plan we had just put in place. It was tough to see Nigel go, but I felt the alternative—operating without a clear and coherent strategy—would be much worse. Even though the situation seemed dire, our sales had actually stabilized and I felt we were on the right track. There would be no turning back, and no trashing our game plan.

Several weeks after Nigel's departure, a few of my managers and I headed out for a night of bowling. We had worked our typical long day, and looked forward to a night of camaraderie. It was dark outside and pouring rain by the time we left. On the way out, I noticed a young Asian man sitting in the lobby.

"Can I help you with something?" I asked.

He smiled and said, "Oh, no, that's okay. I just had an interview, but now my car won't start."

The young man was well dressed and well scrubbed, and I could tell he was fresh out of college.

"Who did you interview with?"

"Steve Moore."

"How did it go?"

"Not so well, I think."

I knew Steve was working overtime, interviewing candidates for product development positions—that's why he hadn't joined us for the outing. Steve was a tough interviewer, so it didn't surprise me that the candidate didn't think he'd done well.

"Well, let's take a look at that car," I said. "What's your name?"

"Tash."

We went outside in the downpour and Tash pointed to his car—an old Honda. He got in and tried to start it, but no luck. My managers and I probed under the hood, but we couldn't fix the problem.

"Tash, I don't think we can help you. Do you need a ride somewhere?"

"No, I already made a call. My ride is on the way. Thanks a lot, though."

"Well, Tash, my name is David Tarver. Give me a call tomorrow. I'm going to follow up on your interview to see if we have something for you."

Ordinarily I wouldn't have bothered to follow up on the first interview of a new college grad, but there was something special about Tash. He seemed bright and energetic and optimistic. He was standing beside his car, stranded and drenched, yet he didn't seem depressed. To the contrary—he seemed positively effervescent. In my mind's eye, I saw Tash as a future star in our organization. He had interviewed with Steve for a product development position, but I had a hunch he might be just the sort of person I needed on the sales force. I made a mental note to follow up with Steve. At home later, I wondered if my hunch about Tash might have been wishful thinking, born of desperation and despair over losing Nigel. I knew nothing of Tash other than what our brief conversation revealed, but there was something almost spiritual about the encounter.

I awoke the next morning thinking about how, when I was a kid, my friends and I would scour the neighborhood for bees. We each carried a jar with a tin lid, and the first to spot a bee sitting on a flowering bush would call out, "Mine!" and capture the insect in his jar. Often, two or more kids would go after the same bee. When that happened, an argument would ensue.

"I saw him first!"

"No you didn't, I did! Get back!"

In the wake of the arguing and jostling, the bee sometimes got away.

I hoped Steve and I wouldn't tussle over Tash.

As soon as I arrived at work, I stopped by Steve's office, told him about my encounter with Tash, and I asked him how the interview had gone.

"Oh yeah, well, he's real good. He's graduating from a really good school, Cooper Union, in New York City. He's got good grades, too, or I wouldn't have brought him in for an interview."

I was disappointed. Steve saw him first. He was Steve's hire.

Steve went on: "The problem is, he doesn't have much practical experience. He's had a few lab projects in school, but besides that he hasn't really designed and built anything significant."

I felt a rush of relief. Steve didn't want him! I didn't care that Tash hadn't built anything. He was a smart guy—Steve confirmed that. He seemed personable and poised, even in a driving rainstorm. The typical engineering candidate lacked interpersonal skills, but Tash seemed to be that rare engineer with an outgoing personality. That, in a nutshell, is what I wanted a sales manager to be. I decided to hire Tash.

I decided something else, too: I was sick of paying sales managers huge salaries only to see them leave a year or two later. I was tired of paying top dollar for "experienced" sales executives who couldn't effectively convey the technical advantages of our products. I decided to build a new sales and marketing organization from scratch. I put off replacing Nigel and the departed regional sales manager. Instead, I assembled a sales and marketing team composed of Tash and four other recent college grads. All five were energetic, smart, and hungry. Each had a technical degree from a top university. None had a whit of experience, but I felt that, in our field, experience was overrated. The world of technology had changed tremendously in just a few years, and was still evolving rapidly. I needed people who could learn fast and adapt quickly.

Drawing on my GMI co-op experience, I designed an indoctrination program that got the new team up to speed quickly. They rotated through assignments in customer support, manufacturing, engineering, and sales. I thought that was the best way for them to get to know the company and our customers, and for the company and customers to get to know them. It was an intense indoctrination, and it required me to spend lots of time with the new team. I worked with them every day, and assured them that if they worked hard and produced results they would soon hold key positions. We quickly developed mutual respect and team spirit. The other employees started referring to the new hires as my "pups." They took the new employees under their wings and helped them get up to speed. Before long, something wonderful took hold—TAS started to feel like a dynamic, hungry company again.

By mid-1993, the restart was clicking. Our hard-driving development engineers were meeting deadlines, and our sales trajectory was positive again. When we ended the year with a decent profit, I looked back and saw that sales had increased each month since Nigel left. Our annual sales nearly matched the previous record high.

Early in 1994, with sales still rising, we introduced our first wireless product, the TAS 4500 Wireless Channel Emulator. We were convinced our newest product faced only one serious competitor, Hewlett-Packard. Fortunately for us, HP no longer seemed to be focusing on their product—it was relatively expensive and sorely outdated. Unfortunately, just as we were introducing our new product, I was astounded to see a new, direct competitor depicted on the cover of *Microwaves and RF* magazine. The manufacturer was a small northern New Jersey company called Wireless Telecom Group (WTG), only an hour away from TAS. I wondered if we'd been spied upon. The competitor's timing and proximity seemed more than coincidental, but I was unable to establish any link between our project and theirs. The appearance of the WTG product served notice that our entry into the wireless business wouldn't be easy, but we had been through that drill before. We were going to have to prepare ourselves for another battle.

Despite the rocky TAS 4500 introduction, by year's end we had achieved a new sales record, $7.5 million. Things were going well, profits and bonuses were good again, and morale was high. The company I loved, the company we'd built from nothing, was back on the right track. I felt happy and relieved, but also completely drained. Oddly enough, the authors of *Hard Drive* had captured the reason for my exhaustion. In their book, they referred to a passage from John Steinbeck's *Cannery Row*.

The things we admire in men, kindness and generosity, openness, honesty, understanding and feeling, are the concomitants of failure in our system. And those traits we detest, sharpness, greed, acquisitiveness, meanness, egotism and self-interest, are the traits of success. And while men admire the quality of the first they love the produce of the second.

Building TAS, and especially turning the company around, had hardened me. I didn't have the sinister edge Steinbeck described, but I had developed a thick skin and a firm hand. I evaluated every decision and every relationship in terms of what was best for the company. I didn't feel I could ever let my guard down. As a result, I lost friends and missed opportunities to connect with people socially. I became formal and robotic, and always wore my game face. Like a professional athlete, though, I could only wear the armor and play the game for so long. I was worn out.

It was time to make good on the promise I made to myself late in 1992 as I was trekking through my personal valley of despair. It was time to move on.

49. Exit Strategy

ON THE EVENING OF SEPTEMBER 18, 1994, I was flying high above the South China Sea. Having just concluded a successful visit to Japan, I was feeling good, and was en route to visit our new distributor in Singapore. Two years after being mired in a sales slump and a morale crisis, TAS hadn't merely turned around—the return to startup mode had us growing strongly again, and customers around the world were buying our products.

After dinner, I settled back in my economy-class seat and prepared to enjoy a glass of wine. Gliding at 30,000 feet, sipping a nice cabernet, I truly felt on top of the world. Anything seemed possible. I savored such moments because they often produced my best ideas about technology and business and life. Usually, the ideas survived the flight and still seemed novel back on the ground. Sometimes, though, the good wine and rarefied atmosphere produced ephemeral ideas that turned out to be little more than daydreams.

The sky outside was pitch-black. The dinner service was over, and the flight attendants were settling down to a well-deserved rest. Most of the other passengers were dozing. I had a whole row of seats to myself. I leaned over and slid my briefcase from under the seat in front of me. It was time.

I placed my briefcase on the adjacent seat and looked at it for a moment. I smiled when I saw the stickers Stacy had placed on it a few years earlier. They gave my briefcase a decidedly un-businesslike appearance, but I never dared remove them. They were a much-needed connection to my daughter and to home.

I glanced around to be sure no one was watching, slowly opened the briefcase, and pulled out a small book. I had removed the dust cover before leaving home because I didn't want anyone to discover what I was reading.

I settled back and opened the book, ready to absorb everything it could tell me about the words on its title page: *The Complete Guide to Selling Your Business,* by Paul Sperry and Beatrice Mitchell.

I smiled again, recalling how I'd acquired the book.

"I'm recommending three books to you," Pamela Carlton said.

At that 1992 lunch meeting in New York City, I was reaching out to Pam, a Yale-educated investment banker, for advice about selling TAS. That was among my company's darkest days, but Pam's advice offered a glimmer of hope. I couldn't have imagined twenty-three years earlier that my girlfriend's little sister would one day play such an important role in my business career.

"One of these books is very technical, one is fairly light, and one is kind of in the middle," Pam said. "I didn't know where you wanted to start, so I figured I'd give you options. If none of these addresses what you're looking for, just let me know."

I returned home and put the books on my den bookshelf, where they remained until two years later, when I was packing for the Singapore trip. I quickly flipped through the three books, and decided to take the "light" one. Back in 1992 I had promised myself I would look into selling the company once things got better. Things had definitely gotten better.

The book was better than I'd expected, full of concise prose and common sense. The authors listed reasons entrepreneurs decided to sell their companies. They described the selling process and included concrete examples. They said a seller should create a market among potential buyers, and should expand the market beyond obvious or well-known suitors. They said the best purchase price often comes from a buyer who is seeking to break into a new market or a new line of business.

The book's tone was friendly and familiar. The authors seemed like normal, helpful people, very different from most financial industry "types" I had encountered. The investment bankers, business brokers, and venture capitalists I contacted in 1992 left a decidedly bad taste in my mouth, and that was one reason I had set aside the whole idea of selling TAS.

Flying toward Singapore, one of those high-altitude, anything-is-possible ideas took hold: Why not see if the book's authors would be interested in helping me sell TAS? They obviously had expertise and a successful track record. Their New York City offices were just an hour from Eatontown. I put the book aside and resolved to call Sperry Mitchell and Co. from my hotel.

It was nearly midnight when I settled into my room at the Westin Plaza. I was still excited about contacting Sperry Mitchell, but I wondered if this was one of those crazy altitude-and-wine-induced ideas that wouldn't fly once I was back on solid ground. Would Paul Sperry and Beatrice Mitchell be the helpful, normal folks they seemed? Would they be interested in helping us sell TAS? Would their compensation requirements be reasonable?

There was only one way to find out. I called New York information to get the Sperry Mitchell number, then placed the call and asked to speak to Paul Sperry or Beatrice Mitchell.

A female voice said, "Your name, please?"

"David Tarver, with Telecom Analysis Systems." I tried to sound my professional best.

"One moment, please."

After a minute of recorded music, a different female voice came on the line.

"Hello. This is Beatrice Mitchell."

I felt a rush of excitement. Beatrice Mitchell was a real person, and Sperry Mitchell was a real company.

"Hello, this is David Tarver. I'm the president of Telecom Analysis Systems, and I just landed in Singapore. On the way here I was reading *The Complete Guide to Selling Your Business*, and I was very impressed."

"Oh, thank you very much," Beatrice said. She seemed almost as excited as I was.

"I've been thinking for some time about selling my company. After reading your book, I decided to call and see if you might be interested in advising us on the sale."

Beatrice asked a few questions about TAS. She wanted to know what business we were in, how long we'd been at it, and what our approximate sales and profits were. She seemed interested, and her tone seemed just as unassuming in person as it did in the book.

"I'd love to talk with you about selling your company," Beatrice said. "When you return to the States, give us a call, and we'll arrange a time for Paul and me to sit down with you."

"That's great," I said. I'll give you a call as soon as I get back."

I sat the phone on its cradle and threw my fist into the air. "Yes!" I shouted to the empty room. I had a feeling this idea was going to work out. I was exhausted from the long flight, and my wine buzz hadn't completely worn off, but I had trouble falling asleep.

As soon as I returned to New Jersey I made the appointment, and drove up to New York City a few days later.

The Sperry Mitchell offices were on Madison Avenue in midtown Manhattan. It was a great address for a financial company, but the building was nothing special and their offices were not at all ostentatious. I noted with interest that one receptionist was Asian and the other was black. The Asian receptionist ushered me into a small conference room. A few minutes later, a tall woman with dark hair entered the room and extended her hand. She had pale skin and reddish cheeks. Her handshake was firm, and her eyes seemed to dance in a mixture of excitement and anticipation.

"Hello. You must be David Tarver.... I'm Beatrice Mitchell."

I returned the greeting and we sat at the table. Beatrice said Paul Sperry would be along in a few minutes.

"How was your trip to Singapore?" Beatrice asked.

"It was good. Our distributor there is really starting to establish our business. He also thinks we have some good opportunities in Malaysia and Thailand." I felt the need to impress Ms. Mitchell with the international nature of our business.

"That's great. I was impressed that you would call us from Singapore. Paul and I have been eager to meet you."

Beatrice was so natural and friendly that I began to expect Paul Sperry would be more like the other investment bankers I had met. When he entered, Paul looked like anything but a Wall Street shark. He wore dark slacks and a white shirt and a rather bland tie, and his shirt was unbuttoned at the neck. Paul clearly had been working hard all morning—his shirt was wrinkled and his armpits stained. I could see he was no "pretty boy" yuppie financier.

"Oh, hello, Paul," Beatrice said. "I'd like you to meet David Tarver."

I liked the fact that Beatrice called me "David" and not "Dave" or "Mr. Tarver."

"Hello, David," Paul said. "I understand you called from Singapore a few days ago."

"Well, I happened to have taken your book on the trip, and was impressed by it. When I got to Singapore, I just got on the phone and called."

"We're glad you did. What can we do for you?"

I told Paul and Beatrice I had started TAS in my basement in 1983 while working at Bell Labs; that I had two cofounders, Steve Moore and Charles Simmons, also ex-Bell Labs engineers; that we had grown the company from nothing; that we'd experienced a downturn after the Gulf War but had turned things around; that TAS was once again strong and growing, and would surpass $7 million in sales for the current year.

They listened intently, and seemed genuinely interested. When I finished describing TAS, they told me about their company. They had both worked at larger firms, but had left to start their own. They specialized in selling small to mid-sized companies. They matter-of-factly informed me, to my surprise, that they were husband and wife.

My initial hunches about Sperry Mitchell and Co. were confirmed. Paul and Beatrice seemed like regular people. They appeared competent, and had no problem working with a company the size of TAS. Still, I had apprehensions based on my experiences two years earlier. I put my concerns on the table, quickly and clearly.

"How do you folks get paid for your services?" I asked.

Paul fielded that one. He didn't seem surprised or defensive.

"We get a commission on the sale of your company. The commission follows the standard Lehman formula. Are you familiar with that?"

I felt I could be honest with Paul. "No," I replied.

"The Lehman formula says we get five percent of the first million of the sale proceeds, four percent of the second million, and so on, down to one percent of everything five million and above."

That sounded good to me—immensely better than Clover Valley Group's straight 5 percent commission. Still, I suspected there might be other charges Paul wasn't telling me about.

"Do you charge any kind of retainer for your services?"

"No, there's no retainer. We get paid only after you sell your company," Paul said.

I couldn't believe what I was hearing. I glanced at Beatrice. She nodded and smiled.

Okay, so Paul and Beatrice were proposing a "Lehman formula" commission and no retainer. It sounded almost too good. There was one other thing, though, and I knew this one would trip them up. I almost didn't want to ask, because things were going so well.

"What about your expenses? Do you get reimbursed for office expenses or travel or anything like that?"

I imagined that a firm like Sperry Mitchell, on Madison Avenue in New York City, would generate lots of reimbursable expenses. I thought Sperry Mitchell's expenses might amount to more than their commission.

Beatrice answered that one. "David, the only payment we get is the commission, and we only get the commission if and when you sell your company. We don't charge a retainer, and we pay our own expenses."

That was exactly the answer I had hoped to hear, so I pressed on. We hadn't discussed references.

"That sounds good," I said, trying my best to disguise my excitement. "Can you give me some references, maybe a company similar to ours that you've sold?"

I knew from their book that Sperry Mitchell had handled the sale of many small companies, but I wasn't sure whether they had ever sold a telecommunications-industry company.

"Of course," Paul said. He turned to Beatrice and said, "What about Cidco? Do you think that would be a good reference for David?"

"Cidco would be perfect," Beatrice said. Then she said to me, "Cidco makes caller ID equipment, and they've done very well. We handled the sale of that company last year, and I'm sure they would be willing to talk to you. We can give you a few others, but that's the first one that comes to mind."

On the outside, I was smiling. On the inside, I was doing somersaults. It wasn't just that the answers Paul and Beatrice gave were so good, or that they seemed so much better than other potential advisers I had talked to. What impressed me most was that they were so casual and comfortable—with each other and with me. I liked the way they discussed the situation on the fly, in front of me. They weren't canned or slick. They were real and effective.

"Thanks," I said. "I look forward to talking with the references, but everything sounds good. Now, do you have any questions for me?"

Paul spoke up: "David, the first question we like to ask a potential client is: Why do you want to sell your company?"

I had anticipated that question, because Paul and Beatrice had addressed it in their book. I had a gut answer and a more polished answer. I gave them the gut answer. I explained that Steve, Charles, and I had worked ridiculously hard for years to build our business. I told them how, in the early days, we would work a full day at Bell Labs and then work in my basement until 2:00 a.m. to build our product prototype. I told them how, through sweat and grief and sheer brainpower, we built the company into a multimillion-dollar entity. I told them how our profits nearly vanished in the 1991–92 recession, and how we then scrambled and turned things around. Then I explained our one big fear: that a bigger, better-funded company would come along and take away everything we had built. Yes, our biggest reason for wanting to sell was fear—fear that we could lose everything if we didn't put the company on a larger, more stable footing.

Paul and Beatrice nodded knowingly as I went through my reasons for selling. I'm sure they had heard it all before, even if not in such straightforward terms. Most business owners would say they wanted to sell to realize the value of their hard work, something they called "shareholder liquidity." They would say that they wanted to acquire a "strategic partner" so they could "capture market opportunities." When you stripped away the fancy terms, though, their reasons often amounted to simple fear.

In my own case, there was one other issue, and I put that one on the table for Paul and Beatrice, too.

"I never saw myself running an engineering business for the rest of my life. I always planned to sell my company at forty and move on. I want to do other things in life, and I can't do those things if I'm tied up around the clock running my business."

Paul and Beatrice didn't seem taken aback. Beatrice asked, "Is that how Steve and Charles feel, too?"

"No," I said. "I'm pretty sure they want to stay with the business, at least for the next few years. Steve and Charles are a few years younger, and they want to keep working for a while."

"That's good," Beatrice said. "Most buyers will want to see continuity in the business."

We spent the next hour or so going over the specific steps we would each need to take to complete a sale of TAS. Most of what we discussed was in their book, but sitting there talking with them brought the subjects to life and made a possible sale much more real. The meeting left me extremely impressed, and convinced me that Paul and Beatrice were the people we should hire to sell TAS. I tried to contain my excitement, but it was obvious that I was sold.

One hurdle remained. I needed to get buy-in from Steve and Charles. I had informed them generally about my desire to sell, and they had generally agreed to the idea. Now it was time to talk specifics and get started. I headed back to New Jersey to share the meeting's results with my cofounders.

The next morning I got together with Steve and Charles. They were eager to hear how it had gone, and I laid it all out. When I told them Paul and Beatrice had asked about our motivations for selling, Steve responded: "Like I always said, I'm doing this because I want to make a lot of money. If selling dope wasn't illegal, I'd do that."

That's what Steve had always said whenever we talked about motivation, but I knew he was just blowing smoke. Steve was the most dedicated, hardest working, most loyal person I'd ever met. Sure, he wanted to make a lot of money, we all did, but Steve was just as dedicated to the adventure—the challenge, the proving of our worth—as I was.

"Well, Steve," I said. "If I'm right about Sperry Mitchell, we should be able to make a lot of money from the sale. I'm convinced that we will all come out of this as millionaires, like I always said."

Charles was more circumspect: "So how much did they think we could get for the business?"

"They were reluctant to get specific, but they thought we could get a lot more than the two to five million Clover Valley Group was talking about," I replied. "The price we get will be determined by the market, and Sperry Mitchell is going to help us maximize the price."

Charles seemed satisfied with that, so I decided to confirm what I told Paul and Beatrice about their plans to stay with the company.

"Another thing Paul and Beatrice asked about was what we planned to do after the sale. I told them my goal was to leave the company, but that you guys planned to stay. Is that still the case?"

"I'd like to keep working at least five to ten more years," Charles said.

"Yeah, we're not old like you, Dave," Steve joked. "I'll keep working as long as I like what I'm doing, but if the buyers turn out to be assholes I'll be out of here in five minutes, and I'll probably kick their ass on the way out."

Steve always had a way with words.

A few days later, Paul and Beatrice visited TAS to see the facilities and to meet Steve and Charles. I intended that the meeting would give us enough information to make a final decision about hiring Sperry Mitchell.

I was surprised when I looked out my office window and saw that Paul and Beatrice had driven down from New York in a Ford Taurus station wagon. No yuppie BMW or Mercedes for these two.

As I expected, Paul and Beatrice got along fine with Steve and

Charles. They made a presentation about the selling process and about their capabilities, then spent an hour or so discussing the process and answering questions. I could see the excitement building in my cofounders as their questions were answered and their concerns addressed. By the time Paul and Beatrice returned to New York, I knew what our verdict would be. We agreed to sign Sperry Mitchell.

As I reflected on our decision that night, I realized that deciding to sell TAS was a huge step, but that for me it was an easy step.

For years, my friends had assumed I could never leave TAS. It was my baby, the company I started in my basement. They knew my self-esteem and the company's performance were inextricably linked. They didn't realize that, for me, TAS was a means to an end. They didn't understand that I had become bored and drained by the constant competitive battles. I had grown weary of proving ourselves to skeptical customers over and over again, of conquering the next product design, of winning the next major client. My life was starting to feel like an endless loop.

My friends couldn't possibly understand the emotional strain I felt. I hated firing people, especially people I liked, because they weren't doing the job. I was tired of lying awake at night worrying that my family's well-being could be destroyed by a new competitor or a new technology or a lawsuit.

I had proved to myself what I had set out to prove and was ready to move on. I needed time to relax and think, time to get to know my kids better, time to plan my next moves in life. I had come to feel that, outside the walls of TAS, life was passing me by. I wanted to get my *time* back. I recalled countless days when, holed up in a conference room for a meeting, I would stare outside wistfully and realize I was missing yet another gorgeous afternoon. I remembered so many trips, whether to Paris or London or Tokyo or Singapore, where I wished I could relax for a few days and experience the place—to meet the people and understand something about their lives. I never felt I could take the time to do those things while running on the TAS treadmill. I felt every minute had to be spent in a productive activity or we might meet our demise.

There are some pursuits, I suppose, one should be willing to carry on for a lifetime. I suppose, too, that there are some businesses one should never consider selling. For me, TAS was neither.

The next morning, I called Sperry Mitchell and informed them that Steve, Charles, and I had decided to hire their company as advisers. We were putting TAS up for sale.

50. Break Some Eggs

I WAS NEVER GIVEN TO WILD PARTYING and drinking on New Year's Eve. All that reveling seemed forced—more obligation than celebration. So it was that on the evening of December 31, 1994, I sat at home reflecting on my career and anticipating a life-changing event. I had only one New Year's resolution for 1995—to sell the company that had defined and consumed my life for more than twelve years.

Paul Sperry and Beatrice Mitchell had completed our "Confidential Memorandum," the document that would inform potential buyers about TAS. The first meeting we held to work on that document provided new insight into our advisors. Beatrice asked if we had a refrigerator. She had recently given birth and wanted to pump and store her breast milk. While she excused herself, I cleared the old sandwiches and moldy fruit from our corporate fridge to make space for her milk. After that interlude, we completed our business. If I needed further confirmation that we had chosen the right advisors, Beatrice's intrepid spirit had just provided it.

By the end of January, Paul and Beatrice were contacting companies to determine the interested parties. The time for meeting potential buyers was near. I wasn't getting cold feet about selling, but I was worried that the process might do serious damage to the company.

My immediate concern was breaking the news to our employees. The TAS team was a close-knit group. Our Eatontown staff of fifty-seven worked in one relatively small building. In a group as tight as ours, it would be impossible to keep a secret as important as the potential sale of the company. In a few weeks, neatly suited executives would be walking around TAS to "kick the tires." How would our employees react? Would our key people think the company was about to collapse? Would they think they were being sold down the river? Would they jump ship? We were on track for another year of record sales and profits, but losing key employees would jeopardize all that.

I was also concerned about how news of our desire to sell the company would play with customers. Would they worry that TAS wouldn't be around to meet their future needs? Would they have concerns about warranty service and technical support? Would news that TAS was for sale dry up orders and make selling the company more difficult?

Mostly I was worried about our confidential information falling into the hands of a competitor. How would we respond if a competitor expressed an interest in buying TAS? What if a potential competitor expressed an interest, received confidential information about our products and customers, then used that information to develop a competing product? Years earlier we'd been stabbed in the back by a supplier who used knowledge about our sales and customers to compete with us. I feared that exposing our most sensitive information to potential competitors could result in a similar or even worse disaster.

Nonetheless, I was determined to move ahead. I decided that, at least as far as employees and customers were concerned, the best approach was to be open about our plans and to frame the sale in positive terms. It would be impossible to prevent employees from finding out about our intention, and it would be wrong to conceal it by lying to them. I didn't want selling TAS to be a furtive process, as if we were doing something underhanded. Our team had come through the roughest times together, and the company was once again strong and growing. The right buyer could put the company in an even stronger position. Such a sale wouldn't only benefit the three principal owners—it would benefit our employees and customers, too.

Despite my desire to be open, I felt it would be crass and dangerous to simply say we were planning to sell the company. I decided to say we were "exploring ways to enhance growth," and that alternatives included private equity, public offering, and outright sale. That sounded a lot better, and it was true. Our primary focus was to sell the company outright, but we were certainly willing to entertain other options.

I felt confident that openness was the right approach to deal with employees and customers, but I remained worried about competitors. I was paranoid that sensitive product information, revealed to a potential buyer, would be used to compete against us. The worst possible result of the process was that we would fail to sell the company, and that revealing our confidential information would generate new competitors. I couldn't get past that possibility on my own, so I decided to share my concerns with Paul Sperry. He was, after all, the expert in such matters.

"David, rest assured that we're not going reveal any sensitive product or customer information to anyone who is not absolutely serious. We won't even send the confidential memorandum to anyone you don't approve in advance. Before anyone gets to look at the really sensitive information, they'll have to sign a letter of intent. Trust me, I've been doing this for a long time, and I've almost never seen a competitor use the process in a nefarious way."

"Almost never?" I said.

"Look, David, if you're going to make an omelet, you have to break some eggs. There's always risk involved in selling a company, but it's our job to help you manage the risk. You've got a great company and a great story, and I think there'll be lots of interest from legitimate buyers. I'm confident we're going to find a buyer for you."

Paul's words didn't completely allay my concerns, but I agreed it was time to "break some eggs." I chuckled as I recalled an even more direct expression my Aunt Mary had often used: "it's time to shit or get off the pot." Our next employee meeting was scheduled for February 4. I decided to announce our plans then.

It was our regular monthly meeting, so our employees weren't expecting momentous news. They were still feeling the glow of their year-end bonuses and holiday vacations. After opening the meeting in the usual way, I paused, took a deep breath, and put up a slide entitled "Expansion and Growth." I said our continued growth would allow us to meet our customers' needs more effectively, and would provide career opportunities and financial rewards. I gave specific examples of new products our customers needed, products we'd been unable to provide for lack of resources. Then I dropped the bomb. I said that to finance our continued growth, we needed to consider three options: finding an equity investor, doing a public stock offering, or selling the company.

I looked around to gauge the response. To my great surprise, I didn't detect concern on anyone's face. I thought maybe I hadn't been explicit enough, so I continued.

"Thanks to the hard work of everybody here, the company is doing great. We want to make sure we put ourselves in position for even more growth and success in the future."

I looked around again. Still no concerned looks, still no questions.

I thought, *don't sell past the sale!* That was the salesman's number-one rule: when you've made the sale, shut your mouth. I couldn't help myself, though. I kept talking.

I stressed that whatever path we chose, the company would continue to succeed and grow, and TAS employees' careers would be enhanced. I said if we succeeded in finding the right investor, everyone would win. I emphasized that if we didn't find a suitable investor, we would keep growing the company using our own resources. I said the company would continue to do well without additional investment, but probably wouldn't grow as fast as it otherwise might. I assured everyone we weren't just looking to cash in on the company's success—we were looking out for everyone's best interests.

I paused again, looked around, and didn't detect any anxiety. I invited questions and comments.

"Do you guys have any specific investors or buyers in mind?"

I was relieved that someone had finally spoken up.

"No. We're at the beginning of the process. We think there'll be a lot of people interested in investing in TAS, but right now it's too early to say."

"How long will it take to find these investors?" another employee asked.

"We think we should know something by the end of the year," I said.

Then one of the younger engineers said, "I think we should take the company public. We could raise a lot of money that way." Several key employees owned small amounts of TAS stock, and they recognized the potential windfall a sale or public offering would bring.

"That's one option we're considering," I said.

I realized that my earlier concerns about the staff's reactions were unfounded. Our employees seemed to think selling the company was a good idea. I could feel the excitement in the room. Their unstated message seemed to be: *What took you so long?*

In the dark days of 1992, when the company was stagnating, the employees wanted to know how we were going to turn things around. Once we had accomplished that, they wanted to know how we were going to keep growing. The idea of selling the company seemed to answer that question and ease their concerns about the future.

I remembered how insulted I was during our team-building meeting in 1992, when some suggested that I bring in "professional management" to run the company. I sensed some residual feeling among the employees that, despite our subsequent success, gaining access to a larger company's resources and management expertise would be a good thing. I realized that, all along, the employees had been seeking career safety and security. The idea that we might become part of a larger company with more resources wasn't scary to them—it was comforting.

I felt tremendous relief. Our employees were onboard, and we could proceed to the next step. I was confident we could "break some eggs" without breaking the company.

51. The Dubious Dance

The cat was out of the bag. Our employees knew we were prepared to sell TAS, and soon our customers and competitors would know, too. It was time to put the selling process into high gear.

Paul and Beatrice had offered more than seventy-five companies the opportunity to review our confidential memorandum. A few were major players in the telecommunications industry, but I had never heard of most. As it turned out, a majority weren't interested, meaning they didn't want to even *consider* buying or investing in TAS. Some reviewed the document but declined to go further. Several companies, after reviewing the memorandum, expressed interest in visiting TAS for a firsthand look.

Our first face-to-face meeting was with a large U.S. telecommunications test equipment company. The meeting took place at the 1995 Communication Networks (CommNet) tradeshow in Washington, D.C. Both we and our potential suitor were exhibiting, which allowed us to meet conveniently and discreetly.

I was working in our booth with Rebecca Fish, my marketing communications manager, when I saw three middle-aged white executives sauntering slowly toward us, talking and pointing in our direction. When they arrived, Rebecca greeted them with a smile. One said, "Good morning. We're looking for David Tarver."

I was standing next to Rebecca. She smiled wryly, raised an upturned palm, and said, "Well, here he is."

I tried to ignore the surprised looks on the faces of our guests.

"Good morning. I'm David Tarver, president of TAS. It's nice to meet you."

"Good morning," one of the visitors said. He didn't extend his hand as he introduced himself and his colleagues. The demeanor of all three changed from neutral to cold—they weren't the least bit friendly.

I hoped our visitors were simply wearing their game faces and were seriously interested in buying TAS. I walked them through our booth and briefly described each of our products, but their sideways glances and lack of questions told me they had no interest. After less than ten minutes, they headed back toward their own booth.

When I returned to New Jersey, Paul Sperry called to say our first nibbler was no longer nibbling. I was disappointed, but not surprised.

The next potential buyer was a Korean company I had never heard of. Paul Sperry told me they had lots of money to spend and had expressed great interest.

Executives from the company visited our Eatontown headquarters, but they clearly didn't understand our products or our business strategy. I wondered if they'd done any homework, and why they bothered to visit. A few days after the meeting, I was shocked to learn they remained interested. The executives said they had only to check with their bosses in South Korea before going forward with a purchase proposal.

I couldn't envision selling TAS to a company that had demonstrated so little understanding of our business. Before the Koreans could get back to us, I told Paul and Beatrice we weren't interested.

A few more promising prospects turned out to be duds. I got discouraged, but Paul assured me that our experience was normal and that we would find a buyer.

The next company Paul brought to us was a venture capital firm. I had always been wary of the investors many called "vulture capitalists," but this group seemed like good people and seemed genuinely interested in TAS. We had a good meeting and a lively discussion. Shortly afterward, they offered nearly $9 million for a 70 percent stake in TAS. Their plan was to continue to grow TAS and later take the company public. They said a lot of the potential payout to Steve, Charles, and me would result from the public offering.

In 1992, I probably would have jumped at that offer, but now I found it barely acceptable. I believed TAS was worth more than they were offering. I also felt that giving up control of the company I had worked long and hard to build was worth a lot more than a nonspecific future payout. I wanted to be paid in full. I discussed the matter with Steve and Charles, and we agreed to table the offer.

I found the results of our selling process discouraging. I couldn't understand why major players such as Hewlett-Packard and Tektronix weren't interested in buying TAS. I began to think we might not find a suitable buyer.

Paul Sperry called to say a British company, Bowthorpe plc, was interested in visiting TAS. I had never heard of Bowthorpe, and doubted anything would come of a visit. I was sure Bowthorpe was another dud, but there was no point in refusing the meeting.

In May, two Bowthorpe executives, Colin McCarthy and Helen Ralston, arrived in Eatontown. McCarthy, a senior executive, headed finance at Bowthorpe. He was immaculately dressed, and a full head of white hair capped his trim frame. He looked like the quintessential British gentleman. Helen Ralston was much younger. She had red hair and a cherubic face, and was McCarthy's deputy.

Steve, Charles, and my CFO, Jim Scott, joined me for the Bowthorpe meeting. The get-acquainted portion was formal but cordial. McCarthy said Bowthorpe was a billion-dollar company that made a wide variety of electrical and electronic products ranging from plastic cable ties to aircraft flight recorders. He said Bowthorpe had nearly one hundred subsidiaries worldwide, a vast network the company had built by acquiring small companies like ours. He said Bowthorpe typically let its subsidiaries run themselves unless they got into trouble.

I gave a brief overview of TAS, but didn't have to say much because McCarthy and Ralston had read our confidential memorandum. Both asked good questions, mostly about our finances. McCarthy asked us to provide some specific financial statements, and we promised to do so. The meeting ended and our guests returned to their hotel. I thought the Bowthorpe delegation was genuinely interested, but I wasn't sure. I didn't sense we had connected, and I couldn't quite see myself working with them.

Jim Scott pulled together the requested financial statements. He suggested we take the papers to our guests' hotel, less than a block away, and invite them downstairs for a drink. We met McCarthy and Ralston in the Eatontown Sheraton bar. To my surprise, both seemed much more relaxed. As we shared information about our interests and our families, I could almost imagine the Bowthorpe execs as colleagues. I liked McCarthy and Ralston, and consequently was beginning to like Bowthorpe. Before we left the bar, McCarthy said Bowthorpe was seriously interested in acquiring TAS, and that their CEO would likely visit soon to see the company for himself.

A week later a patrician gentleman named Dr. John Westhead did exactly that. Paul and Beatrice came down from New York for the meeting. We were all eager to see if we had found our buyer.

Bowthorpe's CEO appeared at least sixty years old. He was tall and stately, if slightly bent, with thinning white hair. Mischief, skepticism, and age lines seemed permanently etched on his face.

As soon as we sat at the conference table, Dr. Westhead asked, "Might I trouble you for a glass of wine?"

The unusual request seemed to put everyone in the room at ease. I thought, *now this is my kind of guy!*

We happened to have a fresh bottle on the premises, so after asking everyone else at the table if they wanted a glass, I went to fetch the wine for Dr. Westhead. I poured myself a glass, too.

The meeting went better than any of us had expected. Despite his apparent eccentricity, Westhead was sharp as a tack. He asked excellent questions about our technology, our marketing, and our business strategy. What impressed me most was that he seemed to bond with Steve and Charles right away, and seemed to identify with their engineering-driven perspective. I liked John Westhead, and I could tell Steve and Charles did, too. After the meeting, the three of us agreed that if Bowthorpe came in with a good offer, we would seriously consider it.

A few days later we got word that Bowthorpe was putting together an offer. Although we didn't know the amount, we were encouraged and hopeful. In early June, we received a preliminary offer from Bowthorpe. To my disappointment, it didn't specify a purchase price. Instead, Bowthorpe offered to pay seven times our 1995 earnings. It seemed to me that, though the Bowthorpe folks claimed they were impressed with TAS, they weren't prepared to put a solid, fixed-price offer on the table. They weren't confident about our business prospects, so they were hedging their bets.

I called Paul Sperry and Beatrice Mitchell to complain about the offer and get their reaction. They listened to me rant for a few minutes, and then Beatrice asked, "How much are you willing to sell the company for?"

I had considered our bottom-line price with Charles and Steve many times, and we had agreed that $15 million would be nice, and $20 million would be great. That's what I told Beatrice.

Beatrice said, "Well, it looks like you're on track for over two and a half million in profit, and seven times that would make the sale price around eighteen million, so you're practically there already!"

"Yeah, but what if something happens and we don't hit our profit target?" I said. "Why can't they just offer us a straight eighteen million and call it a day?" After all the ups and downs and unforeseen events affecting our business, I wanted to eliminate any uncertainty.

Despite my apprehension, after reflecting on Beatrice's question and discussing the offer with Steve and Charles, I realized there was very little possibility we wouldn't make enough profit to result in at least a $15 million selling price. I realized we could heavily impact profits simply by maximizing sales and minimizing expenses. I realized that if we pushed hard for the remainder of the year, the sale could bring even more than we'd hoped. We could do everything possible to maximize our profits and turn Bowthorpe's "variable" offer to our advantage. Steve and Charles agreed, and together we decided that we should give everyone in the company a strong incentive to help maximize profits. We designed a generous bonus plan that would give all employees a share of the year's profits. The bonus would increase dramatically if profits exceeded the $2.6 million target, and would decrease substantially if our profits were below that. We called our scheme the 1995 Bonus Plan.

We told Bowthorpe we were inclined to accept their offer, provided they would accept and fund our bonus plan. We pitched the plan as a win-win-win situation: if we exceeded our 1995 profit target, Steve, Charles, and I would receive fair compensation for selling the company; TAS employees would get great bonuses; and Bowthorpe would get a strong company with an excited, motivated staff.

Bowthorpe agreed, and they accepted a provision that Paul and Beatrice proposed—that Bowthorpe make additional payments to us if our profits for the three years after 1995 exceeded our targets. That provision brought the maximum total purchase price to $30 million.

It seemed that in just a few months Steve, Charles, and I would be wealthy young men, able to chart our futures on our own terms. When we told our employees about the deal, and about the 1995 Bonus Plan, many seemed even more excited than we were. Those who owned TAS stock stood to reap significant additional money from the sale—enough for a new car or a down payment on a home.

The company meeting at which we announced the Bowthorpe deal to employees was the most exciting session we'd ever held. It was a gratifying contrast to the meetings held during the depths of our 1992 downturn. Every TAS employee, from top manager to engineer to manufacturing worker to stockroom clerk, walked out of that meeting ready and eager to send our 1995 profits through the roof.

52. End Game

ON AUGUST 3, 1995 WE SIGNED A FORMAL LETTER OF INTENT to sell TAS to Bowthorpe plc. It was an important milestone, but we were far from the end of the road. The document merely signaled the end of the beginning of the sale process—and the onset of Bowthorpe's due diligence.

The Bowthorpe folks would pore over every piece of information about our business: tax returns, financial statements, customer lists, sales invoices, purchase orders, employee data, and insurance policies. Our CFO, charged with coordinating our side of the effort, endured excruciating pain in this process, but I wasn't worried about the outcome. I knew our business was sound and our record keeping solid.

Then Bowthorpe did something unexpected: They hired a high-powered consulting firm to analyze our business and produce a "red flag analysis" to reveal any potential problems with our products or our market forecasts.

I was concerned because I knew every business faced *potential* problems. If Bowthorpe wanted to find a reason not to buy TAS, they probably could. I feared the consultants would try to prove their worth by identifying a "problem" in order to justify their fee, and that some Bowthorpe executive would then cite that problem as a reason to thwart the deal. I was concerned, but knew I had to cooperate. I didn't want to appear defensive or nervous, or do anything that would give Bowthorpe an excuse to back out.

By late August, Bowthorpe was deep into its due diligence, and the consulting firm's hunt for red flags was well under way. That's when my personal life was rocked by awful news from Flint. The news concerned my grandmother, Mimama, my guardian angel since birth. Mimama was dead.

I had last seen Mimama earlier that summer at Heritage Manor nursing home in Flint. Mimama was one hundred years old. Dementia had greatly decreased her awareness. Dark veins were clearly visible beneath the barely pigmented skin of her arms, and her hair had almost thinned away. Others might have seen her merely as an elderly woman, but to me she was beautiful and timeless.

I bent over Mimama's bed and gently kissed her forehead. She opened her eyes, made a feeble attempt to smile, and then opened her mouth and stuck her tongue out just a bit. I took a butterscotch candy from a bowl on her nightstand and placed it on her tongue. In a weak and garbled voice, she said, "Thank you." I was touched, because her response reminded me that her goodness came from somewhere deep within. Mimama couldn't remember my name, and had been bedridden for years, but she still tried to smile when she saw me. She still said thank you in response to my small gesture of kindness.

I asked Mimama how she was doing, and she kept repeating that she wanted to go home. When I asked why, she said she wanted to see her mother and father. I imagined Mimama sitting in the old farmhouse in Camilla, Georgia, with her industrious father, Timothy, and loving mother, Alice, and her many siblings. At first it seemed odd that she wanted to go back to that place and time. Then I realized that in my own final days I would probably long for our Sixth Street home, my beautiful mother, my father's basement workshop, and Mimama's salmon croquettes and grits. Those images were as indelible and enticing for me as Mimama's childhood home was for her.

I didn't know how to respond to Mimama's request, so I began singing *River of Jordan,* one of her favorite gospel songs. I patted her hand gently as I sang, and she nodded her head as if to keep time. One verse began, "I'm gonna shake hands with my mother," and another began, "I'm gonna shake hands with my father." That was the closest I could come to helping Mimama go home. I felt sad and ashamed because, despite my education and business success and impending wealth, I couldn't do any more for her.

Mimama's death reminded me of how much life I had missed while totally focused on my business. I spent a week in Michigan mourning and reflecting, and then returned to work even more intent on closing the Bowthorpe deal and moving on to new experiences and deeper connections with friends and family.

Despite my personal setback, business continued to go extremely well. Our highly motivated employees were building, selling, and shipping more products than ever. By mid-September it was clear that we would exceed our profit target by a significant margin.

Meanwhile, Bowthorpe's due diligence and the consulting firm's red-flag analysis dragged on into October. Every few days, Bowthorpe raised a concern about some aspect of our business, and we addressed it. In October, the consulting firm made its final report to Bowthorpe, concluding that TAS's business, while solid, could suffer if market trends didn't continue or if our new wireless product failed to gain market acceptance. As I expected, the report didn't recommend a course of action, but I wondered if Bowthorpe might back out.

Meanwhile, my attention was drawn to an event that had nothing to do with Bowthorpe—the Million Man March, scheduled for October 16 in Washington, D.C. The organizers encouraged black men to stand up for each other and their families, and to take responsibility for their homes, their communities, and their economic destiny. Those aims were the heart and soul of the shared dream that led Steve, Charles, and me to leave Bell Labs jobs and launch TAS. I thought seriously about attending, but the Bowthorpe deal was at a fragile stage, and I needed to be on hand and fully focused in case a critical issue arose. I would have time for social activism *after* the sale.

Bowthorpe completed its due diligence at month's end. The merger agreement was nearly finished, and the deal appeared set. Then came a serious last-minute hitch.

Paul Sperry called and said a new player had joined the Bowthorpe side. John Westhead, the CEO who had made the decision to buy TAS, was retiring. His replacement, Nicholas Brookes, wanted to see TAS for himself before signing off on the deal. Paul said he didn't expect Brookes' visit to derail the sale, but I wasn't so sure. I worried that John Westhead's purchase decision might be viewed as an eccentric old man's last hurrah, and that Nicholas Brookes might reverse the decision in order to assert his authority as the new CEO.

A few days later, Brookes arrived at TAS and met with Steve, Charles, and me. The meeting did not go well. Brookes appeared uninterested and negative, and the questions he raised about our business prospects seemed laced with skepticism. Steve and Charles had enjoyed talking with Westhead, but they and Brookes didn't hit it off at all.

I felt terrible. Our sales were great and we were headed for record profits, which would yield a sale price at the high end of our expectations. We were committed to the sale, and our employees were calculating their bonuses and stock sale proceeds. I didn't want to believe that Brookes might back out so late in the game.

After the meeting, Steve, Charles, and I were disappointed and angry, and the uncertainty about Brookes' position was killing us. Desperate to salvage the situation, I decided to try to connect with Brookes on a social level. I was wary of that approach, though, because I assumed that if he and I didn't get along, we could pretty much kiss the deal goodbye. I considered it a risky ploy, but I had to try. I couldn't let the visit end on a sour note.

As I walked Brookes to his limousine, I invited him to have dinner with Diane and me. To my surprise, he accepted. I picked him up at the Sheraton Eatontown and drove him to What's Your Beef in Rumson. Diane drove directly from her office and met us there. It was a fitting place for such a consequential dinner, because What's Your Beef was where Diane and I had our first date, eighteen years earlier.

The dinner atmosphere was the exact opposite of our meeting at TAS. Nick, Diane, and I ate delicious steaks, drank lots of wine, and talked. We shared our backgrounds and talked about our plans and hopes. I assured Nick that Steve, Charles, and I all felt Bowthorpe was the right owner for TAS, and that together we would continue to grow the business. As the evening drew to a close, Nick surprised me by saying he was still "keen" to buy TAS and had been pushing the deal for months. His words and body language said the deal was done. Later that night I told Steve and Charles, and we all breathed a huge sigh of relief.

By mid-November the final merger agreement was drawn, all regulatory approvals had been received, and the closing was scheduled. November 21, 1995, two days before Thanksgiving, was the date we were to complete the deal with Bowthorpe. I thought of it as our Independence Day.

53. Free at Last

SOME FOURTEEN YEARS AFTER STEVE, CHARLES, AND I MET in my basement and began building our first product prototype, I eased my car into a parking space at the Red Bank law offices of Giordano, Halleran and Ciesla. It had been nearly twelve years since we incorporated Telecom Analysis Systems. The excruciating months of due diligence had themselves seemed like years. Now, finally, it was time to close the sale and formally transfer TAS ownership to Bowthorpe plc.

The receptionist at Giordano ushered me into the conference room where the closing was to take place. Giordano had been our law firm almost from the beginning. I had visited their offices many times, but had never noticed this particular room. A huge conference table dominated the space. The top was highly polished marble, and the base looked like fine mahogany. Sumptuous high-backed black leather chairs lined the perimeter of the table. It was an imposing venue worthy of such a momentous occasion.

I had never seen so many neatly arranged papers in one place. Some were in manila folders, others loose. Some folders and papers were arranged vertically in metal organizers; others were stacked horizontally.

Steve and Charles were already there, and both were nattily attired. Steve wore a nice wool sweater and slacks, and Charles sported a new suit. I wore a suit, too. If ever there was an occasion to wear one, this was it. It wasn't every day one entered a room as a workingman and exited as a money-in-the-bank millionaire.

Paul and Beatrice were there, too. I had chosen them to represent us partly because they seemed like regular people. In all our previous meetings, they had dressed like regular people, too. Paul usually wore a shirt and tie, and Beatrice a nondescript business suit. Today was different. I had never seen them so "clean." Paul wore a fabulous dark business suit that exuded power and money. Beatrice was wearing a suit that was at once feminine, elegant, and rich. They looked very much the New York power couple, ready to transact some business, ready to get paid.

The Giordano attorney handling our deal was Phil Forlenza. Phil looked like he always did: serious, professional, and somewhat harried. In the weeks prior to the close, Bowthorpe's legal team had buried Phil in paperwork—the usual blizzard of representations and warranties, plus a few extraordinary items relating to U.S. government approval for the sale of TAS to a foreign entity. I sensed Phil was itching to close the sale so he could move on to his next deal.

Bowthorpe's attorney was a young woman from a big New York law firm. Phil introduced her as Sarah something-or-other. She was dressed in the standard-issue expensive-New-York-female-attorney business suit. It didn't look special, but it probably cost more than a thousand dollars. Sarah's demeanor was formal but matter-of-fact, as if she had done this type of transaction countless times—no big deal. I didn't sense a warm personality. She seemed to be a hired gun, there to do a job and get back to New York.

When I first entered the room, my nerves were tight. I had no second thoughts about the sale, I just wanted it done. I was nervous that a last-minute snafu might unravel the deal. Seeing the huge conference table with the mountain of papers only heightened my anxiety. Gradually, though, my nerves settled. I kept telling myself to relax and enjoy the day. It was a momentous occasion, a graduation of sorts, and an event few people ever experience. It was time to take care of the business at hand, but it was also time to celebrate.

I joked nervously with Steve and Charles, and we made small talk with Paul and Beatrice. Bowthorpe's attorney seemed disinclined to socialize. She seemed more interested in making sure she had a reliable telephone connection to England. Only then did I realize that no executives from Bowthorpe, our new owners, would attend. It looked as if they were going to complete the transaction by remote control. I asked Paul and Beatrice if my observation was correct. They asked the young attorney Sarah, and she confirmed it.

I was disappointed. The closing was one of the most important events in my life, and the people to whom I was selling my company didn't consider it important enough to show up. Paul and Beatrice seemed disappointed, too. We had talked a few days earlier about having a celebration dinner after the closing. That clearly wasn't going to happen. It didn't seem right to celebrate the deal without the buyers. Paul and I agreed we would arrange a proper "closing dinner" sometime when the Bowthorpe folks were in town. That made me feel a little better.

It was time to sign the documents. The attorneys positioned Steve, Charles, and me in a little assembly line along one side of the conference table, and we began signing papers. I autographed each document with a special pen I'd purchased for the occasion, and then passed it along to Steve and Charles. Phil occasionally had to explain the significance of a document, and Paul and Beatrice asked a few questions, but generally things went smoothly. In an hour or so, all the signing was done, and we had only to make sure the money from Bowthorpe had been wired to our bank accounts. I called my Merrill Lynch financial adviser and confirmed that the Bowthorpe funds were indeed there. My cofounders called their advisors and confirmed the same. The deal was officially done. Before celebrating, I wanted to take a moment to commemorate the event. I asked Steve and Charles to pose for a photo with me. We took three, with each of us in turn seated at the conference table signing a document while the other two stood behind. Those were our "graduation photos." I thanked Phil, Paul, and Beatrice for their work, and then Steve, Charles, and I shook hands and congratulated each other on our accomplishment. I couldn't help recalling the day the three of us walked toward the Bell Labs cafeteria, the day I invited them to join me in an improbable business adventure. That day now seemed part of a different lifetime. "I told you guys we would be millionaires," I said, and flashed a big grin. It felt unbelievably satisfying to see that prediction come true.

Our business complete, the parties filtered from the room. Sarah left immediately and unceremoniously to return to New York. Paul and Beatrice also headed back to the city. Phil walked down the hall to his office. Steve, Charles, and I drove separately back to TAS to give our employees and friends the good news.

"GRADUATION"

I was euphoric as I settled into my car. I popped in my latest Stevie Wonder CD and sang along.

> *Rain your love down, won't you rain down your love,*
>
> *Let it drench us like the sun from above....*

As I turned from Newman Springs Road onto Shrewsbury Avenue, I was consumed by thoughts of my father. His intense interest in electronics created a fertile learning environment for me, and I had been intent on making the most of it. I grew up determined to prove it was a new day—that the forces that had frustrated my father's ambitions and limited his career would not stifle me. My father's dreams became my dreams, and his reality became my nightmare. Taken together, they were powerful motivators.

Tears started streaming down my face, and I screamed, "Yeah! Yeah! Ahhhh. Yeah! I did it, Pops! See? I did it!"

I always knew what financial independence meant, but had never felt it. Now I realized that, unlike my father, I could do what I wanted, live where I wanted, raise and educate my kids the way I wanted. I was, to borrow a phrase, a man in full.

I must have been a sight, a black man crawling along in his black

Mercedes, screaming at the top of his lungs and crying at the same time. Luckily, no policeman spotted me, or I might have been arrested, or worse, committed.

Oblivious of my surroundings, I didn't care who was watching. With each scream, I exorcised the doubts and fears and uncertainty that had lurked inside since my Flint childhood. It was as if thirty years of pressure were released in a few moments. I had proved what I had set out to prove: that I, a black man, could establish and build a world-class technology business from scratch. I proved that the many sacrifices my parents made to enable my success were not in vain. I proved that the monumental civil rights struggle waged during my formative years made the difference between my father's bitter frustration and my own sweet elation. So many people worked and endured and marched and suffered and died during that movement, and their images were burned into my young psyche. I proved that those people didn't work and endure and march and suffer and die for nothing.

Their battleground became my proving ground.

54. The Long Goodbye

IT WAS A BALMY, ALMOST PERFECT SUMMER EVENING IN LONDON. The streets were clogged with people hustling home from work or heading to their favorite pubs. I was ensconced in the cocoon-like ambiance of a small, private club called Mosimann's Belfry. It was August 1999, nearly four years after we sold TAS to Bowthorpe, and I was at Mosimann's to meet my boss, Bowthorpe CEO Nicholas Brookes. He had chosen the location for our meeting. I had never heard of the place, but was happy we weren't meeting at the office. I made sure to arrive early so I'd be fully prepared for the encounter. I knew that, whatever the outcome, my life was about to change.

I settled into a high-backed leather chair and sipped a glass of red wine. I listened to the smooth jazz background music and absorbed the scene. These were London's beautiful people, relaxing. I felt like one of them. The wine and the elegant setting were having their intended effect, and before long I drifted into that nether region between reality and dreams, where past, present, and future are often indistinct. As I drifted, I replayed the events of the previous four years.

My mind wandered to the 1995 law-office ceremony at which Steve, Charles, and I closed the deal with Bowthorpe. In the hours following the sale, I was euphoric. I had achieved a major goal, *the* major goal of my career as an engineering and business professional. I hadn't achieved the goal alone, but I alone had dreamed it, conceived it, and led the way to realizing it.

Less than a month after the sale, I got my first taste of the Bowthorpe culture. The occasion was my first official Bowthorpe corporate function, the Presidents' Meeting on Marco Island, Florida. The December 1995 meeting brought together the presidents of Bowthorpe's ninety-seven far-flung subsidiaries. As president of Bowthorpe's latest acquisition I was, of course, expected to attend. Spouses were invited, too, so Diane came along.

The hotel venue was elegant but not extravagant. Bowthorpe's welcome packet indicated an evening reception for company presidents and their spouses. The invitation included a dress code: business suits for the men, "pretty dresses" for the ladies. Diane was highly amused. She was a well-paid AT&T executive, a Stanford-educated mathematician, and she wasn't given to being regarded as ornamental. That evening, rather than wear a "pretty dress," she defiantly wore a yellow pantsuit she'd bought that afternoon in the hotel gift shop.

We entered the ballroom with some trepidation. The room was a sea of white faces, and it felt like foreign territory. I was uneasy—not knowing how or if I would fit in with my new Bowthorpe colleagues. Diane and I spent the evening sauntering gingerly around the ballroom, sipping white wine and eating light hors d'ouevres. We received nothing more than curious glances from many attendees, but some of my new colleagues introduced themselves and we engaged in polite, stilted conversation. Fortunately, the focus of the reception wasn't the black president of Bowthorpe's latest acquisition and his Chinese American wife. The spotlight was on Dr. John Westhead, the outgoing CEO who'd made the initial decision to purchase TAS. For me, talking to Dr. Westhead, wishing him well, and hearing him reiterate his complete confidence in his decision to buy TAS was the high point of the evening.

I kept reminding myself not to worry about the Bowthorpe corporate culture or about Dr. Westhead's departure. I told myself none of that mattered because I didn't plan to stay with Bowthorpe any longer than necessary. My personal agenda was to focus on TAS, continue to grow the company, maximize our incentive payments, and leave at the end of the three-year "earn-out" period. I returned from Marco Island even more intent on doing just that.

A few weeks after the presidents' meeting, I received a surprise invitation from Nick Brookes. He wanted me to fly to the U.K. to meet with his management team and talk about further telecommunications business opportunities. I accepted the invitation, despite my determination to avoid any distractions that could be detrimental to TAS. I decided that, as long as I was a Bowthorpe employee, I would try my best to be helpful and supportive. I was grateful to Bowthorpe for making me a wealthy man, and I wanted to justify their belief in TAS. I wanted to help Nick and Bowthorpe achieve success, as long as it didn't interfere with maximizing TAS's business. I decided to use the meeting at Bowthorpe headquarters to present a strategy for building a much larger telecommunications business, a strategy that utilized the business model we'd established at TAS.

I took an evening flight from Newark and arrived at London's Gatwick Airport early the next morning. Nick Brookes' driver, Angus, ushered me into the back seat of Nick's big Mercedes and we set off. I expected to be driven to a nearby hotel so I could freshen up and prepare for the meeting, but Angus drove me straight to Nick's home—an impressive country estate called Wolver's Hall. Nick greeted me warmly and introduced me to his wife Maria. He then showed me to a guest bedroom and invited me to sleep off the jet lag. I gladly accepted.

After my nap, Maria served a delicious lunch, and then Nick and I headed to the office. I was rested and ready, much more so than if I'd spent the morning checking into the airport hotel and trying to grab a few hours' sleep. Nick's hospitality was totally unexpected, but it was an effective gesture. By graciously opening his home to me, Nick made me feel like a valued member of his team.

Bowthorpe headquarters was near Gatwick Airport, amid a cluster of post-World War II industrial buildings. The building wasn't fancy, but it wasn't unpleasant, either. Nick gave me a brief tour during which he explained that Bowthorpe's priority was to invest in the businesses it owned, not in fancy corporate headquarters. I was impressed.

After the tour, Nick gathered his team of six top executives in a conference room. He gave me a rousing introduction, and then I launched into my presentation. I gave the executives a TAS overview, and then talked about our business direction and strategy. I identified other companies Bowthorpe might acquire to further develop its telecommunications business. The presentation was warmly received. The executives were astounded by TAS's growth and profit margins. They certainly seemed receptive to buying similar companies.

A month or so later, Nick and I traveled to Hawaii to visit one of the acquisition targets I had identified. During the trip, Nick said he had decided to assemble a group of telecommunications companies, and that he wanted me to run the group. I was surprised and flattered, but I had no desire to stay with Bowthorpe. I was committed to doing everything I could to build the business during my three-year transition period, but after that, I was gone. After experiencing the freedom and control of running my own company, and then the freedom of financial independence, the last thing I wanted was another corporate job. Anyway, the telecommunications "group" at that time still consisted of just one company, TAS, so Nick's offer was premature.

By the end of 1996, TAS sales had grown another 50 percent, and Nick was even more anxious to create a telecommunications group. At my second presidents' meeting, this time in London, I was still the only black face in the room, but I was viewed as a star. The 1996 Bowthorpe annual report featured a large photo of me standing behind several TAS products. Bowthorpe was actually showcasing TAS and me to its investors! My stature had risen mightily since that first, uneasy meeting on Marco Island. I had earned the respect of the other company presidents, and had established budding friendships with some.

In 1997, with my help, Bowthorpe started acquiring more telecommunications companies. We bought the Hawaiian company Nick and I visited, and we bought a U.K.-based GPS test system manufacturer identified by another Bowthorpe executive. That same year, TAS sales grew by more than 60 percent, to over $25 million. The telecommunications equipment market was exploding, and that fueled Bowthorpe's desire to gobble up more companies.

Following the 1997 acquisitions, the telecommunications group consisted of three substantial companies, and it needed a leader. I still didn't want the job because I was determined to leave at the end of the earn-out period, but I did want to help Nick find someone to fill the position. I reached out to an executive at another successful telecommunications test equipment company, a fellow I'd known since I was a young engineer at Bell Labs. When I first met him, he was a young sales engineer for one of our key vendors. He had impressed me back then, and I had followed his career ever since.

The fellow I recruited agreed to join Bowthorpe, as "Group President—Telecoms," in early 1998. Nick Brookes was glad to have a leader for the group, but he still wasn't satisfied. He wanted me to run the group

TAS WAS FEATURED IN
1996 BOWTHORPE ANNUAL REPORT.

David Tarver, President of Telecom Analysis Systems Inc (TAS), with telecoms testing equipment made by the company at its base in Eatontown, New Jersey. TAS equipment allows developers, manufacturers and evaluators of telecomms equipment to test products over a wide range of realistic network conditions. TAS provides market-leading solutions for testing analogue modems, ISDN terminal adapters and wireless base and mobile station equipment.

along with my recruit, and the new group president favored the idea. I said I would accept the offer if Bowthorpe would commit to paying the TAS incentive payment in full. Given TAS's exploding sales and profits, that wasn't much of a stretch. There was little possibility we wouldn't qualify for the maximum payment. In July 1998 I became co-president of Bowthorpe's telecommunications business unit.

I still felt no desire to stay with Bowthorpe past the earn-out period, but I had several good reasons for accepting the offer. I wanted to see what it was like to serve near the top of a large, global company. I wanted to learn more about corporate acquisitions. I wanted to create the best possible business environment for my TAS cofounders and former employees. I wasn't giving up anything—I could still leave whenever I wanted, but I reasoned that it would be better to leave as telecommunications group co-president than as president of a relatively small business unit, TAS. There was also a slim possibility that I would enjoy my new role and decide to stay.

My first task following the promotion was to choose my successor at TAS. Steve and Charles were the obvious candidates. Steve was totally committed to the company and was passionate about his work, but he didn't want the job. He had little interest in corporate politics and constant travel, and he felt his product development skills were indispensable. I couldn't argue with his reasoning. Like Steve, Charles had been with the company since our days in my basement, and he *did* want the job. I appointed Charles to succeed me as president of TAS.

As telecommunications group co-president, my top priority was to identify, evaluate, and negotiate further acquisitions. In that regard, my co-president and I achieved considerable success. In 1999, we added three significant, successful companies to the group. By the middle of 1999, telecoms group sales approached $250 million, and market value was probably in excess of $2 billion. Nick Brookes was ecstatic. He decided to remake all of Bowthorpe by selling off the other divisions and acquiring more telecommunications companies. To emphasize the company's new focus, he decided to change the name from Bowthorpe to Spirent Communications. Less than four years after acquiring TAS, Nick was prepared to completely revamp Bowthorpe, a successful sixty-year-old company, change its identity, and adopt the TAS business model for the entire corporation.

Despite our success at acquisitions, I became concerned that the telecommunications business was overheating, and the price required to acquire companies was getting too high. One pricey outfit was a California company we visited in the spring of 1999. Technically, the company was a good fit for the group, but I saw warning signs. Most of the company's engineers were located off-site. The headquarters building seemed deserted on the day we visited, except for the company president and his fiancée, the marketing director. The president was forecasting a seemingly unrealistic 100 percent sales increase, from $9 million to $18 million, in the current year. The asking price was startling: $140 million. I advised strongly against buying the company.

A few weeks later, I learned that Nick was still seriously considering the purchase because, as he indicated to financial analysts, he was "keen to make larger acquisitions." I flew to London specifically to meet Nick and head off the acquisition, and though I succeeded, I remained concerned that our telecom bubble might burst.

I should have been excited about the transformation of Bowthorpe, excited about my prospects for advancement, excited about our success in acquiring companies, and about my position as leader of a rapidly growing high-tech business. I should have been excited, but I wasn't. I felt that I'd already proven everything I had set out to prove, and that it was time to move on. I wasn't after more money or lofty titles. I wanted freedom. I wanted time. I wanted to try new things.

On the other hand, I wasn't stupid, so I wasn't about to make any rash decisions. I strongly considered the possibility that I might simply be burnt out and in need of a long vacation. The strain of growing and then selling TAS, and the subsequent intense wave of acquisitions, had taken a toll on my psyche. Starting in June 1999, I took a three-month leave from Bowthorpe to recharge my batteries and take a look at my situation from a remote, relaxed perspective.

Removed from the day-to-day grind, I was able to think more clearly about the direction of the new telecommunications group. I realized that important strategic decisions needed to be made, and that those decisions couldn't wait until after my leave of absence. I organized a strategic planning session for me, my co-president, and Nick, but I was determined not to go anywhere near the office. I arranged for the session to take place in Montego Bay, Jamaica, and hired my old friend Clifton Smith to facilitate. One issue I planned to address was group leadership. I hoped to convince Nick that it was time for the telecommunications business to have one leader—that it was unnecessary and unproductive to have two presidents. I hoped to convince him that it was time for my co-president to run the business on his own, and that I should leave the company as I had planned to do.

We held the strategic planning meeting at Half Moon Resort. It was an intense session, a classic Clifton Smith performance. In that remote, intoxicating setting, Cliff got us to take an open and honest look at our business, the marketplace, and ourselves. We found that we agreed on the broad strategy for the telecommunications business, and in the end, Nick realized that the group needed to have one leader. However, in keeping with the spirit of openness and honesty, Nick said he wanted that one leader to be me, not the group president I had recruited. I was flattered by that revelation, but also embarrassed. I felt badly for my co-president, but I owed Nick fair consideration of his offer. I asked him to give me a few weeks to think it over.

The perks associated with being the sole "Group President—Telecoms" were clear. Huge salary, bonus, and stock options; bases of operations in New York, London, LA, and Tokyo; frequent visits to Hawaii to check on the operations there; a seat on the Bowthorpe board of directors; possibly succeeding Nick as Bowthorpe CEO.

On the other hand, those weren't my goals. I wanted my time back; I wanted to explore and pursue other interests; I wanted to reconnect with my family and my community. I had recruited someone else to run the group precisely because I had to move on. I knew that if I stayed, it would be for money and power and glamour. I didn't have the desire.

Still, the offer was tempting. As telecoms group president, I could guarantee great opportunities for Steve, Charles, and my other TAS colleagues—and nothing said I had to keep the job forever. I could try it for a few years, and if I didn't find it to my liking, I could exit as the leader of a large and powerful business.

When the time came to share my decision, I called Nick to arrange a meeting. This wasn't a matter to be discussed by telephone. I called my co-president to let him know what I was going to do, and then boarded a plane for the overnight flight to England.

The morning of my arrival, I checked into the Landmark London hotel. I'd intended to take a long nap, but slept intermittently. After a few hours, I decided to get up and out of the hotel. I walked over to the Marks and Spencer department store for lunch. Later, I caught a taxi to Hyde Park and took a long walk. It was a gorgeous afternoon, and the sights and sounds and smells reminded me how much I enjoyed being in London.

I arrived at Mosimann's Belfry forty-five minutes or so before Nick and I were to meet, and was pleasantly surprised. The place seemed more like a comfortable sitting room than a commercial establishment. I was impressed by the elegant and relaxing atmosphere, and could easily envision making it one of my London hangouts. In contrast to the din of activity outside, the Mosimann's interior was dimly lit and calm. A gorgeous hostess greeted me as if I were a close friend and walked me to a quiet corner of an upstairs room. Seated in a throne-like chair, listening to soft music, sipping wine, I felt like royalty.

Just when I was at my mellowest, savoring the wine and the jazz and the lovely hostesses, Nick arrived. We made small talk for a few minutes—about business, about recent acquisitions, about how exciting it was that we had built such a great "telecoms" business so quickly.

Nick seemed perturbed and impatient, though. He didn't have the look of someone who was going to give or receive good news. He looked at me earnestly and said, "You've decided what you're going to do?"

I felt a tinge of anxiety. In that moment, I realized fully the impact of my decision.

"Yes," I said. "I'm leaving."

"So, what are you going to do now?"

"Try to enjoy my life."

Nick seemed disgusted and on the verge of anger. I understood. He probably felt jerked around by my leave of absence, by the way I had summoned him down to Jamaica for a strategic planning session, and by my insistence on this face-to-face meeting. Why would I do all that if I planned to leave?

The fact was that my feelings had not changed since before we sold TAS. I wanted badly to change the course of my life—that was why I'd been ready to sell the company. What Nick perceived as being "jerked around" was just my spending the time and effort to carefully assess the situation and make the right choices.

The day after my meeting with Nick, I returned to New Jersey and shared my decision with Steve and Charles. They weren't at all surprised—they knew I had been "tuned out" for weeks. Steve had called my condition "short-timer's disease," and his diagnosis was correct.

I didn't even have the energy to clean out my office.

I quickly became persona non grata with Nick. I offered to serve on the Bowthorpe board, but he didn't take me up on that. I did manage to advise TAS for the better part of a year, and that smoothed the transition to Charles's leadership.

At the beginning of 2000, Bowthorpe changed its name to Spirent to reflect its new telecommunications technology focus. Spirent continued the acquisition binge, at higher prices. Spirent purchased the company I had begged Nick not to buy. Indeed, Spirent didn't pay $140 million—the announced purchase price was $400 million! In 2001, Spirent wrote off the entire cost of that acquisition. Spirent made other large acquisitions, and then the tech bubble burst. When I left the company, the stock price stood at roughly seven pounds. Two years later, it was seven pence. I took no joy in that, because I had come to like and respect my Bowthorpe/Spirent colleagues. I only wished I could have joined the board of directors or otherwise advised the company.

Fortunately the company formerly known as Telecom Analysis Systems has continued to thrive, powered by Spirent's resources, a sound business strategy, Charles's and Steve's committed and capable leadership, and our fortuitous decision in 1992 to enter the wireless business. At this writing the company has well over one hundred employees and annual revenues approaching $100 million—not bad for a company started by three young African American men who began their careers not knowing what was possible.

Epilogue

Proving Ground. When I was coming of age in Flint, the cradle of the U.S. auto industry, those two words sparked my imagination. The proving ground was a place where automobile designers took their latest creations to evaluate performance and wring out imperfections. The words evoked a romantic image: a pristine new car being subjected to the vagaries of a simulated "real world." Each proving-ground test would either confirm or invalidate an aspect of the car's design. Each failure would induce improvements until a solid product emerged.

The proving-ground concept made a substantial impression on me. My winning Flint Area Science Fair projects all hinged upon subjecting something to a "proving ground" I created. My first subjects were single-cell organisms—paramecia and amoeba. Later, I moved on to transistors. Years later, I applied that same proving-ground concept to telecommunication devices. At TAS we designed and manufactured products that subjected sophisticated telecommunications gear—first modems, then cell phones—to an *electronic* "proving ground."

Perhaps most significantly, the proving ground is also a metaphor for the struggle my African American contemporaries and I engaged in as we came of age and, in the midst of a civil rights revolution, embarked upon our careers. The gates of opportunity were open to us at last, but our success was far from assured. Even the most confident among us were prone to moments of doubt. As black professionals in newly accessible fields, we were entering a world in which nearly everyone we dealt with was white, where few people who looked like us had ever competed. We were, for the most part, willing and ready to be tested, and above all we were anxious to prove that we could survive and thrive in this new and exciting world.

Steve Moore, Charles Simmons, and I saw our business venture in the context of this larger civil rights struggle. We were out to prove something not just about ourselves but also about all African Americans. Greater financial success was only one motivation for leaving comfortable and prestigious jobs at Bell Labs and going to work in my basement, and I doubt that that motivation alone would have sustained us. The historic significance of our venture enabled us to persevere when less-motivated teams might have faltered. Telecom Analysis Systems was our *personal* proving ground, and the fact that we emerged from it stronger, more confident, and financially secure is a huge source of pride.

My childhood science and electronics experiences left deep and lasting impressions on me, too. The failure of my first science fair project was one of the most indelible. Losing because I didn't apply the scientific method seared the lesson onto my brain in a way no textbook ever could.

When I left Bell Labs and started TAS, I was still thinking about the scientific method. I viewed our business venture as an experiment, the outcome of which was far from certain. I had no role models—I wasn't aware of another black person who had started a successful high-tech company. TAS was not a traditional black-owned business. We existed in a white business universe, and that fact alone spawned several questions. Would white engineers buy our products? Would white engineers work for us? Would white bank officials lend us money? Would white lawyers, accountants, and other professionals represent us fairly and enthusiastically? If the answer to any of those questions was no, our business was probably doomed. I wanted to reap the rewards of success, but more than that, I wanted to know if we *could* succeed. If we failed, I wanted to know why. Drawing on my childhood lesson, I applied the scientific method to our venture.

Question: Can three young black men, educated at the best schools and trained at the world's foremost research and development institution, create and run a successful high-tech company?

Hypothesis: Social, cultural, and political conditions have changed significantly, primarily as a result of the civil rights movement. It is indeed possible for three young black men to create and run a successful high-tech company.

Experiment: Create a company (TAS) and give it a go. Don't accept any special assistance along the way—that would contaminate the experiment. Do it "the old-fashioned way."

Conclusion: Our "experiment" consumed more than twelve years of our lives and required exceedingly hard work, but in the end, we verified the hypothesis. Along the way, we gained invaluable experience in business and human relationships. We achieved financial independence. We created a satisfying and productive work environment, and we respected and rewarded people based upon their contributions. The sale to Bowthorpe confirmed the value of our business, and our vision was further validated when our new owners made TAS the kernel of a much larger enterprise.

We accomplished our business goals, but not because racism had somehow evaporated in the wake of all that 1960s consciousness raising. Race was an ever-present factor in our business relationships, but the signs were rarely, if ever, unambiguous. In fact, I was often struck by a kind of "race uncertainty principle." In these pages, I have described cases in which I was convinced someone was racist, yet they bought our products or otherwise supported our company. Did that mean my assessment was wrong? I've described other cases in which a white person treated me unfairly, but only a hefty dose of truth serum could extract a motive. Should I assume the motive was racism?

Racism definitely remains with us, but to focus too intently on it is to be defeated by it. Racism turned out to be only one of many forces governing the behavior of people we dealt with. Like gravity, it was weak but ubiquitous. In most person-to-person or even business-to-business instances, it could be overcome. If we believed otherwise, there would have been little point in embarking upon our venture.

I was surprised to find that people outside the U.S., particularly those in Asia, seemed less resistant to the idea of a black technology entrepreneur. In the case of my Asian colleagues, that is perhaps because nearly all U.S. entrepreneurs they dealt with were of a different race. Our Asian customers didn't seem to buy into white superiority *or* black inferiority.

The most encouraging thing I learned was that most people were rational, and based their business decisions on economic self-interest. That was true of customers, suppliers, professional advisers, and employees. No one was going to buy from us, or work for us, simply because we were bright and charming—we had to present a compelling rationale. In the long run, the need to justify our products and our company made us better business people. Though we sometimes encountered people who behaved irrationally, i.e., contrary to their own economic interest, they were in the minority, and their irrationality was a fairly reliable sign that racism was afoot.

More than a few white colleagues went out of their way to help our business. Some offered expert advice, and others purchased and enthusiastically recommended our products. I don't think any of them would have helped us if they believed we produced an inferior product. At the same time, I think they were heartened to see three black engineers, definite underdogs, trying to build a successful business. To them, our venture was fresh, new, exciting. On the other hand, a few whites actively avoided doing business with us, and, given the opportunity, would have impeded our progress. Because we were independent entrepreneurs, those people had little power over us. We simply focused our time and energy elsewhere.

I cannot overstate how difficult it was to build our company. We worked ridiculously long hours, solved vexing technical problems, and dealt with a wide range of people and situations, all while trying to maintain some semblance of normal family life. If we had known from the start how difficult our venture was going to be, we might not have attempted it. Still, for every setback we experienced, we were blessed with some unexpected piece of good fortune. At times I felt that something beyond our understanding propelled us toward our goal. I am certain God was with us.

The changes to my life after departing Bowthorpe were huge, abrupt—and welcome. In August 1999 I was the leader of a rapidly growing global technology business. In September I was the volunteer director of a tiny Red Bank community center. In August my annual compensation was half a million dollars. In September, it was zero. In August, the high point of my day was my morning staff teleconference. In September, I looked forward each day to serving afternoon snack to grade school children.

I didn't mind the career change one bit, and didn't for a moment regret leaving Bowthorpe. I was excited and reinvigorated because I was finally embarking on the last phase of the plan I had dreamed of in high school and formalized in college. I was embarking on my self-defined, self-described anthropology studies. I hoped what I learned would enable me to one day make a contribution to the world beyond the field of engineering.

My new place of employment was the Count Basie Learning Center, named in honor of the late, world-famous jazz musician. Red Bank was Count Basie's hometown, and he was a shining example of a local kid, a black kid, who had distinguished himself on the world stage. The Count Basie center was a grass-roots organization created to address a serious problem—Red Bank kids' declining academic performance. The center was supposed to give kids tutoring, mentoring, and cultural enrichment. I started serving on the board of directors in 1997, while still with Bowthorpe. When I left the company, I was Count Basie Learning Center board president, and I looked forward to helping make it an exemplary institution. Barely a month later, the center's director abruptly resigned, and I had to step in and run the place.

I dove into managing the programs, hiring staff, sprucing up the building, and recruiting a new director. As I became more engaged, I noticed that several third-grade clients were poor readers, and were incapable of performing simple addition and subtraction. Exploring the situation further, I was surprised to find that many, if not most, Red Bank students were substandard academic performers. I was also surprised to learn that although whites comprised roughly half of Red Bank's school-age population, the public schools were 85 percent black and Hispanic. Most white Red Bank students were attending private or parochial schools. A group of disgruntled white parents had recently started a charter school in town, and that initiative spawned deep divisions. Vehement arguments often erupted between pro- and anti-charter-school factions. Despite these troubles, many Red Bank residents were under the impression that the Count Basie Learning Center, which was serving fewer than thirty students per day, was somehow cleaning up Red Bank's academic mess.

Aching to gain further insight, I met with the Red Bank school superintendent and a local pastor, both black men. The superintendent had been on the job for several years, and many people around town, particularly in the black community, were dissatisfied with his performance. The pastor was a former school board member who had repeatedly expressed concern that the district lacked a strategic plan. During our meeting, he repeated that assertion. The superintendent replied that any strategic plan had to be a *community* plan, not merely a school district plan. I thought the superintendent was passing the buck, so I decided to call him on it, and asked, "If I bring the community together to create a plan, will you support the effort and participate?" Of course he had to answer in the affirmative, before a witness who was a man of the cloth.

I left the meeting with a new hypothesis and a new cause. The hypothesis: "If all segments of the Red Bank community work together to devise and execute a plan to address the academic, social, and cultural needs of Red Bank children, academic results will improve." The unstated cause: "Give Red Bank children the kinds of opportunities my friends and I had when we were growing up in Flint." Red Bank was an ideal laboratory for such an experiment. It was a 1.7-square-mile town with a population of 12,000 and fewer than 1,000 public school students. It was a middle-class community, not a poverty-stricken one. Red Bank had improved markedly since my early days in New Jersey. It had a thriving downtown and a vibrant cultural scene. In fact, it was known throughout the region as "Hip City." Red Bank was home to several community service organizations such as the YMCA, Catholic Charities, the Salvation Army, and the Visiting Nurse Association. Moreover, it was surrounded by some of the richest communities in the country, and many residents of those communities were willing to financially support opportunities for Red Bank kids.

I turned to my old friend and adviser Clifton Smith to help organize the community. We spent the better part of a year establishing support for a not-for-profit organization we dubbed the Red Bank Education and Development Initiative (RBEDI). We engaged all segments of the community to identify the most pressing problems and devise effective solutions. We secured the support of Red Bank's civic, academic, religious, and community service leadership, and presented the project to the public at a Molly Pitcher Inn reception. That first community event took place on September 20, 2001—just nine days after the terrorist attacks that destroyed the World Trade Center.

In June 2002, RBEDI produced a series of findings and recommendations, and Red Bank finally had its "community plan." Our work was the catalyst for many changes in the schools, including a new superintendent, new teachers, and a documented school improvement plan. We induced community organizations to provide more and better services to Red Bank children. Ultimately, we witnessed dramatic improvement in the academic performance of Red Bank's kids. In 2001, the year RBEDI started, less than 25 percent of Red Bank students passed the New Jersey eighth grade performance assessment. By 2004, that number had risen to nearly 60 percent.

My Red Bank "experiment" confirmed that when a community comes together and provides great education and enrichment opportunities—the kinds of opportunities my Flint peers and I took for granted—most kids can achieve academic success. My earlier TAS "experiment" confirmed that, with appropriate education, experience, and opportunities, an African American person, indeed any person, can compete at the highest levels of global business and entrepreneurship. Those findings are encouraging, yet when I look at the situation confronting African Americans today, I am often discouraged and disappointed. Many legal barriers to black progress were dismantled in the 1950s and '60s, but new barriers have arisen. They comprise a familiar litany: white flight/bright flight from urban centers; rapid displacement and/or export of unskilled and low-skill jobs; the illegal drug epidemic and associated high black male incarceration rate; wealth transfer from poor to rich rather than from rich to poor; dramatic increase in the cost of higher education; less access to affordable medical care. All of these factors disproportionately affect people on the lowest rungs of the economic ladder, and that is where most black, brown, and red people still find themselves.

The city of Flint I knew in my childhood is barely recognizable today. The public schools, once among the nation's best, are now among its worst. The schools I attended—Walker Elementary, Whittier Junior High, and Central High—are now closed.

Drugs and gun violence were rare during my days in Flint. Sure, we had our cough-syrup addicts and drunks, and we heard about the occasional stabbing. Today, hard drugs are readily available, and Flint's murder rate is among the highest in the USA.

General Motors, the company that introduced me to engineering and manufacturing and business, has all but abandoned Flint. AC Spark Plug, the place where I received valuable engineering co-op training, is gone. In its place lies a vast, barren field. The same fate has befallen most of the other once-mighty GM facilities.

The changes I am referring to extend far beyond Flint and Michigan—they are national. They extend even to my beloved Bell Labs. The place where I began my professional career is no longer the shining beacon of technology and opportunity it was when I arrived in New Jersey. The amazing Holmdel facility, the place where I came of age as a technical professional, has been completely abandoned. Wild animals now roam the grounds.

Companies are not seeking to hire African Americans the way they were when I graduated from college. Recent graduates are finding that "affirmative action" is a dirty word, and "equal opportunity" has been replaced by "it's not what you know, it's who you know."

All of these changes occurred since the 1960s and '70s, those magical days when everything seemed possible, when it was a true blessing to be young, gifted, and black. Is the timing of the changes mere coincidence?

I often wonder, *what happened to us?* My success, and the success of my black baby-boomer peers, sometimes seems an aberration, the product of one brief, shining moment in American history. Looking back, it's clear that the civil rights movement, which unleashed so much progress and productivity, also unleashed reactionary forces. Those forces were already evident in 1955 when my folks moved onto Sixth Street. That's why our white neighbors were moving out. The reactionary forces accelerated in the 1960s, and hit warp speed in the '70s and '80s. I now realize that the almost ideal upbringing I had in Flint was an exception even then. Today, my childhood experience borders on urban myth.

Opportunity for all is a bedrock principle in this country. It is what *makes* us exceptional. Our nation won't be recognizable much longer if we regress further into camps of rich and poor, educated and illiterate, white and black and brown. We can't do away with the tools that foster social mobility just because the people on the bottom rungs of the social ladder are disproportionately nonwhite. We must be about enhancing, rather than destroying, our sense of community, as difficult as that may be.

At times I feel the situation is dire—not just for African Americans, but for the entire country. Such thoughts weigh heavily after my first decade of anthropological studies. Still, I also think at times about my ancestors, and how far we have come since their days on this earth. That comparison is one of my greatest sources of renewed strength and energy. The questions arising from the comparison are easy to answer.

Does the world look bleaker today than it did for my great-great-great-grandmother Cassie, as her first rapist slave master shipped her and her infant son off to her second rapist slave master?

Do any of us have a tougher job than my great-grandfather Timothy Titus Catchings, newly freed from slavery and working diligently, against all odds, to establish the first black-owned farm—the most advanced farm—in Mitchell County, Georgia?

Are conditions today worse than they were for my grandmother Elizabeth Bernice Hayden, severely burned as a child, then beaten and abused as an adult as she struggled to hold her family together?

Do young people today face choices as difficult as the one my mother faced, severely reprimanded for using a "white" bathroom and compelled to seek employment in a distant state?

Do any of us harbor more frustration than did my father, who suppressed his career ambitions while he pursued his passion in a musty basement workshop?

When viewed in the context of my ancestors' struggles, the challenges African Americans face today seem at least manageable. My forebears were sustained by their faith, family, determination, and skills. That combination has worked since the beginning of recorded history, and regardless of how dire the situation appears, I am convinced it still works today.

My story illustrates, I believe, important and useful facts about technology, race, and business in the wake of the mid-twentieth-century social and technological upheaval. That's why I chose to write this memoir. Doing so has necessarily introduced you to my personal life, and that requires bringing you up to date on some significant changes since my TAS and Bowthorpe days.

Diane and I separated in 2000 and divorced in 2001. Though we were married for nearly twenty-two years, we couldn't bridge the emotional chasm that grew between us as I single-mindedly pursued my career dreams. Though our marriage didn't survive, our friendship did, and I am grateful for that.

In 2003 I proudly accepted recognition as a distinguished alumnus of Flint Central High School. The reception was held at the swanky Flint Golf Club, which I had never entered before that night. In my speech, I noted the high points of my eclectic high school experience, including the rather abrupt end of my basketball career, and managed to extract both laughter and tears from the audience. That evening, more than anything else, helped me mentally shelve the disappointment I had carried since the 1971 senior awards assembly.

In 2004 I married Kishna Arien Sharif at the Half Moon Resort in Montego Bay, Jamaica. It was the same place where, almost exactly five years earlier, I hosted my Bowthorpe colleagues at a pivotal strategic planning session. Kishna and I are kindred spirits, both born in Michigan, both graduates of Flint Central High School, although years separated her school days from mine. Before meeting Kishna, I was physically ailing and emotionally depleted. Now I am able to laugh, and to dream, again. Kishna has saved my life, figuratively many times, literally at least once. One hundred and ten friends, family members, and former colleagues attended our wedding, and we had the time of our lives. Though it was a bittersweet occasion for them, my daughter Stacy was maid of honor, and my son Aaron was best man.

My sister Bernice died in 2005. I will never forget how, ailing, wheelchair-bound, and on dialysis, she made the trip to Jamaica for our wedding. In the weeks before her death, I was rehearsing the role of *The Wiz* at the Count Basie Theatre in Red Bank, and I remember Bernice smiling and making funny faces as I danced and sang some of my numbers in her hospital room. A few weeks later she was gone, but her smiles and those last happy memories remain. My sister's death devastated my mother, and for the first time since my early days in New Jersey, I seriously considered moving back to Michigan.

In 2007 Stacy graduated from New York University's Stern School of Business, and Aaron graduated from Red Bank Regional High School. There was no longer any compelling reason for Kishna and me to remain in New Jersey. On August 9, 2007, thirty-one years to the day after I started my Bell Labs career, we finished packing our belongings and drove to our new home in Birmingham, Michigan. That fall, Aaron began his college career at the University of Michigan in Ann Arbor.

In 2008, Kishna and I were blessed with a daughter: Nadiyah Louise Tarver.

Nearly four years have passed since my return to Michigan. I have spent most of that time raising Nadiyah, hanging out with Kishna and my mom and occasionally my brother Fred, allowing my anthropology studies to lead me to various observation posts, and writing this book. I write these closing words on the eve, literally, of Aaron's graduation from the University of Michigan. He will begin working for a California technology company this fall. Stacy, having eschewed a Wall Street career, is now a marketing professional in New York City. Nadiyah will begin preschool this fall. Even at this early stage, she likes to be called "Doctor Nadiyah."

The world my children face is certainly a mixed bag. On one hand, there is the legacy of racism. On the other hand, a *black* man has been elected president of the United States. On one hand, there is the litany of social and cultural problems I described. On the other hand, technology is advancing more rapidly than ever, creating opportunities and solutions I could have not have imagined at their age. As my children traverse their own proving ground, I hope they will find a way to use the tools available in their era to improve their world. I hope also that they will equip *their* children with the skills, determination, and faith my ancestors bequeathed to me.

Acknowledgments

WHERE DO I BEGIN? I have been laboring over this book for more than ten years, and until very recently its contents have been pretty closely held. Writing it has truly been a labor of love, but as with many love affairs, the relationship has contained elements of frustration, consternation, and, well, hate. I won't dwell on those, though. Let's just say I'm happy the project is finished.

More than a few people assisted in the completion of this book. First and foremost, my wife Kishna demonstrated saintly patience in the face of my obsession with this work. She provided much encouragement and advice, and at times she offered needed and welcome distraction.

Marion Landew of the New York University School of Continuing and Professional Studies guided me as I put my story on paper for the first time, and she assured me that the story needed to be told.

My aunt Lillian Hayden and cousins Cecil Catchings, Evelyn Ruhlak, and Charlotte Ruhlak serve as family historians. Without them, it would have been impossible to capture the story of our family's early days in Georgia.

Yvonne Grayson read several manuscript drafts and performed insightful reviews. Her input was critical in shaping the tone of the book.

My mother Claudia Louise Tarver, always the dedicated one, was one of the first people to read the completed manuscript end-to-end. If no one else reads the book, I will still consider it a worthwhile endeavor, because she did.

My brother Fred Tarver, aided by his near-photographic memory, sharpened (and in some cases corrected) my recollections of our Flint childhood. His wife Pat also read and commented on the manuscript. Other readers include: Reggie Barnett, Charles Hughes, Hassan Sharif, Mihoko and Craig Mierzwa, Karen Davis, and Tracey Tarver. My daughter Stacy and son Aaron also read and commented on parts of the manuscript, and Stacy suggested the approach we used for the cover design.

My TAS co-founder Steve Moore meticulously reviewed the parts of the story that dealt with our business venture, and it was fun recalling the more dramatic elements of that story with him. I am forever indebted to Steve and to Charles Simmons. Without their dedication and loyalty, our business success would not have been possible, and my story might have ended quite differently.

Mineo Yamamoto hosted Kishna and me on a 2006 visit to Japan where I was able to retrace my steps and recall the perspective-altering events that took place there. During that visit, we enjoyed one of the more memorable evenings of our lives, and I was reminded why Mr. Yamamoto was such a special person and valued friend.

Laura Berman of the *Detroit News* was immensely helpful in steering me to people who helped bring this project down the home stretch. One of those people, Bill Haney, was a godsend.

Cynthia Frank, Joe Shaw, and the expert staff at Cypress House edited the book and provided logistical guidance. Jacinta and Ken Calcut of Image Graphics and Design did their usual expert job on the book and cover design, and were a joy to work with. Shana Milkie produced the very useful index.

I offer my humble and heartfelt thanks to all the above people. I also thank all of those who assisted me on my lifelong journey through technology, race, and business. The world turned out better for me than it did for my father because of people like you.

W. David Tarver
November, 2011

Author's note regarding pseudonyms:

I have employed a limited number of pseudonyms in this book. I have done this for several reasons, but principally where 1) I thought the use of the actual name might cause undue embarrassment or consternation, 2) using the actual name might violate the letter or spirit of a confidentiality agreement, or 3) I simply couldn't remember the actual name. The use of pseudonyms in no way diminishes the veracity of the events described.

The following names are pseudonyms:

Janice, Kathy, Kim, Debbie, Gopal, Dipa, Adilah, Mr. Jones, Kayla, Sarah, Geoff Harris, Neil Hodges, Raymond Grant, Dr. Donald Pierce, Mr. White, Mr. Rodriguez, Laverne, Bob, Alice, Jenny, Bill Benson, Frank Weldon, Michael Chen, Bob Simpson, Peter Hammond, Barry Richards, George Wilson, Clover Valley Group.

INDEX

Note: Page numbers in *italics* refer to photos. WDT refers to William David Tarver.

A

AC Spark Plug division
    Advanced Development Lab assignment, 89–90
    vs. Bell Labs, 166
    farewells to, 100–101
    first assignment, 86–87
    Mike Ebel friendship, 88–89
    production control assignment, 90–96
    razing of building, 450
    Tech Club meetings, 87–88
Adilah (U-M student), 126, 127–128
Adkins, Kedrick, 250
AEA Technology, Inc.
    Concord Data Systems order, 283–284
    Interface '86 tradeshow presence, 307, 309, 313, 314
    MB Electronique distribution of products, 338
    Rockwell showdown, 324–330
    TAS 1010 comparison, 264–267, 300
    as TAS competitor, 332, 340, 363
    WDT examination of simulator, 246–247
Affirmative Action Center at Bell Labs (AAC), 188–189, 190
African American-white relations. *See* race relations and racism experiences
Al (Bell Labs supervisor), 227, 228, 230, 231–233, 234–235
Alexander, Ammie, 8, 9
Alexander, Cassandra ("Cassie"), 7–9, *8*, 451
Alexander, Henry, 8, 9
Alexander, Rick, 279–281
Alice (shipping and receiving worker), 319
"Am I Black Enough for You" (song), 123
Amy (Diane's friend), 217
Anfuso, Frank, 260
Angus (driver), 434
Arp synthesizers, 201
Asbury Park, New Jersey riots (1970), 174
Askew, James Newton, 8–9
AT&T, Inc.
    Graphics Workstation project, 234, 235–236
    industrial design of products, 268
    modem industry involvement, 294
    SARTS project, 193
    telecommunications revolution, 226
AT&T Consumer Products, 237, 242, 252
AT&T Information Systems, 241–242
Austin, Peggy, 274–275
autograph books incident, 26

B

Baker's Keyboard Lounge (Detroit, Michigan), 197, 198
Barnett, Oscar, 100
Barnett, Reggie
    autograph books incident, 26–27
    "Buffalo Soldier" song, 330
    graduation celebration, 79
    at Notre Dame, 82
    Reggie's Rule, 120, 126

senior awards, 77–78
sports activities, *73*, 78, 120
basketball participation
  at Bell Labs, 165, 171, 209, 210
  at school, 72–73, *73*, 75, 76–77
Bell, William David, 12
Bell Communications Research (BellCore), 321
Bell Ringers Club, 25
Bell Telephone Laboratories
  Charles Simmons at, 247–248, 275–277
  decline in, 450
  Diane Sheng employment, 209–210, 217–218
  Donald Pierce employment, 295
  "Giant Transistor" water tower, *161*, 163
  government projects, 169
  modem industry involvement, 226, 246, 294, 295, 302
  prestige of, 355, 357
  product design, 268
  product development vs. research projects, 229, 236
  purchases of TAS products, 272–274, 295, 302
  Steve Moore at, 247–248, 270
Bell Telephone Laboratories, WDT experiences
  Affirmative Action Center, 188–189, 190
  basketball games, 175–176, 209, 210
  changes in company, 252–253
  decision to accept offer, 156–160
  first days of work, 162–172
  first resignation, 213–214
  Graphics Workstation project, 226–236, 237, 238, 240, 241, 367
  Grill Room gatherings, 173–175
  interviews at, 157–158
  Jim Ingle promotion, 219–224
  meeting fellow black employees, 172–175
  performance review participation, 228–229, 230
  promotion to group supervisor, *224*, 224–225
  recruiting at U-M, 216
  resignation and last days, 251–254
  SARTS project, 184–185, 193, 223–224, 247–248
  team building intervention, 231–233
  Touch-Tel project, 237–242, 251, 252
Bement, Spencer, 131–135, 142–148
Benson, Bill, 328
Berkman, Bill, 344–348
Big Storm (1967), 42–44
Birmingham, Alabama church bombing (1963), 30–31
Black Action Movement (BAM), 124
black-white relations. *See* race relations and racism experiences
Blocker, George, 83
Bob (TAS manufacturing manager), 319, 321
Bolinsky, Jimmy, 26
Borum, John, 287
Borum, Vera, 287
bottle collecting business, 38–40
Bowthorpe plc
  closing of TAS purchase, 1–2, 427–430
  due diligence and final agreement, 422–426
  offer for TAS, 418–421
  presidents' meetings, 433, 435
  Spirent Communications name change, 437, 441
  strategic planning session, 438–439
  telecommunications business unit, 434–439
Bradley, Frank, 207–208
Bradley 2A/2B Impairment Simulator
  vs. AEA simulator, 246, 247
  as inspiration for TAS, 206–208, 225, 248
  vs. TAS 1010, 262, 263, 265, 266, 282, 283
Bradley Telecom, 207
Brookes, Maria, 434
Brookes, Nicholas
  Bowthorpe name change, 437
  Bowthorpe telecommunications unit expansion, 434–436
  strategic planning session, 438–439
  visit to TAS, 424–425
  WDT departure from Bowthorpe, 432, 439, 440
Brown, Alice, 9
Brown, Earl, 170–171, 175

Brown, Mrs. (counselor), 78–79
Brown, Randy, 69
Bruni, Giordano, 331
"Buffalo Soldier" (song), 330
busing issues, 94–95

C

calculators, 83–84
Caldwell, Darthulia, 12
Caldwell, Harold, 70
Callahan, George, 167–169, 171–172, 173
Cannery Row (Steinbeck), 398–399
Carlos (student), 126
Carlton, Al ("Fat Finger"), 57
Carlton, Gay
    breakup with WDT, 103
    Denice Davis party, 57–62
    high school relationship, 76, 77, 368, 386
    MIT application and attendance, 80, 81, 97
    WDT consideration of MIT, 97, 98, 99, 100, 101–102
Carlton, Pamela, 57, 58, 386–387, 401
Catchings, Francis Emory, 9
Catchings, Timothy Titus, 9, 10, 16, 451
Cathers, Rick, 365, 366
cell phone industry, 239, 375–376, 443
Chandler, Neville, 238, 239–240, 241, 242
channel simulators. *See* AEA Technology, Inc.; Bradley 2A/2B Impairment Simulator; TAS 1010 Channel Simulator
Charles (neighbor), 29–30
Chen, Mike, 325–330
chipsets, 324
Chris (TAS production supervisor), 385
Cidco, 406
civil rights movement
    corporations pushed to hire blacks, 174–175
    demand for increased black enrollment at U-M, 124–126
    Flint open housing ordinance sleep-in, 45–48
    integration at GMI, 82
    proving ground metaphor for, 443–444
    reactionary forces to, 450–451
    TAS success aided by, 431, 445
    WDT awareness of, 30–31
    See also race relations and racism experiences
Clover Valley Group, 387–392
Clyde (classmate), 27–28
college years. *See* General Motors Institute (GMI); University of Michigan
Colquitt, Nettie, 172–175, 187, 188, 210–211, 214
Colton, John
    basketball participation, 165, 175–176
    diversity in group, 167, 171, 175, 227, 247
    Jim Ingle promotion, 219–224
    "nigger in the woodpile" incident, 184–191
    promotion of, 219, 221
    tone generator project, 194–195
    WDT interactions, 157–158, 164–167, 213–214
CommNet tradeshows, 331, 416
Commtec Associates, 292
The Complete Guide to Selling Your Business (Sperry and Mitchell), 401–402
Concord Data Systems, 281–284, 364
Confidential Memorandum for TAS sale, 411, 412, 413, 416
Count Basie Learning Center, 447–448
Crawley, Chris, 74–76
Cross, Martha, 25, 30
Crow, David, 333–334, 336

D

Dargue, Frank, 185
data analyzers, 309–310, 314, 365
Davis, Darwin, 58, 368–369, 386
Davis, Denice
    class officer elections, 73, 74
    sweet-sixteen party, 57, 58, 59, 60, 61
Davis, Elizabeth Bernice Tarver, 12, *14*, 204, 453
Davis, Evans, 108, 158, 204, 206
Debbie (assembly line worker), 95
Detroit, Michigan
    presence of black middle class, 179
    riots of 1967, 45, 51
digital modems, 373
Dinkins, David, 368, 369, 370
Dipa (roommate), 122–123, 128, 129
direct frequency synthesis (DFS), 194–195
Dones, Jenny, 73, 74
drum major post, 72, 76

Dutcher, Phil, 205

E
Ebel, Mike, 88–89
electronic music interest. *See* music synthesizer project
Engel, Joel, 239
entrepreneurial ambitions
    abandoned ideas, 204
    childhood activities, 38–44
    decision on Bell Labs vs. PhD, 155–160
    ECE 472 course grade effects, 148
    Healthcare Information Service venture, 204–206
    as life goal, 1–2, 89, 150, 218, 254, 322–323
    music synthesizer project, 104–106, 151–154, 166, 196–202, 207, 208
    Small Business Policy course, 149–150
    See also Telecom Analysis Systems, Inc. (TAS)
Environmental Research Institute of Michigan (ERIM), 131, 149
Equitable Black Achievement Awards, 368
Erickson, Jim, 381

F
Farr, Ian, 331–336, 337, 363
Fish, Rebecca, 416
Flint, Michigan
    black middle class presence, 179
    boom days, 13–14, 15
    decline in, 450
    open housing ordinance sleep-in, 45–48
    urban renewal plan, 51–52
Flint Area Science Fair. *See* science fair entries
Flint Central High School, 71, 72, 93, 452
    See also high school years
Flint Journal
    ham radio classified ad, 34
    open housing ordinance election coverage, 47
    science fair coverage, 20, 69, *70*
    WDT paper route, 41–44
Flint River(s), 16

Flint Spokesman, 69
focus group, Touch-Tel project, 240–241
Ford Motor Company, 155–156
Forlenza, Phil, 428, 429, 430
Fran (Bell Labs supervisor), 228, 231–233
France prospecting visit, 336–340
Frank Stank (classmate), 18, 20, 21
French, Bert, 258–260
Frenkiel, Dick, 238–239, 242, 253
Fukumoto, Mr. (department head), 360–361

G
Gangnath, Clifford, 86, 100–101
Garcia, Jose, 169
Gates, Bill, 381–382, 384
General Instrument D AY-t-1013 Universal Asynchronous Receiver/Transmitter, 153–154
General Motors
    boom days, 13–14, 15
    decline in, 450
    fifty-millionth vehicle parade, 5
    GMI relationship, 81, 89
    instrument panel assembly line, 91–96
General Motors Institute (GMI)
    AC Spark Plug division experience, 86–96, 100–101, 166, 450
    acceptance to, 80
    first year, 81–85
    second year, 103, 104
Genovese, Tony, 169–170
"Giant Transistor" water tower, *161*, 163
Giguerre, William ("Gigs"), 184, 185
Giordano, Halleran and Ciesla (law office), 2, 427, 428
GMI. *See* General Motors Institute (GMI)
Goodman, David, 375
Gopal (roommate), 122–123, 128, 129
"graduation" photo, 429, *430*
Grady Colored School of Nursing (Atlanta, Georgia), 12
Graham, Bill
    Concord Data Systems sales call, 281–284
    Phoenix Microsystems flier, 310–311
    TAS sales representation agreement, 279, 280
    WDT refusal to do business in South Africa, 363–364, 370

INDEX

Grant, Raymond
  advice on Hayes management visit, 299–300, 302, 304–305
  Hayes presentation by WDT, 296
  TAS 1010 evaluation, 297–298, 303
Graphics Workstation project
  assignment to, 226–228
  failure of, 234–236, 237
  intervention on, 231–233, 367
  vs. Touch-Tel project, 238, 240, 241
  WDT concerns with, 229–230
Grill Room gatherings, 173–175
Gulf War (1991), 372

H

ham radio operation, 33–37
Hamilton, Theresa, *23*, 23–24, 26
Hammond, Peter, 387–392
Hampton, Shirley, 25, 55, 73–74
handheld calculators, 83–84
Hans (AEA sales rep), 309, 331
Hard Drive (Wallace and Erickson), 381–382, 384, 385, 394, 395, 398
Harris, Geoff, 285–291
Harvard University, 80
Hauser, Emilie, 25
Hayden, Claudia Louise. *See* Tarver, Claudia Louise Hayden
Hayden, Elizabeth Bernice Catchings ("Mimama")
  biographical details, 9–11, 451
  death of, 423–424
  illness and recovery, 48–50, 51
  Louise and Fred Tarver move to Flint, 13
  MIT consideration by WDT, 102
  on MLK assassination, 48
  at Quinn Chapel A.M.E. Church, *46*
  visit to New Jersey, 182–183
  WDT relationship, 6–7, 48–50, 51, 423–424
Hayden, Emerald Earl, 10, 11
Hayden, Mary Elizabeth (later, Watson), 10, 11, 54, 98
Hayden, William Wise, Sr., 10, 11, 13
Hayden, William Wise, Jr., 10, 11, 197–198, 199
Hayes Microcomputer Products, Inc.
  demonstration visit in Georgia, 292–297
  management visit to TAS, 299–304

  placement of orders, 297–299, 304–305
Healthcare Information Service venture, 204–206
Heath, Captain (plantation owner), 9
Heick, Bob, 203–204, 206
Henry Dreyfus Associates, 268
Hewlett-Packard
  Hayes visit to, 299
  HP 35 calculator, 83–84
  Interface '86 tradeshow, 308
  lack of interest in TAS, 418
  measuring instruments for Bradley Box, 207
  as standard for TAS, 268, 269, 344
  as TAS competitor, 313, 398
  WDT experience with instruments, 193, 206
high school years
  basketball participation, 72–73, *73*, 75, 76–77
  class officer elections, 73–74
  marching band participation, 72, 76
  science fair entries, 64–70, *70*, 76, 77
  senior awards ceremony, 71, 77–79
Hirsch, Don
  Graphics Workstation project, 230–231, 232, 233, 234
  Touch-Tel project, 237–238, 239
  WDT resignation, 253
Hodges, Neil
  TAS 1010 evaluation, 297–299, 300, 303
  WDT visits to Hayes, 293–296, 304–305
Hosmer, Larue Tone, 150
Humphrey, Hubert, 50
Hure, Jean-Michel, 337–338
Hurley Hospital (Flint, Michigan)
  Dody treatment at, 55
  Fred Tarver, Sr. treatment at, 107
  Healthcare Information Service venture, 204–206
  Louise Tarver employment, 13, 14, 48, 114, 204
  Mimama treatment at, 48

I

IBM Office Products, 156
impairment simulators. *See* AEA Technology, Inc.; Bradley 2A/2B

Impairment Simulator; TAS 1010 Channel Simulator
Infinet, 310
Ingle, Jim
   Frank Bradley lunch, 207–208
   promotion of, 219–224
   sales leads, 272
   Scientific Devices connection, 279, 280
   TAS 1010 evaluation, 262–267, 282, 325
instrument panel assembly line, 91–96
Intel Corporation, 90, 193
Interface '86 tradeshow, 306–314, 316, 331

J

Jamaica trips
   Bowthorpe strategic planning session, 438–439
   family trip, 7, 15–16
   marriage to Kishna, 452, 453
James Wickstead Design, 268–269
Janice (girlfriend), 55–56
Japanese business connections. *See* Toyo Corporation Japan
Jenny (software engineer), 320–321
Jones, Mr. (Hayes executive), 301–304
Jones, Mr. (housing specialist), 177–178

K

Kaleidoscope talent show, 77
Kathy (assembly line worker), 92–94, 96
Kayla (housing consultant), 179–181
Kennedy, John Fitzgerald, 30, 31
Kennedy, Robert F. ("Bobby"), 48, 50–51
Kerscher, Bill, 90, 101
Kettering University. *See* General Motors Institute (GMI)
Kim (assembly line worker), 94–95
Kimbrough, Clarence, 46
King, Martin Luther, Jr. (MLK), 30, 31, 48, 50–51
Kirk (friend), 45–47
Kobayashi, Miss (secretary), 352, 356, 359
Kurzweil synthesizers, 201, 202

L

Lariz, Mondy, 324–330, 371, 372
Larry (Bell Labs supervisor), 228, 231–233

LaVerne (office manager), 318–319, 321
Lawrence, Victor
   advice on new products, 374–375
   order from TAS, 277–278
   performance review rankings, 229, 230
Lehman formula, 405
Little Silver, New Jersey, WDT integration of, 243–245
Long Branch, New Jersey, decline in, 181
Lucky Goldstar Company, 242

M

Mandela, Nelson, 364, 368–370
marching band participation, 72, 76
Mayo, Mel, 309–310, 311–312
MB Electronique, 337–338, 340
McAfee, Leo ("No-Plea McAfee"), 138–141, 150–151, 154
McCann, Les, 198–200
McCarthy, Colin, 418–419
McCormick, Richard, 205–206
McCoy, Adora, 54–55
McCree, Floyd, 45, 47, 48
McDermott, Les, 152–153
McDonald's parking lot police confrontation, 183
McLaughlin, Pete, 279–281, 284
Mehmet (roommate), 121–122, 128, 129
Merrill, F. Gordon, 184–192
microprocessors, early, 90, 193
Microsoft Corporation, 381–382, 384
MidLantic Bank ("The Hungry Bankers"), 260–261
Migliorini, Gianni, 331
Mike (classmate), 26–28
Million Man March (1995), 424
Minority Engineering Programs Office (MEPO), 125–126, 144
MIT (Massachusetts Institute of Technology)
   Gay Carlton attendance, 80, 81, 97
   WDT consideration of, 97–103
Mitcham, Maymie, 49
Mitcham, Reverend, 49–50
Mitchell, Beatrice
   as author, 401
   Bowthorpe offer, 420, 421
   call from Singapore, 402–403
   hiring of as TAS advisor, 410
   meeting with WDT, 403–407

potential TAS suitors, 411, 416, 417
sale of TAS, 428, 429, 430
visit to TAS, 408–409
modem industry
   Bell Labs involvement, 226, 246, 294, 295, 302
   Concord Data Systems involvement, 281–283
   data analyzer role, 309–310
   digital modem development, 373
   Hayes involvement, 292–294, 295, 297, 300–302
   Japanese companies, 342, 354–356, 359–360, 362
   Phoenix Microsystems involvement, 310, 313–314
   Rockwell involvement, 324, 371
   Sharp Corporation involvement, 342
   U.K. companies, 334
   See also Telecom Analysis Systems, Inc. (TAS)
Molly Pitcher Inn, 165, 177, 180, 449
monophonic synthesizers, 201
Monterio, Anne, 125–126, 144
Moog, Robert, 106
Moog concert, 104–106
Moog synthesizers, 105–106, 201, 202
Moore, Kimberly, 270
Moore, Steve
   Bell Labs resignation, 270
   Bowthorpe meetings and offer, 418–421, 425–426
   Charles Simmons meeting, 275–277
   civil rights context for TAS, 444
   continued leadership at Spirent, 441
   early TAS years, 256, 269, 273–274, 279, 292
   equity share in TAS, 270–272
   growth period for TAS, 367, 373
   Hayes management visit, 301–304
   hiring of Sperry Mitchell, 404, 406, 407–409, 410
   Interface '86 tradeshow, 307, 308, 310, 312, 313–314
   marketing staff difficulties, 285, 286, 289, 290
   Mount Airy retreat, 365, 366
   as neighbor of Diane Sheng, 218
   Nigel Wright resignation, 394
   personnel woes, 315, 316–317, 318–319, 320, 321–323
   photos of, *255*, *430*
   prototype development, 249–250, 251, 265
   recruitment by WDT, 247–249
   rejection of TAS president position, 437
   return to startup mode, 382–385
   sale of TAS, 392, 427–430, 432
   sales concerns, 279
   on sharing design information, 343, 346
   Tash hire, 395–397
   team-building session, 377, 380
   Toyo relationship, 347, 362
   WDT decision to turn company around, 393
   WDT departure from Bowthorpe, 439, 440, 441
   wireless technology research, 375–376
Moore, Walter, *73*
"More Today Than Yesterday" (song), 62
Morgan, Dennis, 227, 228, 230–231, 234
Morse code, 33, 34, 35, 36, 37, 375
Mosimann's Belfry, 432, 439–440
Mott, Charles Stewart, 15, 46
Mulally, Mr. (special education teacher), 17, 21
music synthesizer project
   Directed Study project, 151–154
   end of dream for business, 196–202
   Moog concert inspiration, 104–106
   similarity to Bradley Box, 207, 208
   similarity to RTS-5, 166

N

NAACP, 189, 191
NASA Lewis Research Center, 156
Nesbit, Leroy, 32–33
newspaper route, 41–44, *44*
Nice, John, 333, 334–335, 336
Nick (friend), 151
"nigger in the woodpile" incident, 184–192
Nixon, Charlie
   Hayes account management, 304, 305
   Interface '86 attendance, 307, 312
   WDT visit to Hayes, 292–297
Nixon, Richard, 51

O

Operation Desert Storm (1991), 372

Organizational Mirror exercise, 378–379
Ottesen, Lloyd, 170, 175

P

Pang, Leesing, 162, 176, 209
Parker, Paul, 159
Peaks, Cathy, 25
Perry, Dewayne, 240, 242
Phi Eta Psi fraternity, 83
Phoebe-Putney Hospital (Albany, Georgia), 12, 13
Phoenix Datacom Limited, 331–332
Phoenix Microsystems, 309–310, 311–314, 331, 365
photocells, 18, 19, 21
Pieczara, Fran, 182, 183
Pieczara, Stan, 182, 183, 211
Pierce, Donald, 295, 296
polyphonic synthesizers, 201
Pontiac Central High School basketball game, 76–77
Powell, Mary, 250–251
proving ground metaphor, 431, 443–444

Q

Quinn Chapel A.M.E. Church, 46, 111

R

race relations and racism experiences
  assembly line discussions, 94–96
  Bell Labs interview, 157
  Bowthorpe presidents' meetings, 433, 435
  Chai Yeh achievements, 137
  class officer elections, 73
  Concord Data Systems sales call, 283, 284
  Diane Sheng family disapproval, 215, 216, 217
  diverse childhood friendships, 25–26
  diversity at Bell Labs, 167, 171, 174–175
  English dialects, 334, 335
  few black college students, 82, 124–126
  fishing trip incident, 112–113
  at GMI, 82–83, 84–85
  Grill Room gatherings, 173–175
  Hayes interactions, 295, 296, 302–303, 305
  hospital restroom incident, 3, 13, 16, 451
  ID requirement if no suspicion of crime, 244
  Jim Ingle promotion, 220
  Leo McAfee interactions, 138, 139, 140, 141
  Little Silver house police stop, 243–245
  McDonald's parking lot police confrontation, 183
  middle-class vs. poor black neighborhoods, 179, 180
  new barriers to progress, 449–452
  "nigger in the woodpile" incident, 184–192
  performance review rankings, 228–229, 230
  personnel woes at TAS, 321, 322
  potential white girlfriend, 75–76
  "professional managers" comment, 379, 380
  "race uncertainty principle", 445–446
  radio experimentation, 34, 36, 37
  Raymond Grant interactions, 296, 298, 299–300, 304–305
  Rockwell interactions, 325, 326, 330
  roller skating parties, 25
  SBA loan banker, 259
  science fair awards, 69–70
  senior awards ceremony, 78–79
  South Africa, refusal to do business in, 363–364
  special education overrepresentation of blacks, 17
  Spencer Bement bias, 132, 135, 144, 145, 147–148, 154
  team building intervention, 231–233
  Toyo seminar, 354–357, 360, 361
  UCLA job retraction, 215
  University Towers roommates, 122, 129–130
  Victor Lawrence treatment, 229, 230, 278
  WDT reflections on, 16, 30–31, 444–446, 453–454
  See also civil rights movement
"race uncertainty principle", 445–446
radio experimentation, 32–37
Rag Chewer's Club, 36, 37
Ralston, Helen, 418–419

Rand Corporation, 212–213, 214, 215
Rao (engineer), 273
Raymond (friend), 57, 58, 59
Reaves, Johnny, 31, 39–40
Red Bank Education and Development Initiative (RBEDI), 449
Red Bank Mini-Mall office, 275, 300, 302, 303
Red Bank, New Jersey
　decline in academics, 447–449
　improvements in, 448
　vs. Little Silver, 244
　WDT observations on, 177, 179
Reggie's Twenty-Five Percent Rule, 120, 126
Remote Test System 5 (RTS-5), 165–166, 175, 184–185, 193, 203–204, 206
Richards, Barry, 387
"River of Jordan" (song), 423
Robart, Bruce, 72
Robertson Stephens and Co., 387, 388
robot science fair entry, 19–22, 25
Rockwell Japan, 357
Rockwell Semiconductor Products, 324–330, 371–372
Rodriguez, Mr. (Hayes executive), 301–304
roller skating parties, 25
ROM (Read Only Memory) devices, 90
Ross, Wally, 28
RTS-5 (Remote Test System 5), 165–166, 175, 184–185, 193, 203–204, 206
Russell, Campanella ("Campy"), 76–77

S
Saarinen, Eero, 163
Salazar, Andy, 310
Saltman, Mrs. (teacher), 30–31
SARTS (Switched Access Remote Test System), 184–185, 193, 223–224, 247–248
Sato, Steve, 169
Schripsema, Richard, 205, 206
Schwartzwalder, Karl, 86, 87
science fair entries
　fourth grade, 17–22, 25, 444
　high school, 64–70, 70, 76, 77, 443
　junior high, 63–64
Scientific Devices, Inc., 279–281
scientific method

Red Bank schools application, 448, 449
science fair application, 17–18, 22, 63, 64, 65, 66, 444
TAS application, 444–445, 449
Scott, Jim, 365, 418, 419
segregation
　busing issues, 94–95
　Flint open housing ordinance sleep-in, 45–48
　hospital restroom incident, 3, 13, 16, 451
　Little Silver lack of black residents, 244
　Red Bank schools, 447
　See also race relations and racism experiences
Sereys, Guy, 337–340
Shand Electronics, 34, 37, 66
Sharif, Kishna Arien (later, Tarver), 452, 453
Sharp Corporation, 341–344, 357
Shaw, Mr. (teacher), 67
Shaw brothers, 38
Sheehan, John, 234
Sheng, Diane Dwan
　basketball participation, 209
　Bell Labs work, 209–210, 217–218
　Bowthorpe events, 425, 433
　calls home from WDT, 335–336, 339, 350, 353
　children's births, 287–288, 368
　China trips, 215–216, 348
　divorce, 452
　early relationship with WDT, 209–212
　engagement and wedding, 214–218
　Equitable Black Achievement Awards attendance, 368
　financing of TAS, 256, 260
　Little Silver, NJ house, 243–245
　Stanford University studies, 209–210, 212
　Steve Moore dinner, 270
　UCLA job offer and retraction, 212, 213, 215
Sheng, Suzanne, 215–216, 218
Simmons, Charles
　Bell Labs resignation, 277
　Bowthorpe meetings and offer, 418–421, 425–426
　civil rights context for TAS, 444

concerns about leaving Bell Labs, 271, 275–277
continued leadership at Spirent, 441
early TAS years, 256, 269, 273–274, 279, 292
equity share in TAS, 270–272
growth period for TAS, 367, 373
Hayes management visit, 301–304
hiring of Sperry Mitchell, 404, 406, 407–409, 410
Interface '86 tradeshow, 307, 310, 313–314
marketing staff difficulties, 285, 286, 290
Mount Airy retreat, 365, 366
Nigel Wright resignation, 394
personnel woes, 315, 316–317, 319, 320, 321–323
photos of, *255, 430*
prototype development, 249–250, 251, 265
recruitment by WDT, 247–249
return to startup mode, 382–385
sale of TAS, 392, 427–430, 432
sales concerns, 279
on sharing design information, 343, 346
succession as TAS president, 437
team-building session, 377, 380
Toyo relationship, 347, 362
WDT departure from Bowthorpe, 439, 440, 441
wireless technology research, 375–376
Simmons, Marlene, 275, 365
Simon, Becky, 25
Simpson, Bob, 329
"Sing a Simple Song" (song), 123
slide rules, 83–84
Small Business Administration loan programs, 258–261
Small Business Policy course, 149–150
Smith, Clifton L.
Bell Labs intervention, 231–233
Bowthorpe strategic planning session, 438–439
Red Bank Education and Development Initiative, 448–449
TAS team-building events, 365–367, 377–378, 380–381, 385
Sobolewski, John, 279–281

South Africa, WDT refusal to do business in, 363–364
Spack, Ed, 167, 187, 219–220, 221–223
Spaniola, Jim, 87
Sperry, Paul
as author, 401
Bowthorpe offer, 420, 421, 424
call from Singapore, 402
hiring of as TAS advisor, 410
meeting with WDT, 403–407
potential TAS suitors, 411, 413, 416, 417
sale of TAS, 428, 429, 430
visit to TAS, 408–409
Sperry Mitchell and Co., 402, 403, 404, 405, 408, 409, 410
Spirent Communications, 437, 441
Spotsville, Mr. (neighbor), 68
Stannard, Jan, 68–69
Steinbeck, John, 398–399
Stokes, Dot, 172–175, 185, 187
Swindell, Mr. (music teacher), 25
Switched Access Remote Test System (SARTS), 184–185, 193, 223–224, 247–248
synthesizer project. *See* music synthesizer project

T
Talking Book (album), 106
Tang, Myra, 268–269
Tarv Acoustic Synthesis (TAS), 106
Tarver, Aaron Michael, 368, 452, 453
Tarver, Claudia Louise Hayden
and Aurora McCoy, 54, 55
birth and childhood of, 10
children's births, 12, 14
hospital restroom incident, 3, 13, 16, 451
marital relationship, 11, 111–112, 113–114
MIT consideration by WDT, 98, 102
move to Flint, 13–14
move to Sugar Hill neighborhood, 51–52
nursing studies and work, 12–13, 14, 48, 114, 204
photos of, *46, 107*
science fair awards attendance, 68, 69
sleep-in permission for WDT, 46–47

visits with WDT, 182–183, 453
Tarver, Elizabeth Bernice (later, Davis), 12, *14*, 204, 453
Tarver, Fred Douglas, Sr.
  death and funeral of, 107–111, 114, 115
  Denice Davis party argument, 60
  electronics workshop in basement, 3, 16, 33, 114–115, 116, 243, 451
  family life, 12, 14
  on GMI, 81
  maintenance technician work, 33, 111, 114, 115
  marital relationship, 11, 111–112, 113–114
  military service, 12
  MIT consideration by WDT, 102–103
  move to Flint, 13–14
  move to Sugar Hill neighborhood, 51–52
  photo of, *107*
  radio involvement, 33–34, 37
  science fair incident, 67–68, 69
  WDT reflections on, 37, 108–109, 111–115, 430, 451
Tarver, Fred Douglas, Jr.
  as best man, 218
  birth of, 12
  fishing trips, 112
  McDonald's parking lot police confrontation, 183
  parental arguments, 114
  photo of, *14*
  radio experimentation, 32–33
  science fair assistance, 18, 19, 21, 22
  theremin device construction, 105
  Tiny introduction to WDT, 53
  visits with WDT, 182–183, 453
  WDT fight with Charles, 29–30
Tarver, Ida, 12
Tarver, Kishna Arien Sharif, 452, 453
Tarver, Nadiyah Louise, 453
Tarver, Stacy Sheng, 287–288, 368, 400–401, 452, 453
Tarver, William David
  AC Spark Plug work, 86–96, 100–101, 166, 450
  announcement of sale to employees, 413–415
  "anthropological studies" goal, 2, 7, 150, 392, 447, 451, 453

  basketball participation, 72–73, *73*, 75, 76–77, 165, 171, 209, 210
  Bell Labs work (*see* Bell Telephone Laboratories, WDT experiences)
  birth of, 14
  Bowthorpe due diligence and final agreement, 422–426
  Bowthorpe group co-president position, 436–439
  Bowthorpe meetings and offer, 418–421
  Bowthorpe presidents' meetings, 433, 435
  Bowthorpe strategic planning session, 438–439
  Bowthorpe telecommunications business unit involvement, 434–439
  burnout periods, 399, 409, 438
  childhood entrepreneurial activities, 38–44
  childhood romances, 53–62
  children's births, 287–288, 368, 453
  closing of TAS sale, 1–2, 427–431
  college applications, 80–81
  decision to turn company around, 392–393
  deterioration at TAS, 371–376
  divorce and remarriage, 452
  early childhood memories, 15
  early relationship with Diane, 209–212
  electronic music interest, 151–154
  engagement and wedding, 214–218
  entrepreneurial ambitions (*see* entrepreneurial ambitions)
  factors in success, 2–3, 443–446
  family history, 7–16, 451–452
  fifth grade experience, 30–31
  financing of TAS, 256–261
  first attempt at selling TAS, 386–392
  first year of TAS, 268–278
  fourth grade experience, 23–30
  France prospecting visit, 336–340
  Georgia visit, 6–7
  GMI acceptance and studies, 80–85, 97, 101, 102, 103
  Graphics Workstation project, 226–236, 237, 238, 240, 241, 367
  Hard Drive strategy, 377–385, 394
  Hayes management visit, 298–304
  Hayes presentation, 292–296

Healthcare Information Service
  venture, 203, 204–206
hiring of Sperry Mitchell, 403–410
hopes for children's future, 453–454
inspiration from simulator
  examinations, 206–208, 246–247
Interface '86 tradeshow, 306–314
Jamaica trips, 7, 15–16, 438–439, 452, 453
Jim Ingle assessment of TAS 1010, 262–267
leave of absence from Bowthorpe, 438
leaving Bell Labs, 252–254
leaving Bowthorpe, 439–441
Little Silver police stop, 243–245
Mandela reception, 368–370
marching band participation, 72, 76
marketing staff difficulties, 285–289
Mimama relationship, 48–50, 51, 423–424
MIT consideration, 97–103
MOOG concert, 104–106
motivation for selling TAS, 406–410
Mount Airy retreat, 365–367
move to Sugar Hill neighborhood, 51–52
music synthesizer project, 104–106, 151–154, 166, 196–202, 207, 208
New Jersey living situations, 177–183, 243–245
Nicholas Brookes meeting, 432, 439–440
"nigger in the woodpile" incident, 184–192
open housing ordinance sleep-in, 45–48
personnel woes, 285–289, 315–323
PhD vs. Bell Labs decision, 155–160
photos of, *14, 46, 73, 255, 361, 430, 436*
post-Bowthorpe life, 446–449
potential TAS suitor meetings, 416–418
promotion at Bell Labs, 219–225
prototype and business plan
  development, 247–251
proving ground metaphor, 431, 443–444
radio experimentation, 32–37
Rand Corporation job offer, 212–214
reflections on father, 37, 108–109, 111–115, 430, 451

Rockwell showdown, 324–330
sales indoctrination program, 397–398
science fair entries, 17–22, 25, 63–70, *70*, 76, 77, 443, 444
Scientific Devices sales representation, 279–284
Singapore trip, 400–403
South Africa, refusal to do business in, 363–364
Touch-Tel project, 237–242, 251, 252
Toyo relationship (*see* Toyo Corporation Japan)
turnaround at TAS, 394–399
U.K. prospecting visit, 331–336
U-M studies (*see* University of Michigan)
worries about selling TAS, 411–413
TAS (Tarv Acoustic Synthesis), 106
TAS (Telecom Analysis Systems). *See* Telecom Analysis Systems, Inc. (TAS)
TAS 1010 Channel Simulator
  Concord Data Systems sales call, 281–284
  data analyzer addition to, 310
  displaying of, for financing, 256–258
  first orders, 272–274, 276, 277–278, 374
  Hayes orders, 293, 296–297, 298, 300, 304
  improvements to, 365
  production units, 268–270, *269*, 272
  prototype development, 249–250, 251
  replacement for, 365
  testing of, 262–267
  theft of demo unit, 296–297
  Toyo orders, 341–342, 343, 348
TAS 4500 Wireless Channel Emulator, 398
Tash (TAS sales manager), 395–397
TASKIT software, 310, 373
Tauson, Rick, 272–273
Tech Club meetings, 87–88
Technical Associate position (TA), 170
Telecom Analysis Systems, Inc. (TAS)
  administrative assistant hire, 274–275
  announcement of sale to employees, 413–415
  bonus check distribution, 392–393
  bonus plans, 420–421
  Bowthorpe due diligence and final agreement, 422–426

Index

Bowthorpe meetings and offer, 418–421
business plan, 250–251
cash crunch, 277–278
closing on sale of, 1–2, 427–431
Concord Data Systems sales call, 281–284
decision to turn company around, 393
Eatontown headquarters, 367
financing of, 256–261
first attempt at selling, 386–393
first sales, 272–274, 276, 277–278, 374
France prospecting visit, 336–340
Hayes relationship, 292–305
incorporation of, 270–272
Interface '86 tradeshow, 306–314, 316, 331
low employee morale, 376, 377–381, 386
move to Birch Avenue, 306
new product development, 365, 367, 374–376
personnel woes, 285–291, 315–323
Phoenix Microsystems relationship, 309–310, 311–314
post-acquisition success, 441
potential suitors, 416–418
preparations for sale of, 411–415
reasons for selling, 406–407
Red Bank Mini-Mall office, 275, 300, 302, 303
return to startup mode, 382–385, 394, 395
Rockwell relationship, 324–330, 371–372
sales indoctrination program, 397–398
sales milestones, 278, 306, 332, 367, 368, 398
Scientific Devices sales representation, 279–284
scientific method applied to, 444–445
Sharp proxy order, 341–344
softening sales, 372–376, 386, 415
Steve Moore and Charles Simmons join, 247–249
TAS 1010 Channel Simulator (*see* TAS 1010 Channel Simulator)
team-building sessions, 365–367, 377–381, 415
Toyo orders, 341–344, 348

Toyo seminar preparation and rehearsal, 350–356
Toyo seminar presentation, 356–360
Toyo visit to TAS, 347–348
turnaround of, 394–399
U.K. prospecting visit, 331–336
uniqueness of product, 301, 302, 322–323
wireless technology involvement, 375–376, 377, 398, 441
teleconferencing industry, 226, 234–236
telephone channel simulators. *See* AEA Technology, Inc.; Bradley 2A/2B Impairment Simulator; TAS 1010 Channel Simulator
Texas Instruments Japan, 357
theremins, 105
Thompson, Allen, 101
Thompson, Mike ("Dookey-Hookey"), 21, 29–30, 72–73
Thompson, Rich, 237, 238, 239–240, 241
Thompson, Tommy, 234, 235
Tiny (friend), 53–54
Tom (Bell Labs supervisor), 227–228, 231–233
Tom (TAS manufacturing manager), 307, 315–316, 317, 321
Tom, Mike, 169, 170
tone generator project, 193–195
touch-screen telephone/data terminal, 241–242
Touch-Tel project, 237–242, 251, 252
Toyo Corporation Japan
 Bill Berkman call, 344–347
 orders to TAS, 341–344, 348, 362
 seminar preparation and rehearsal, 350–356
 seminar presentation, 356–360
 visit to TAS, 347–348
 WDT arrival in Japan, 348–352
 WDT dinner with management, 360–361
transistors
 "Giant Transistor" water tower, *161*, 163
 heat effects science fair entry, 64–66
 radiation effects science fair entry, 66–70
 radio experimentation, 32–37
Turner, Cliff, 77

Tyree, Dorian, 83–84

U

UART (universal asynchronous receiver transmitter), 151–154
United Kingdom prospecting visit, 331–336
University of California - Los Angeles (UCLA), 212, 215
University of Illinois, 159
University of Michigan
    Aaron Tarver attendance, 453
    acceptance to, 80–81
    anticipation for, 119–120
    Bioelectric Sciences Lab research assistantship, 118–119
    contrast to GMI, 123–124
    Directed Study project, 151–154
    ECE 312 course, 138–141
    ECE 380 course, 137–138
    ECE 471 course, 131–136
    ECE 472 course, 142–148
    few black college students at, 124–126
    first semester classes, 108, 109–110, 115–116, 123–124
    foreign student friends, 126–128
    graduate studies, 149–154
    lice infestation, 128–129
    Minority Engineering Program Office, 125–126
    radiation source for science fair entry, 67
    recruiting for Bell Labs, 216
    University Towers living situation, 120–123, 128–130

V

Volz, Richard A., 108, 109–110, 115–116

W

wagon, newspaper delivery, 42, *44*
Wallace, James, 381
Waller, Fred, 46
Walling, Ann, 164
Watson, Aurora, 54–55
Watson, Mary Elizabeth Hayden, 10, 11, 54, 98
Watson, Odest, 11, 54
WDT. *See* Tarver, William David
Weaver, Rex, 46
Weldon, Frank, 325–330
West Long Branch, New Jersey, police confrontation, 183
Western Electric, 193, 216, 226, 234
Westhead, John, 419, 424, 425, 433
What's Your Beef restaurant, 211, 425
White, Mr. (Hayes executive), 301–304
white-black relations. *See* race relations and racism experiences
Wickstead, James, 268–269
Wilborn, Eugene, 77
Williams, William J., 118, 119, 131–135, 143, 144
Wilson, George, 387–389
Wilson, Rhonda, 25
wireless technology industry, 375–376, 377, 398, 441
Wireless Telecom Group, 398
Wonder, Stevie, 106, 118, 196–197, 430
Wood, Raymond, 83
Wright, Nigel, 383, 394–395, 397

Y

Yamaha synthesizers, 201, 202, 342
Yamamoto, Mineo
    seminar preparation and rehearsal, 350–352, 353–356, *361*
    seminar presentation, 356–360, 362
    Sharp Corporation proxy order, 341–344
    visit to TAS, 346, 347–348
    WDT arrival in Japan, 349–352
Yeager, Bob, 167, 188, 189–190
Yeh, Chai, 137–138
Yumoto, Mr. (department head), 360–361

Z

Zaman, John, 249
Zamorski, Maureen
    Hayes visit, 301, 304
    hiring of, 275
    Interface '86 tradeshow, 307, 308, 309, 312, 313
    resignation of, 316, 317, 321

Permissions Acknowledgments

Regent Music Corp.: lyric excerpt from *More Today Than Yesterday*. Words by Patrick Upton. Copyright © 1969 by Regent Music Corp. Reprinted by permission of Regent Music Corp. International copyright secured. All rights reserved.

Alfred Publishing Company, Inc.: lyric excerpt from *Am I Black Enough for You?* Words and music by Kenneth Gamble and Leon Huff. Copyright © 1972 (Renewed) by Warner-Tamerlane Publishing Corp. All rights reserved. Used by permission.

Alfred Publishing Company, Inc.: lyric excerpt from *Stand*. Words and music by Sylvester Stewart. Copyright © 1970 (Renewed) Mijac Music. All rights administered by Warner-Tamerlane Publishing Corp. All rights reserved. Used by permission.

Alfred Publishing Company, Inc.: lyric excerpt from *Sing a Simple Song*. Words and music by Sylvester Stewart. Copyright © 1968 (Renewed) Mijac Music (BMI). All rights administered by Warner-Tamerlane Publishing Corp. (BMI). All rights reserved. Used by permission.

Alfred Publishing Company, Inc.: lyric excerpt from *Don't Call Me Nigger, Whitey*. Words and music by Sylvester Stewart. Copyright © 1969 (Renewed) Mijac Music (BMI). All rights administered by Warner-Tamerlane Publishing Corp. All rights reserved. Used by permission.

Alfred Publishing Company, Inc.: lyric excerpt from *Everybody is a Star*. Words and music by Sylvester Stewart. Copyright © 1970 (Renewed) Mijac Music (BMI). All rights administered by Warner-Tamerlane Publishing Corp. All rights reserved. Used by permission.

Hip Hill Music Publishers: Lyrics from the song composition, Buffalo Soldier, used by permission of Hip Hill Music Publishing. Songwriters: David M. Barnes, Margaret Lewis, Mira Smith.

About the Author

DAVID TARVER was born and raised in Flint, Michigan. He received bachelor's (1975) and master's (1976) degrees in electrical engineering from the University of Michigan in Ann Arbor, then went to work for AT&T Bell Laboratories in Holmdel, New Jersey. Tarver left Bell Labs in 1983 to start Telecom Analysis Systems (TAS) in his basement with colleagues Steve Moore and Charles Simmons. In 1995, he engineered the sale of TAS to Bowthorpe (now Spirent) plc for $30 million. From 1996-99, Tarver spearheaded development of a Spirent telecommunications test equipment business that had sales of over $250 million and a market value in excess of $2 billion. He left Spirent as president of the telecom equipment business unit at the end of 1999 to pursue community service and family interests. In 2001, Tarver founded the Red Bank Education and Development Initiative (Red Bank, New Jersey). The community-based not-for-profit catalyzed dramatic improvements in academic performance and opportunities for Red Bank children. He has also served on the National Advisory Committee for the University of Michigan (U-M) College of Engineering, the U-M Alumni Association board of directors, the Red Bank (NJ) Board of Education, the National Commission on NAEP 12th Grade Assessment and Reporting, and several other civic and not-for-profit organization boards. In 2007, he returned to Michigan, where he resides with wife Kishna Sharif Tarver and daughter Nadiyah Louise. *Proving Ground* is his first book.